Plant Morphology and Function, Geometric Morphometrics, and Modelling: Decoding the Mathematical Secrets of Plants

Plant Morphology and Function, Geometric Morphometrics, and Modelling: Decoding the Mathematical Secrets of Plants

Editors

Jie Gao
Weiwei Huang
Johan Gielis
Peijian Shi

MDPI

Basel • Beijing • Wuhan • Barcelona • Belgrade • Novi Sad • Cluj • Manchester

Editors
Jie Gao
College of Life Science
Xinjiang Normal University
Urumqi
China

Weiwei Huang
Bamboo Research Institute
Nanjing Forestry University
Nanjing
China

Johan Gielis
Department of Biosciences Engineering
University of Antwerp
Antwerp
Belgium

Peijian Shi
Bamboo Research Institute
Nanjing Forestry University
Nanjing
China

Editorial Office
MDPI
St. Alban-Anlage 66
4052 Basel, Switzerland

This is a reprint of articles from the Special Issue published online in the open access journal *Plants* (ISSN 2223-7747) (available at: https://www.mdpi.com/journal/plants/special_issues/Morphology_Geometry).

For citation purposes, cite each article independently as indicated on the article page online and as indicated below:

Lastname, A.A.; Lastname, B.B. Article Title. *Journal Name* **Year**, *Volume Number*, Page Range.

ISBN 978-3-0365-9422-4 (Hbk)
ISBN 978-3-0365-9423-1 (PDF)
doi.org/10.3390/books978-3-0365-9423-1

Contents

About the Editors

Jie Gao

Professor Jie Gao, born in 1991, is a distinguished scholar and a leading voice in the fields of global change ecology, forest management, and conservation science. He holds a doctoral degree and has completed a prestigious Boya Postdoctoral Fellowship at Peking University, where he also fulfills the vital role of a doctoral supervisor. As a member of the American Scientist Honorary Society, Professor Gao stands out for his academic prowess and contributions to the scientific community. His work has been published in over 20 articles across esteemed international journals including 'Nature Climate Change', 'Journal of Integrative Plant Biology', and 'FEMS Microbiology Ecology', to name a few. These publications highlight his extensive research and expertise in addressing some of the most pressing environmental issues of our time. Professor Gao's scholarly focus encompasses a holistic approach to understanding and mitigating the effects of global environmental changes, with a particular emphasis on forestry operations and the preservation of biodiversity. His commitment to research excellence not only advances the field of ecological science but also lays the groundwork for practical strategies in forest management and biodiversity conservation that can be applied on a global scale.

Weiwei Huang

Weiwei Huang is an associate professor at Nanjing Forestry University, China. She obtained her PhD in Forestry from the University of Copenhagen, Denmark. She mainly works on dendrochronology, dendroecology, tree physiology, and wood anatomy. Her research work focuses on exploring the forest tree/bamboo species and their ability to cope with climate change. She was awarded the Heilmann honorary award in 2018 for her research analyzing Danish forest tree species and their ability to cope with future climate.

Johan Gielis

Dr. Johan Gielis is a renowned scientist acclaimed for his significant contributions to the field of mathematical biology. With a career marked by a profound interest in the geometry of nature, he has been celebrated for the development of the Gielis Formula, a mathematical equation that describes a wide variety of natural and geometric forms. His interdisciplinary work bridges the gap between mathematical theories and biological applications, offering insights into the patterns and forms that underpin the structure of living organisms.

At the core of Dr. Gielis' research is the notion that nature's complexity can be understood through the lens of mathematical simplicity. He has applied his equation to explain phenomena ranging from the morphology of plants to the geometry of sea stars, highlighting the universal applicability of geometric principles. As a lecturer and researcher, he continues to inspire students and peers alike with his innovative approach to deciphering the codes of life through mathematics. Dr. Gielis' work not only deepens our understanding of the natural world but also paves the way for advancements in fields such as biomimetics and computational biology.

Peijian Shi

Dr. Peijian Shi is an associate professor of Nanjing Forestry University. He obtained his PhD from the Institute of Zoology, Chinese Academy of Sciences, which is a top research institution in ecology and zoology. His research interests are remarkably diverse, spanning plant morphology, insect physiology and ecology, and social management science. Dr. Peijian Shi has published more than 120 papers in some international peer-reviewed journals, a testament to his active engagement with the scientific community. Dr. Shi is a vocal advocate for academic freedom. His numerous studies have significantly contributed to the advancement of botany and ecology in China, influencing both the academic landscape and the practical application of ecological principles within social management contexts. Dr. Shi's dedication to exploring the complex interplay between living organisms and their environments, coupled with his commitment to societal issues, positions him as a pivotal figure in the progression of ecological studies and their implementation in the real world. His work not only furthers academic dialogue but also invites a broader discussion on the role of academia in societal governance and policy making.

Preface

In this artfully curated Special Issue of *Plants*, we unfold the intricate tapestry of plant morphology, a field where the architecture of life meets the rigor of mathematics. The subject at hand delves into the profound relationship between the structural features of plants and their adeptness at navigating the environmental labyrinth, revealing a fascinating narrative of survival and adaptability.

The scope of our inquiry extends beyond the mere contours of leaves and branches, touching upon the very essence of botanical function. The symmetrical ballet of plant shapes—leaves, fruits, seeds, roots, and trunks—choreographs a story of life that, while seemingly aligned with geometric perfection, dances vivaciously with variability. Even within the same species, the stage of development dresses plants in a wardrobe of different morphological characteristics.

Our aim is to quantify these characteristics with mathematical elegance, utilizing geometric and statistical models as our compass. This pursuit is not solely academic; it seeks to arm botanists, ecologists, and mathematicians with the tools to interpret the evolutionary and developmental narrative of flora. The purpose, deeply rooted in scientific curiosity, is to enhance our collective understanding of plant strategies for thriving in a dynamic world.

The authors, a coalition of minds from diverse disciplines, have convened in these pages, offering their insights like seeds hoping to germinate new avenues of thought. Their work is an ode to the unity of natural forms, inspired by the universal geometric equation proposed a decade ago by Dr. Johan Gielis.

To our discerning audience—researchers and enthusiasts alike—this Special Issue addresses you in the spirit of exploration and understanding. As editors, we extend our deepest gratitude to those who have watered the seeds of this endeavor with their support, wisdom, and critique. Together, we stand at the cusp of a deeper comprehension of the green mysteries that cloak our planet, poised to reveal the hidden rhymes and reasons of plant form and function.

Jie Gao, Weiwei Huang, Johan Gielis, and Peijian Shi
Editors

plants

MDPI

Editorial

Plant Morphology and Function, Geometric Morphometrics, and Modelling: Decoding the Mathematical Secrets of Plants

Jie Gao [1,*], Weiwei Huang [2], Johan Gielis [3] and Peijian Shi [2]

1 College of Life Sciences, Xinjiang Normal University, Urumqi 830054, China
2 Bamboo Research Institute & College of Ecology and Environment, Nanjing Forestry University, Nanjing 210037, China; wh@njfu.edu.cn (W.H.); pjshi@njfu.edu.cn (P.S.)
3 Department of Biosciences Engineering, University of Antwerp, B-2020 Antwerp, Belgium; johan.gielis@uantwerpen.be
* Correspondence: jiegao@pku.edu.cn

Citation: Gao, J.; Huang, W.; Gielis, J.; Shi, P. Plant Morphology and Function, Geometric Morphometrics, and Modelling: Decoding the Mathematical Secrets of Plants. *Plants* **2023**, *12*, 3724. https://doi.org/10.3390/plants12213724

Received: 19 October 2023
Accepted: 23 October 2023
Published: 30 October 2023

Functional plant traits include a plant's phenotypic morphology, nutrient element characteristics, and physiological and biochemical features, reflecting the survival strategies of plants in response to environmental changes. In this special issue, we aim to uncover the environmental adaptation mechanisms of plant functionality by studying the morphology and functions of plant organs including leaves, fruits, and seeds, providing a theoretical basis to understand the impact of global changes on plant growth and development.

Different species of *Silene* seeds with distinct morphologies exhibit varying harmonic numbers when analyzed using the elliptical Fourier transform (EFT) model [1]. Smoother seeds have fewer harmonic numbers, while wrinkled, spiky, and papillate seeds require more harmonic numbers. The Gielis equation and the modified Brière equation have shown remarkable validities in describing plant leaf morphology [2–4]. The morphological characteristics of bamboo leaves are closely related to canopy management practices [5], and different canopy management practices result in significant differences in leaf morphology and the trade-off relationships among various functional traits.

Environmental factors play a crucial role in shaping plant functional traits. Climate factors, such as temperature and precipitation, significantly influence the temporal variation in nutrient elements of leaf litter in the Ailao Mountains of China, effectively improving nutrient utilization efficiency and shortening turnover cycles [6]. Leaf nutrient content in forests is mainly influenced by soil nutrients and climate factors [7,8]. Different ecosystems also exhibit significant differences in soil nutrient content, with forest soil having higher total nitrogen content than grassland soil, while the carbon-to-nitrogen ratio in forest soil is lower than in grassland soil [9]. Controlled factors for functional leaf traits vary with geographic locations. Leaf area, carbon-to-nitrogen ratio, carbon-to-phosphorus ratio, nitrogen-to-phosphorus ratio, phosphorus content, and nitrogen isotope content in Chinese forests show significant correlations with latitude and longitude. Leaf characteristics in the southern regions are mainly influenced by climatic factors, while those in the northern regions are primarily affected by soil factors [10]. Additionally, varying degrees of light exposure also play a significant role in plant functional traits [11]. Apart from environmental factors, plant hormones also have a significant impact on plant functional traits. Exogenous sucrose and gibberellin significantly increase the internode length and total internode number of bamboo, significantly contributing to increased plant height. Gibberellin more significantly affects internode length, while sucrose increases the total internode number [12]. In addition, functional leaf traits can differ significantly between hybrid plants and their parents [13]. Functional plant traits play an essential role in exploring vegetation productivity and mitigating air pollution. Urban green spaces composed of various functional trait plants help to reduce air pollution levels [14] and serve as an important factor affecting urban ecosystem productivity [15,16].

Author Contributions: J.G. (Jie Gao): conceptualization, methodology, and investigation. J.G. (Jie Gao), W.H., J.G. (Johan Gielis) and P.S.: formal analysis. J.G. (Jie Gao): writing—original draft. All authors have read and agreed to the published version of the manuscript.

Funding: This work was supported by the Natural Science Foundation of Xinjiang Uygur Autonomous Region (No. 2022D01A213), Fundamental Research Funds for Universities in Xinjiang (No. XJEDU2023P071), Innovation and Entrepreneurship Training Program for College Students in 2023 (No. S202310762004), Xinjiang Normal University Landmark Achievements Cultivation Project (No. XJNUBS2301), Xinjiang Graduate Innovation and Entrepreneurship Project and Tianchi Talent Program.

Conflicts of Interest: The authors declare no conflict of interest.

References

1. Cervantes, E.; Rodríguez-Lorenzo, J.L.; Martín-Gómez, J.J.; Tocino, Á. Curvature Analysis of Seed Silhouettes in *Silene* L. *Plants* **2023**, *12*, 2439. [CrossRef] [PubMed]
2. Yao, W.; Niinemets, Ü.; Yao, W.; Gielis, J.; Schrader, J.; Yu, K.; Shi, P. Comparison of Two Simplified Versions of the Gielis Equation for Describing the Shape of Bamboo Leaves. *Plants* **2022**, *11*, 3058. [CrossRef] [PubMed]
3. Wang, L.; Miao, Q.; Niinemets, Ü.; Gielis, J.; Shi, P. Quantifying the Variation in the Geometries of the Outer Rims of Corolla Tubes of *Vinca major* L. *Plants* **2022**, *11*, 1987. [CrossRef] [PubMed]
4. Jin, J.; Quinn, B.K.; Shi, P. The Modified Brière Equation and Its Applications. *Plants* **2022**, *11*, 1769. [CrossRef] [PubMed]
5. Fang, D.; Xian, J.; Chen, G.; Zhang, Y.; Qin, H.; Fu, X.; Lin, L.; Ai, Y.; Yang, Z.; Xu, X.; et al. Rapid Adaptation of Chimonobambusa opienensis Leaves to Crown–Thinning in Giant Panda Ecological Corridor, Niba Mountain. *Plants* **2023**, *12*, 2109. [CrossRef] [PubMed]
6. Dai, S.; Wei, T.; Tang, J.; Xu, Z.; Gong, H. Temporal Changes in Litterfall and Nutrient Cycling from 2005–2015 in an Evergreen Broad-Leaved Forest in the Ailao Mountains, China. *Plants* **2023**, *12*, 1277. [CrossRef] [PubMed]
7. Gao, J.; Wang, J.; Li, Y. Effects of Soil Nutrients on Plant Nutrient Traits in Natural Pinus tabuliformis Forests. *Plants* **2023**, *12*, 735. [CrossRef] [PubMed]
8. Wang, X.; Wang, J.; Zhang, L.; Lv, C.; Liu, L.; Zhao, H.; Gao, J. Climatic Factors Determine the Distribution Patterns of Leaf Nutrient Traits at Large Scales. *Plants* **2022**, *11*, 2171. [CrossRef] [PubMed]
9. Zhao, J.; Xiang, K.; Wu, Z.; Du, Z. Varying Responses of Vegetation Greenness to the Diurnal Warming across the Global. *Plants* **2022**, *11*, 2648. [CrossRef] [PubMed]
10. Zhou, X.; Xin, J.; Huang, X.; Li, H.; Li, F.; Song, W. Linking Leaf Functional Traits with Soil and Climate Factors in Forest Ecosystems in China. *Plants* **2022**, *11*, 3545. [CrossRef] [PubMed]
11. Huang, W.; Ding, Y.; Wang, S.; Song, C.; Wang, F. Growth and Development Responses of the Rhizome-Root System in *Pleioblastus pygmaeus* to Light Intensity. *Plants* **2022**, *11*, 2204. [CrossRef] [PubMed]
12. Wu, C.; Bai, Y.; Cao, Z.; Xu, J.; Xie, Y.; Zheng, H.; Jiang, J.; Mu, C.; Cheng, W.; Fang, H.; et al. Plasticity in the Morphology of Growing Bamboo: A Bayesian Analysis of Exogenous Treatment Effects on Plant Height, Internode Length, and Internode Numbers. *Plants* **2023**, *12*, 1713. [CrossRef] [PubMed]
13. Zheng, X.; Niklas, K.J.; Ratkowsky, D.A.; Jiao, Y.; Ding, H.; Shi, P. Comparison of Leaf Shape between a Photinia Hybrid and One of Its Parents. *Plants* **2022**, *11*, 2370. [CrossRef] [PubMed]
14. Wang, C.; Guo, M.; Jin, J.; Yang, Y.; Ren, Y.; Wang, Y.; Cao, J. Does the Spatial Pattern of *Plants* and Green Space Affect Air Pollutant Concentrations? Evidence from 37 Garden Cities in China. *Plants* **2022**, *11*, 2847. [CrossRef] [PubMed]
15. Wang, J.; Li, Y.; Gao, J. Time Effects of Global Change on Forest Productivity in China from 2001 to 2017. *Plants* **2023**, *12*, 1404. [CrossRef] [PubMed]
16. Du, Z.; Liu, X.; Wu, Z.; Zhang, H.; Zhao, J. Responses of Forest Net Primary Productivity to Climatic Factors in China during 1982–2015. *Plants* **2022**, *11*, 2932. [CrossRef] [PubMed]

plants

MDPI

Article

Effects of Soil Nutrients on Plant Nutrient Traits in Natural *Pinus tabuliformis* Forests

Jie Gao [1,2,*], Jiangfeng Wang [1] and Yanhong Li [1,*]

1 College of Life Sciences, Xinjiang Normal University, Urumqi 830054, China
2 Key Laboratory of Earth Surface Processes of Ministry of Education, College of Urban and Environmental Sciences, Peking University, Beijing 100871, China
* Correspondence: jiegao72@gmail.com or jiegao@pku.edu.cn (J.G.); liyh1330824@163.com (Y.L.)

Abstract: In light of global warming, the interaction between plant nutrient traits and soil nutrients is still unclear. Plant nutrient traits (e.g., N and P) and their stoichiometric relationships (N/P ratio) are essential for plant growth and reproduction. However, the specific role of soil nutrients in driving variation in plant nutrient traits remains poorly understood. Fifty natural *Pinus tabuliformis* forests were used as the research object to clarify the interaction between plant nutrient traits and soil nutrients. We show that: (1) The N_{mass}, P_{mass} and N/P ratios of leaves were significantly higher than those of roots. The N/P ratio of both leaves and roots was less than 14. (2) Leaf nutrient traits showed diverse relationship patterns with root nutrient traits throughout the growing period. Significant changes were found in root nutrient PC2 (the second principal component of root nutrient traits) and leaf nutrient PC1 (the first principal component of leaf traits), and non-significant changes were found in other relationships between leaf and root traits ($p > 0.05$). Root nutrient traits explained 36.4% of the variance in leaf nutrient traits. (3) With the increase in soil nutrient PC2 (related to N), leaf PC2 (related to N) showed a significant trend of first decreasing and then increasing ($p < 0.05$). Only the soil N_{mass} was significantly correlated with the leaf N_{mass} ($p < 0.05$), which demonstrated that the growth and survival of *Pinus tabuliformis* forests were mainly affected by N-limitation.

Keywords: plant–soil interaction; leaf nutrient; N-limitation; *Pinus tabuliformis*

Citation: Gao, J.; Wang, J.; Li, Y. Effects of Soil Nutrients on Plant Nutrient Traits in Natural *Pinus tabuliformis* Forests. *Plants* **2023**, *12*, 735. https://doi.org/10.3390/plants12040735

Academic Editor: Yasutomo Hoshika

Received: 4 January 2023
Revised: 2 February 2023
Accepted: 4 February 2023
Published: 7 February 2023

1. Introduction

Nutrient traits refer to traits related to nutrient characteristics (e.g., N_{mass} and P_{mass}), reflecting the survival strategies of plants in response to global warming, and are widely used in ecology [1–3]. Nitrogen (N) and phosphorus (P) are basic components of plant genetic material and nutrients, and their stoichiometric relationships significantly influence the process of plant growth and reproduction [4,5]. As a major limiting element, N is a fundamental component of enzymes [4]. Meanwhile, phosphorus drives the generation and maintenance of proteins and is also limiting in most environments [6]. The absence of nitrogen and phosphorus in leaves will affect the formation of chlorophyll, further reduce the productivity of forest communities and regulate carbon cycling [3]. The ratio of nitrogen to phosphorus (N/P) in plants reflects environmental factors, especially the nutrient supply of soil-to-plant growth [7,8]. It can clarify which elements restrict the plant's productivity. In other words, in a habitat where P is scarce and N is relatively abundant, plant N/P is relatively high, while in the habitat where N is scarce and P is relatively rich, plant N/P is relatively low and plant P content is significantly increased [3]. Soil nutrient limitation not only affects the nutrient structure of species but also affects the composition of community species and the direction of community succession [9].

Some studies have found that climatic factors may have a certain impact on plant nutrient traits [10]. However, as the direct living environment of plants, soil provides the necessary water and nutrients for plants to survive [11]. Therefore, soil nutrient factors may play a more critical role in shaping plant nutrient trait differences. However, the specific interaction between soil nutrients and plant nutrients remains unclear.

Complex feedback regulation mechanisms exist between plant nutrients and soil nutrients [12]. For example, litter in nutrient-rich soils will introduce higher nutrients, which release large amounts of nutrients after decay, thus maintaining a higher soil fertility level. In nutrient-poor soils, plants naturally produce less litter, and their decay progresses very slowly, which further leads to soil barrenness [4,11]. N and P are important components of soil nutrients, and it is unclear whether their performance is consistent with that of plant nutrient feedback and which nutrient trait plays a more important role in plant–soil nutrient feedback.

Plants with different life forms occupy different plant–soil nutrient feedback [13]. The nutrient profiles of leaves vary significantly between different life forms due to differences in their survival strategies [2]. This plant response feedback on soil nutrient supply reflects a nutrient trade-off in plant growth and development and reflects the survival strategies adopted by plants in coping with survival pressure [14]. Based on global data, Wright et al. [15] found that the leaf nitrogen and phosphorus contents of shrubs were significantly higher than those of trees. Plants with long-living leaves have low N and P contents [2] and, thus, tree species can adapt to a living environment with low nutritional status.

In recent years, an increasing number of studies have attempted to explore the correlations between traits of different organs from the perspective of plant functional traits. Relevant studies not only help to understand the mechanisms of interaction between plant traits [4] and the utilization and allocation of resources during plant growth [16] but also have important significance in further predicting the response of plants to environmental changes. Previous studies on plant nutrient traits (e.g., N, P) often focused on aboveground organs, and few studies were conducted on roots [17]. Exploring the difference in nutrient traits between aboveground and underground organs can help us to better understand the nutrient allocation strategies of plants as well as the plant–soil nutrient feedback [18].

Songshan Nature Reserve preserves the only natural *Pinus tabuliformis* forest in North China. *Pinus tabuliformis* plays an extremely important role in resisting wind and sand, conserving water sources and purifying air, etc. The shrub species are *Syinga reticulata* var. *Mandshurica, Corylus mandshurica* and *Euonymus verrucosus*. *Pinus tabuliformis* is the constructive species in Songshan Nature Reserve. We aim to explore the relative effects of soil nutrients on plant nutrient traits based on the plant nutrient trait data (e.g., N_{mass}, P_{mass}) of roots and leaves and soil nutrient data collected from 50 natural *Pinus tabuliformis* forests. We proposed the following hypotheses: (1) There exists a significant difference between roots and leaves in nutrient traits (e.g., N_{mass}, P_{mass}). (2) Soil nutrient factors are better at explaining the variation in root nutrient traits than leaf nutrient traits.

2. Results

The N_{mass} (Figure 1A), P_{mass} (Figure 1B) and N/P ratios (Figure 1C) of leaves were significantly higher than those of roots (Table 1). We also found that the phosphorus content of trees was significantly higher than that of shrubs. However, there was no significant difference ($p > 0.05$) in nitrogen content between trees and shrubs (Figure S1).

The first two principal components (PC1 = 56.97%; PC2 = 41.54%) can explain 98.51% of the soil nutrient variation. The N/P ratio and PC1 showed a positive correlation. However, the N_{mass}, P_{mass} and PC1 showed a negative correlation (Figure 2A). The first principal component mainly represents the components related to nutrient restriction. The second principal component mainly represents the components related to nitrogen.

The first two principal components (PC1 = 53.61%; PC2 = 44.22%) can explain 97.83% of the plant nutrient variation. Both the nutrient function traits of roots and the nutrient traits of leaves were positively correlated with the second principal component (Figure 2B). The first principal component mainly represents the components related to nutrient restriction. The second principal component mainly represents the components related to nitrogen.

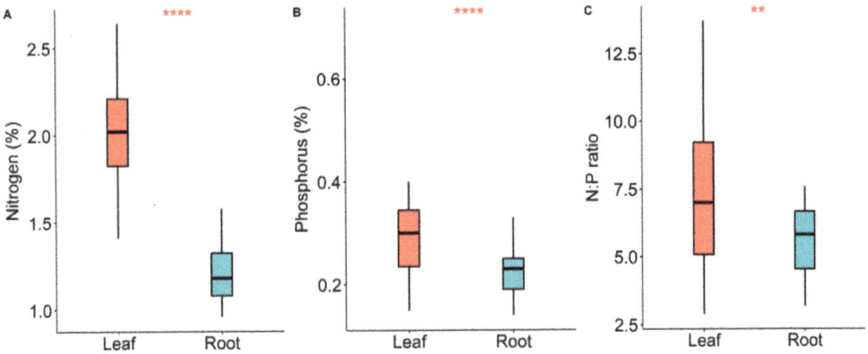

Figure 1. A comparison of the differences in the N_{mass} (**A**), P_{mass} (**B**), and N/P (**C**) of different organs (leaf and root). * represents $p < 0.05$, ** represents $p < 0.01$, **** represents $p < 0.0001$.

Table 1. Variations in the N_{mass}, P_{mass} and N/P ratio of leaves, roots and soil.

	Nutrient Characteristics	Mean Value (g/g)	Coefficient of Variation (%)	Max–Min (g/g)
Leaf	N_{mass}	2.03	3.81	1.23
	P_{mass}	0.32	6.02	0.56
	N:P	7.38	10.59	5.82
Root	N_{mass}	1.19	2.15	0.62
	P_{mass}	0.22	1.23	0.20
	N:P	5.6	29.56	4.42
Soil	N_{mass}	3.22	41.72	5.38
	P_{mass}	0.43	7.74	0.62
	N:P	8.46	20.05	6.4

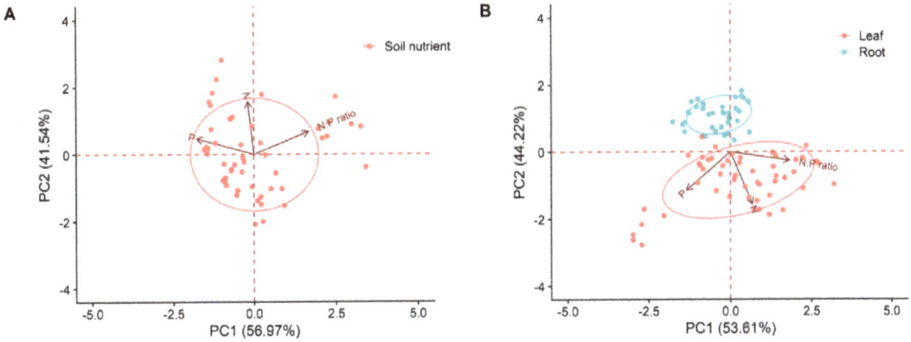

Figure 2. Principal component analysis (PCA) of nutrient characteristics of (**A**) soil and (**B**) leaves and roots. Nutrient characteristics included the N_{mass}, P_{mass} and N/P ratio.

From the results of the generalized additive models (GAMs), it is clear that the investigated leaf traits showed diverse relationship patterns with root nutrient traits throughout the growing period (Table 2; Figure 3). Significant dynamic changes were found in the root nutrient PC2 and leaf PC1, and non-significant dynamic changes were found in other relationships between the leaf and root traits. Root nutrient traits explained 36.4% of the variance in leaf nutrient traits (Table 2; Figure 4).

The soil nutrient factors had non-significant effects on the root nutrient traits (Table 2; Figure 5). The soil nutrient factors had significant nonlinear effects on the leaf nutrient traits ($p < 0.05$). With the increase in soil nutrient PC2 (related to N), leaf PC2 (related to N) showed a significant trend of first decreasing and then increasing ($p < 0.05$). Generally speaking, soil nutrients explained 25% of the variance in the leaf nutrient traits.

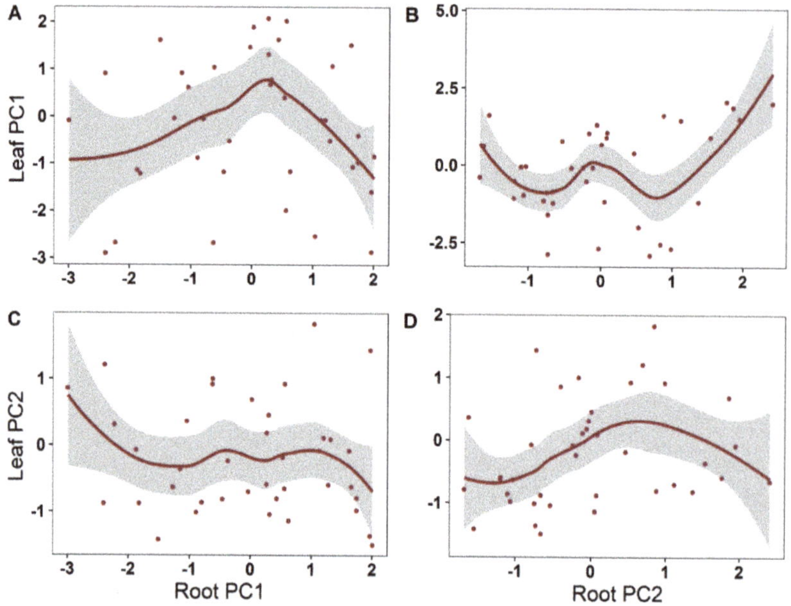

Figure 3. The plots of the GAMs smooth function indicating the effects of root nutrient traits on leaf nutrient traits. (**A**): root PC1 vs. leaf PC1; (**B**): root PC2 vs. leaf PC1; (**C**): root PC1 vs. leaf PC2; (**D**): root PC2 vs. leaf PC2.

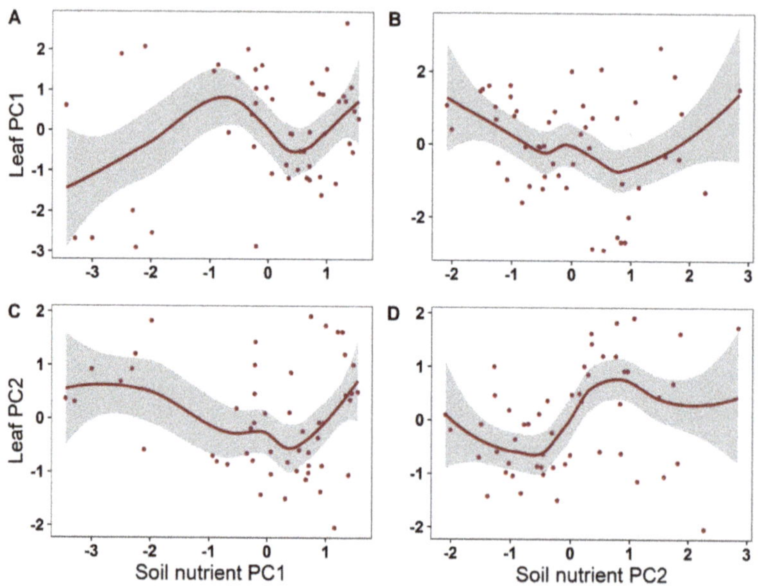

Figure 4. The plots of the GAMs smooth function indicating the effects of soil nutrient factors on leaf nutrient traits. (**A**): soil nutrient PC1 vs. leaf PC1; (**B**): soil nutrient PC2 vs. leaf PC1; (**C**): soil nutrient PC1 vs. leaf PC2; (**D**): soil nutrient PC2 vs. leaf PC2.

Table 2. Results of the generalized additive models (GAMs) explaining the influence of soil nutrient and root nutrient traits on leaf and root nutrient traits. * $p < 0.05$.

Parameters	Independent Variable	Degrees of Freedom	F Value	Pr (>∣t∣)	R^2_{adj}
Leaf PC1	Soil PC1	3.328	1.496	0.227	0.163
	Soil PC2	1.970	1.452	0.217	
Leaf PC2	Soil PC1	1.432	0.498	0.6806	0.25
	Soil PC2	5.484	2.282	0.0475 *	
Root PC1	Soil PC1	1.797	1.729	0.190	0.108
	Soil PC2	1.641	0.790	0.451	
Root PC2	Soil PC1	3.086	2.757	0.0511	0.181
	Soil PC2	1.000	0.231	0.6340	
Leaf PC1	Root PC1	2.843	1.939	0.1352	0.364
	Root PC2	4.418	2.802	0.0337 *	
Leaf PC2	Root PC1	1.000	0.392	0.536	0.0861
	Root PC2	1.927	1.740	0.153	

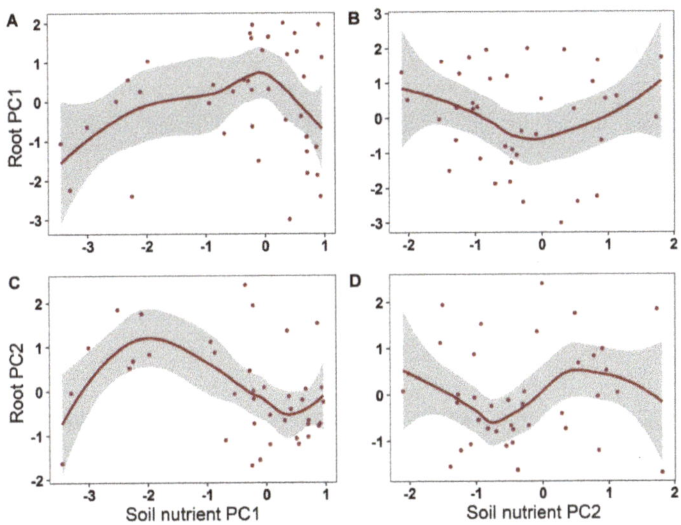

Figure 5. The plots of the GAMs smooth function indicating the effects of soil nutrient factors on root nutrient traits. (**A**): soil nutrient PC1 vs. root PC1; (**B**): soil nutrient PC2 vs. root PC1; (**C**): soil nutrient PC1 vs. root PC2; (**D**): soil nutrient PC2 vs. root PC2.

3. Discussion

We found that the N_{mass} and P_{mass} of leaves were significantly higher than those of roots. Leaves are the main organ for photosynthesis, and N_{mass} and P_{mass} are closely related to protein synthesis [2]. Plant roots cannot participate in photosynthesis due to lack of chlorophyll. Therefore the N_{mass} and P_{mass} of plant roots are significantly lower than those of leaves [19–25].

We also found that the P_{mass} of tree leaves was significantly higher than that of shrub leaves. Plants of different life forms have unique niches and different resource utilization strategies for light, temperature and water under environmental pressure [2,25–32]. However, there was no significant difference in N_{mass} between tree and shrub leaves. The "community construction theory" based on nutrient traits explains that, in a local community, competition may lead to divergence between traits, but habitat screening may lead to the convergence of traits [33]. Often, the habitat selection effect causes different species to form more consistent characteristics so as to adapt to the same environment [3].

There exists a tradeoff between the nutrient traits of different plant organs [15]. N_{mass} and P_{mass}, in plant leaves, are mainly used for photosynthesis, while those of roots are mainly used for underground ecological processes so as to adapt to adverse environments. Therefore, when plants absorb nutrients from the soil, they will balance the nutrients according to their environment [34]. When plants are in an environment with sufficient resources (sufficient light, water, and heat), they will use more nutrients for photosynthesis to maximize resource utilization and facilitate plant growth and reproduction [26]. When plants are affected by osmotic stress, they will use more nutrients in underground processes (e.g., rooting) to avoid the threat [9].

In the past century, aboveground ecology has attracted extensive attention. However, the ecological links between aboveground and underground components remain unclear [2,3]. This knowledge gap hampers our ability to understand and predict the comprehensive responses of an ecosystem to environmental stresses [2,3]. Increasing evidence emphasizes that the importance of strong interactions between aboveground and underground components in regulating ecosystem multifunctionality and responses to global change [19–26].

Plants and soil participate in the global material cycle together, existing in a close relationship [27]. Plants absorb nitrogen and phosphorus from the soil through their roots and return them to the soil in the form of litter [20]. Therefore, there is a feedback relationship between soil and plant nutrients [3]. The aboveground element characteristics of plants are usually related to the soil nutrient content. As the main substrate for plant growth, soil contains organic matter, nitrate nitrogen and ammonium nitrogen, which are decomposed to continuously provide essential nutrients for the normal physiological activities of plants. This enables the soil and plant to achieve and maintain a balanced element ratio through the dynamic exchange of nutrient supply and demand [28–32]. We found that only soil N_{mass} was significantly correlated with leaf N_{mass}, offering evidence that the growth and development of *Pinus tabuliformis* forests were mainly limited by the supply of soil N_{mass}, and P_{mass} was not the key element limiting factor.

4. Materials and Methods

4.1. Study Area

The longitude and latitude range of Songshan Nature Reserve is 115°43′44″ E–115°50′22″ E, 40°29′9″ N–40°33′35″ N. The annual average temperature is 8.5 °C, the highest temperature in the hottest month is 39 °C, and the lowest temperature in the coldest month is −27.3 °C. The annual average duration of sunshine is 2836.3 h, the annual average frost-free period is approximately 150 days, the annual average rainfall is 493 mm, and the annual average evaporation is 1770 mm. The reserve has the second highest peak in Beijing, with a maximum altitude of 2198.39 m. Most mountains measure between 600 and 1600 m. There are three types of soil connected with the elevation changes: brown forest soil, mountain brown soil and mountain meadow soil. Mountain meadow soil is mainly distributed under shrub vegetation above an altitude of 1800 m. The reserve is rich in animal and plant resources, including 713 species of higher wild vascular plants, more than 300 species of medicinal plants and 158 species (subspecies) of birds. Fifty representative plots (30 m × 30 m) were established in natural *Pinus tabuliformis* forests. The average elevation is 800 m. The maximum altitude is 875 m, and the minimum altitude is 770 m. The soil type is brown forest soil.

4.2. Nutrient Trait Data

More than 10 mature and well-developed *Pinus tabuliformis* trees were selected from each plot to collect fresh (one-ye ar-old) needles and twigs. The collected samples were mixed evenly and placed into paper file bags. We selected roots with diameters greater than 2 mm for our research. The contents of N (%) and P (%) in the leaves were determined after sterilization at 105 °C, drying at 60 °C and mechanical grinding. The average value of each sample was taken to calculate the average contents of N (%) and P (%). Soil samples of the surface layer (0–20 cm) were collected under the selected tree. The soil samples were mixed fully and evenly. After air-drying in the laboratory, impurities were removed and the contents of N (g/kg) and P (g/kg) were determined after grinding and screening with 0.25 mm mesh. Roots of *Pinus tabuliformis* collection were carried out by a root-tracking method. The main roots of the sampled *Pinus tabuliformis* were found first, and the fine roots on the main roots were sequentially exposed by gradually removing sediment downward in the direction of the main roots. Roots containing at least five grades were cut off with pruning scissors, and these fine root samples were placed in self-sealing bags for preservation through temporary freezing. After taking the root samples back to the laboratory, the soil attached to the root samples was washed with water. The samples were graded according to the root order. The measured fine roots were placed in an oven at 60 °C for 72 h to maintain their weight. Root samples were kept for testing after crushing and screening with a 2 mm sieve. The total N content was determined by Kjeldahl determination, the leaf P content was determined by the molybdenum antimony anti-colorimetric method, and the soil total phosphorus was determined by the alkali fusion-Mo-Sb anti-spectrophotometric method [19–21,35]. The calculation method for the N_{mass} and P_{mass} of the leaves and roots of shrub species is the same as that for *Pinus tabuliformis*.

4.3. Data Analysis

Principal component analysis (PCA) was used to reduce the dimensions of the plant nutrient traits, root nutrient and soil nutrient factors and was conducted within the R environment using the "vegan" package.

Generalized additive models (GAMs) were used to evaluate the effects of soil nutrient factors on leaf and root nutrient traits. This approach utilizes both parametric and non-parametric components to reduce the model risks inherent to linear models [22]. The model can be summarized as:

$$g(E(Yi)) = \beta_0 + S_1(x_i) + S_2(x_i) + e_i \quad (1)$$

where g is a link function, $E(Y_i)$ is the estimate for the responsible variable Y_i, S_1 is the smooth function of x_i for different light treatments, and S_2 is the smooth function of x_i throughout the investigation time. x_i (i = 1, 2, 3, ..., 12) are the explanatory variables, and they are the number of new rhizomes, new rhizome length, new rhizome diameter, etc. β_0 is the constant term and e_i is the error term. All calculations were conducted within the R environment using the "mgcv" package.

Supplementary Materials: The following supporting information can be downloaded at: https://www.mdpi.com/article/10.3390/plants12040735/s1, Figure S1: A comparison of the differences in the Nmass (A), Pmass (B), and N/P (C) at different life forms (Tree and Shrub), **** represents $p < 0.0001$.

Author Contributions: Conceptualization, J.G.; experiment implementation, J.G.; validation, J.G.; formal analysis, J.G.; writing—original draft preparation, J.G.; writing—review and editing, J.G., J.W. and Y.L.; visualization, J.G.; project administration, J.G. and Y.L.; funding acquisition, J.G. and Y.L. All authors have read and agreed to the published version of the manuscript.

Funding: The study was supported by the Xinjiang Normal University Landmark Achievements Cultivation Project, China (grant number: no number), the Scientific Research Program of Colleges and Universities in Xinjiang (no. XJEDU2021I023) and the General Program in Xinjiang (no. 2022D01A213).

Institutional Review Board Statement: Not applicable.

Informed Consent Statement: Not applicable.

Data Availability Statement: Not applicable.

Conflicts of Interest: The authors declare no conflict of interest. The funders had no role in the design of the study; in the collection, analyses, or interpretation of data; in the writing of the manuscript; or in the decision to publish the results.

References

1. Ma, J.; Niklas, K.J.; Liu, L.; Fang, Z.D.; Li, Y.R.; Shi, P.J. Tree size influences leaf shape but does not affect the proportional relationship between leaf area and the product of length and width. *Front. Plant Sci.* **2022**, *13*, 850203.
2. Wang, X.; Wang, R.; Gao, J. Precipitation and soil nutrients determine the spatial variability of grassland productivity at large scales in China. *Front. Plant Sci.* **2022**, *13*, 996313. [CrossRef] [PubMed]
3. Wang, X.; Wang, J.; Zhang, L.; Lv, C.; Liu, L.; Zhao, H.; Gao, J. Climatic Factors Determine the Distribution Patterns of Leaf Nutrient Traits at Large Scales. *Plants* **2022**, *11*, 2171. [CrossRef]
4. Gong, H.; Li, Y.; Yu, T.; Zhang, S.; Gao, J.; Zhang, S.; Sun, D. Soil and climate effects on leaf nitrogen and phosphorus stoichiometry along elevational gradients. *Glob. Ecol. Conserv.* **2020**, *23*, e01138. [CrossRef]
5. Huang, C.; Xu, Y.; Zang, R. Variation patterns of functional trait moments along geographical gradients and their environmental determinants in the subtropical evergreen broadleaved forests. *Front. Plant Sci.* **2021**, *12*, 1414.
6. Han, W.X.; Fang, J.Y.; Guo, D.L.; Zhang, Y. Leaf nitrogen and phosphorus stoichiometry across 753 terrestrial plant species in China. *New Phytol.* **2005**, *168*, 377–385. [CrossRef]
7. Aerts, R.; Chapin, F.S. The mineral nutrition of wild plants revisited: A re-evaluation of processes and patterns. *Adv. Ecol. Res.* **2000**, *30*, 1–67.
8. Güsewell, S. N:P ratios in terrestrial plants: Variation and functional significance. *New Phytol.* **2004**, *164*, 243–266. [CrossRef]
9. Chen, X.; Chen, H.Y.H.; Searle, E.B.; Chen, C.; Reich, P.B. Negative to positive shifts in diversity effects on soil nitrogen over time. *Nat. Sustain.* **2020**, *4*, 225–232. [CrossRef]
10. Taylor, P.; Asner, G.; Dahlin, K.; Anderson, C.; Knapp, D.; Martin, R.; Mascaro, J.; Chazdon, R.; Cole, R.; Wanek, W. LandscapeScale Controls on Aboveground Forest Carbon Stocks on the Osa Peninsula, Costa Rica. *PLoS ONE* **2015**, *10*, e0126748.
11. Ordoñez, J.C.; Bodegom, P.M.V.; Witte, J.P.M.; Wright, I.J.; Reich, P.B.; Aerts, R. A global study of relationships between leaf traits, climate and soil measures of nutrient fertility. *Glob. Ecol. Biogeogr.* **2009**, *18*, 137–149. [CrossRef]
12. Berendse, F. Litter decomposability—A neglected component of plant fitness. *J. Ecol.* **1994**, *82*, 187–190. [CrossRef]
13. Booth, M.S.; Stark, J.M.; Rastetter, E. Controls on nitrogen cycling in terrestrial ecosystems: A synthetic analysis of literature data. *Ecol. Monogr.* **2005**, *75*, 139–157. [CrossRef]
14. Wright, I.J.; Reich, P.B.; Cornelissen, J.H.C.; Falster, D.S.; Garnier, E.; Hikosaka, K.; Lamont, B.B.; Lee, W.; Oleksyn, J.; Osada, N.; et al. Assessing the generality of global leaf trait relationships. *New Phytol.* **2005**, *166*, 485–496. [CrossRef]
15. Wright, I.J.; Reich, P.B.; Westoby, M.; Ackerly, D.D.; Baruch, Z.; Bongers, F.; Cavender-Bares, J.; Chapin, T.; Cornelissen, J.H.C.; Diemer, M.; et al. The worldwide leaf economics spectrum. *Nature* **2004**, *428*, 821–827. [CrossRef]
16. Koerselman, W.; Meuleman, A.F.M. The vegetation N:P ratio: A new tool to detect the nature of nutrient limitation. *J. Appl. Ecol.* **1996**, *33*, 1441–1450. [CrossRef]
17. Valverde-Barrantes, O.J.; Horning, A.L.; Smemo, K.A.; Blackwood, C.B. Phylogenetically structured traits in root systems influence arbuscular mycorrhizal colonization in woody angiosperms. *Plant Soil* **2016**, *404*, 1–12. [CrossRef]
18. Hobbie, S.E. Effects of plant species on nutrient cycling. *Trends Ecol. Evol.* **1992**, *7*, 336–339.
19. Gong, H.D.; Yao, F.G.; Gao, J. Succession of a broad-leaved Korean pine mixed forest: Functional plant trait composition. *Glob. Ecol. Conserv.* **2020**, *22*, e00950. [CrossRef]
20. Gong, H.D.; CUI, Q.J.; Gao, J. Latitudinal, soil and climate effects on key leaf traits in northeastern China. *Glob. Ecol. Conserv.* **2020**, *22*, e00904. [CrossRef]
21. Gao, J.; Song, Z.P.; Liu, Y.H. Response mechanisms of leaf nutrients of endangered plant (Acer catalpifolium) to environmental factors varied at different growth stages. *Glob. Ecol. Conserv.* **2019**, *17*, e00521. [CrossRef]
22. Huang, W.; Ding, Y.; Wang, S.; Song, C.; Wang, F. Growth and development responses of the rhizome-root system in Pleioblastus pygmaeus to light intensity. *Plants* **2022**, *11*, 2204. [CrossRef] [PubMed]
23. Elser, J.J.; Dobberfuhl, D.R.; MacKay, N.A.; Schampel, J.H. Organism size, life history, and N:P stoichiometry. *BioScience* **1996**, *46*, 674–684. [CrossRef]
24. He, J.S.; Han, X.G. Ecological stoichiometry: Searching for unifying principles from individuals to ecosystems. *Chin. J. Plant Ecol.* **2010**, *34*, 2–6.
25. Guo, D.; Xia, M.; Wei, X.; Chang, W.; Liu, Y.; Wang, Z. Anatomical traits associated with absorption and mycorrhizal colonization are linked to root branch order in twenty-three Chinese temperate tree species. *New Phytol.* **2008**, *3*, 180. [CrossRef] [PubMed]
26. Wang, J.; Wang, X.; Ji, Y.; Gao, J. Climate factors determine the utilization strategy of forest plant resources at large scales. *Front. Plant Sci.* **2022**, *13*, 990441. [CrossRef] [PubMed]
27. Aerts, R.; De Caluwe, H.; Beltman, B. Is the relation between nutrient supply and biodiversity CO2 limited by the type of nutrient limitation. *Oikos* **2003**, *101*, 489–498. [CrossRef]
28. Elser, J.J.; Fagan, W.F.; Kerkhoff, A.J.; Swenson, N.G.; Enquist, B.J. Biological stoichiometry of plant production: Metabolism, scaling and ecological response to global change. *New Phytol.* **2010**, *186*, 593–608. [CrossRef] [PubMed]
29. Sweeney, C.J.; Vries, F.T.; Dongen, B.E. Root traits explain rhizosphere fungal community composition among temperate grassland plant species. *New Phytol.* **2021**, *229*, 1492–1507. [CrossRef]
30. Potarzycki, J. Effect of magnesium or zinc supplementation at the background of nitrogen rate on nitrogen management by maize canopy cultivated in monoculture. *Plant Soil Environ.* **2011**, *57*, 19–25. [CrossRef]

31. Shen, X.; Yang, F.; Xiao, C.W.; Zhou, Y. Increased contribution of root exudates to soil carbon input during grassland degradation. *Soil Biol. Biochem.* **2020**, *146*, 107817. [CrossRef]
32. Liu, C.; Sack, L.; Li, Y.; He, N. Contrasting adaptation and optimization of stomatal traits across communities at continental-scale. *J. Exp. Bot.* **2022**, *73*, 6405–6416. [CrossRef] [PubMed]
33. Fukano, Y.; Guo, W.; Uchida, K.; Tachiki, Y. Contemporary adaptive divergence of plant competitive traits in urban and rural populations and its implication for weed management. *J. Ecol.* **2020**, *108*, 2521–2530. [CrossRef]
34. Reichardt, K.; Timm, L.C. How Plants Absorb Nutrients from the Soil. In *Soil, Plant and Atmosphere*; Springer: Cham, Switzerland, 2020.
35. Chen, Y.S.; Hou, M.T.; Dan, M.A.; Han, X.H.; Zhang, R.Y.; Zhang, X.Z. *Determination of Total Phosphorus in Soil by Alkali Fusion-Mo-Sb Anti Spectrophotometric Method*; China Standardization: Beijing, China, 2018.

plants

MDPI

Article

Curvature Analysis of Seed Silhouettes in *Silene* L.

Emilio Cervantes [1,*], José Luis Rodríguez-Lorenzo [2], José Javier Martín-Gómez [1] and Ángel Tocino [3]

1 Instituto de Recursos Naturales y Agrobiología (Consejo Superior de Investigaciones Científicas), Cordel de Merinas 40, 37008 Salamanca, Spain; jjavier.martin@irnasa.csic.es
2 Plant Developmental Genetics, Institute of Biophysics v.v.i, Academy of Sciences of the Czech Republic, Královopolská 135, 612 65 Brno, Czech Republic; rodriguez@ibp.cz
3 Departamento de Matemáticas, Facultad de Ciencias, Universidad de Salamanca, Plaza de la Merced 1-4, 37008 Salamanca, Spain; bacon@usal.es
* Correspondence: emilio.cervantes@irnasa.csic.es

Abstract: The application of seed morphology to descriptive systematics requires methods for shape analysis and quantification. The complexity of lateral and dorsal views of seeds of *Silene* species is investigated here by the application of the Elliptic Fourier Transform (EFT) to representative seeds of four morphological types: smooth, rugose, echinate and papillose. The silhouettes of seed images in the lateral and dorsal views are converted to trigonometric functions, whose graphical representations reproduce them with different levels of accuracy depending on the number of harmonics. A general definition of seed shape in *Silene* species is obtained by equations based on 40 points and 20 harmonics, while the detailed representation of individual tubercles in each seed image requires between 100 and 200 points and 60–80 harmonics depending on their number and complexity. Smooth-type seeds are accurately represented with a low number of harmonics, while rugose, echinate and papillose seeds require a higher number. Fourier equations provide information about tubercle number and distribution and allow the analysis of curvature. Further estimation of curvature values in individual tubercles reveals differences between seeds, with higher values of curvature in *S. latifolia*, representative of echinate seeds, and lower in *S. chlorifolia* with rugose seeds.

Keywords: Bézier curve; complexity; curvature; development; Elliptic Fourier Transform; models; morphology; seed; systematics

Citation: Cervantes, E.; Rodríguez-Lorenzo, J.L.; Martín-Gómez, J.J.; Tocino, Á. Curvature Analysis of Seed Silhouettes in *Silene* L. *Plants* **2023**, *12*, 2439. https://doi.org/10.3390/plants12132439

Academic Editors: Jie Gao, Weiwei Huang, Johan Gielis and Peijian Shi

Received: 17 May 2023
Revised: 13 June 2023
Accepted: 23 June 2023
Published: 25 June 2023

1. Introduction

The Caryophyllaceae Juss. comprises ca. 100 genera and 3000 species of herbs and small shrubs [1,2] of a cosmopolitan distribution and characterized by a peripheral position of the embryo in the developing seed [3], anatropous to campylotropous ovules [4], with an interesting diversity in seed shape [5–14].

Morphological analysis of seeds in the Caryophyllaceae focuses on two aspects: overall seed shape and detailed seed surface structure. Cardioid-derived models have been applied to the quantification of overall shape in lateral views of the seed [5–7], while ellipse-based models fit well to dorsal views of the seed in many species [8]. The application of models to seed shape quantification contributes to the identification of useful characters for taxonomy. Seeds of *Silene* L. subg. *Behenantha* conform better to a cardioid than those of *Silene* subg. *Silene* [5]. Seed images of species of sect. *Melandrium* fit better to models derived from a modified cardioid closed in the hilum region, whereas seeds of sect. *Silene* fit better to open models [9].

In relation to the seed surface, and based on their geometric properties, *Silene* seeds were divided into four groups: smooth, rugose, echinate and papillose [10,11]. Smooth seeds are defined by the absence of visible protuberances and this type was already described by other authors working with *Silene* [12–14] or related genera, such as *Arenaria* L. [15–17], *Minuartia* L. [18] and *Moehringia* L. [19]. In *Silene*, species with smooth seeds belong to S. subg. *Silene* sec. *Silene* such as *S. apetala* Willd., *S. borderei* Jord., *S. colorata* Poir., *S. damascena* Boiss. and Gaill., *S. diversifolia* Ott, *S. legionensis* Lag., *S. micropetala* Lag., *S. nicaeensis* All., *S. pomelii* Batt. subsp. *adusta* (Ball) Maire, *S. secundiflora* Ott, *S. vivianii* Steud.), other sections in S. subg. *Silene* such as *S. crassipes* Fenzl (sec. *Lasiocalycinae*), *S. colpophylla* Wrigley and *S. ramosissima* Desf. (sec. *Siphonomorpha*), and more rarely to S. subg. *Behenantha*, such as *S. baccifera* Roth (sec. *Cuccubalus*) and *S. littorea* Brot. (Sec. *Psammophilae*) [10,11].

Smooth seeds are characterized by high values of circularity and solidity in their lateral views, while, in dorsal views, higher values of circularity are shared by echinate and rugose seeds [10]. Papillose seeds are characterized by the lowest values of circularity and solidity in both lateral and dorsal views. Species included in this group are *S. holzmani* Heldr. ex Boiss. (sec. *Behenantha*), *S. laciniata* Cav. (sec. *Physolychnis*), *S. magellanica* (Desr.) Bocquet (sec. *Physolychnis*) and *S. perlmanii* W.L.Wagner, D.R.Herbst and Sohmer (sec. *Sclerophyllae*) [10].

Due to the large number of species, as well as infraspecific variation, it is important to quantitatively define seed surface structure properties and tubercle curvature in species of *Silene* and other genera in the Caryophyllaceae. The application of the Elliptic Fourier Transform (EFT) to seed images can provide new methods and models for seed shape description and quantification [20–22]. Once the seed silhouettes are represented by elemental functions, it is possible to calculate the corresponding curvature values. Curvature of a plane curve is a descriptive measure of shape that measures the rate at which the tangent line turns per unit distance moved along the curve. Departing from Bézier curves representing the root silhouettes, curvature was measured in *Arabidopsis* Heinh in Hall and Heinh (Brassicaceae) roots showing reduced values in ethylene-insensitive mutants (*etr1-1* and *ein2-1*) [23], as well as under hydrogen peroxide treatment [24]. In addition, curvature analysis allowed researchers to differentiate morphotypes in wheat kernels [25] and to define three groups of seeds in cultivated grapevine (*Vitis vinifera* L., Vitaceae) [26].

Closed plane curves based on EFT reproduced the seed silhouettes of representative *Silene* species [20]. The number of harmonics required to obtain curves reproducing the silhouettes provides an idea of the complexity of the seed surface and can be related to the four described types [10,11]. Curvature analysis based on EFT curves provides information about the geometry of the tubercles [20]. The analysis of seed surface structure has been applied to seeds of four *Silene* species: *S. colorata* Poir., *S. chlorifolia* Sm., *S. latifolia* Poir. and *S. perlmanii* W.L.Wagner, D.R.Herbst and Sohmer, representative of smooth, rugose, echinate and papillose seeds, respectively [10,11]. First, EFT curves are described for the lateral and dorsal views of seeds, and curvature analysis is performed on the EFT curves. Curvature analysis based on Bézier curves is also applied to the individual tubercles in the seeds of *S. chlorifolia* and *S. latifolia* providing an example for comparison of tubercle shape between seeds, populations or species. The analysis of curvature based on the combination of both EFT and Bézier curves provides a solid basis for the description and comparison of seed surface structure in *Silene*.

2. Results

2.1. General Morphological Aspects of the Seeds

Table 1 contains a summary of the morphological characteristics for the lateral and dorsal views of the seeds used in this work. In the lateral view, the seeds of *S. perlmanii* had the smallest area and lowest values of circularity and solidity, as was reported for the papillose-type seeds [10,11]. In the dorsal view, the highest values of solidity corresponded to the seeds of *S chlorifolia* and *S. latifolia*, representing the groups of rugose and echinate seeds, respectively. The high values of solidity and relatively low coefficients of variation are indicative of relatively stable morphological conditions in the seed populations.

Table 1. Morphological characteristics of the lateral and dorsal views of the seed populations subject of this work. A = area (mm^2); P = perimeter (mm); L = length (mm); W = width (mm); AR = aspect ratio; C = circularity; R = roundness; S = solidity. Different superscript letters indicate significant differences between files for the measurement indicated. The coefficients of variation are indicated between parentheses.

Lateral View								
Species	**A**	**P**	**L**	**W**	**AR**	**C**	**R**	**S**
S. chlorifolia	1.10 ^{c} (6.9)	4.32 ^{b} (3.4)	1.31 ^{c} (3.3)	1.07 ^{c} (4.9)	1.23 ^{b} (5.0)	0.74 ^{c} (3.9)	0.81 ^{b} (4.9)	0.96 ^{b} (0.7)
S. colorata	1.65 ^{d} (6.6)	5.27 ^{c} (3.2)	1.57 ^{d} (3.9)	1.34 ^{d} (3.1)	1.17 ^{a} (2.7)	0.75 ^{c} (1.8)	0.86 ^{c} (2.7)	0.96 ^{b} (0.6)
S.latifolia	1.01 ^{b} (2.0)	4.26 $^{a\,b}$ (2.6)	1.26 ^{b} (2.0)	1.03 ^{b} (1.7)	1.23 ^{b} (3.1)	0.70 ^{b} (4.2)	0.82 ^{b} (3.2)	0.96 ^{b} (0.4)
S. perlmanii	0.54 ^{a} (12.2)	4.07 ^{a} (11.0)	0.94 ^{a} (6.4)	0.73 ^{a} (6.6)	1.28 ^{c} (4.7)	0.42 ^{a} (16.3)	0.79 ^{a} (5.0)	0.88 ^{a} (2.5)
Dorsal View								
Species	**A**	**P**	**L**	**W**	**AR**	**C**	**R**	**S**
S. chlorifolia	0.94 ^{b} (10.4)	4.08 ^{b} (5.0)	1.37 ^{c} (4.8)	0.88 ^{c} (6.8)	1.56 ^{b} (5.9)	0.71 ^{c} (3.5)	0.64 ^{b} (5.7)	0.96 ^{c} (1.4)
S. colorata	0.53 ^{a} (18.7)	4.03 ^{b} (6.1)	1.36 ^{c} (5.0)	0.50 ^{a} (16.9)	2.82 ^{c} (15.3)	0.41 ^{a} (13.1)	0.36 ^{a} (16.1)	0.81 ^{a} (4.0)
S.latifolia	0.89 ^{b} (3.2)	3.95 ^{b} (1.9)	1.29 ^{b} (2.2)	0.88 ^{c} (2.9)	1.46 ^{ab} (4.0)	0.71 ^{c} (3.9)	0.68 ^{c} (3.9)	0.96 ^{c} (0.6)
S. perlmanii	0.49 ^{a} (13.4)	3.52 ^{a} (9.1)	0.92 ^{a} (7.4)	0.68 ^{b} (7.0)	1.34 ^{a} (4.8)	0.50 ^{b} (11.6)	0.75 ^{d} (4.9)	0.90 ^{b} (1.7)

2.2. *Elliptic Fourier Transform and Curvature: General Aspects*

Closed curves resulting from EFT analysis of seed contours reproduced the lateral and dorsal views of seed silhouettes [20]. The EFT curves for eight images representing the lateral and dorsal views of *Silene* seeds are presented. Similarity between the curve and the image silhouette is recognized by the coincidence between seed surface and EFT curve. Once a similarity was reached, Fourier curves representing the seed images were the subject for curvature analysis (Method 1: curvature based on EFT [20]; see supplementary data). In the case of seeds with tubercles, these and the protuberances of the EFT curve coincide. Curvature values were estimated for each EFT curve—including all the tubercles in a single analysis, and, later, individually for representative tubercles (Method 2: Curvature on individual tubercles, based on Bézier curves [23–26]). Both methods of curvature measurement, based, respectively, on EFT and Bézier curves, were applied to seed images of the three tuberculate species (*S. chlorifolia*, *S. latifolia* and *S. perlmanii*), while Method 1 alone was enough to describe the surface of *S. colorata* seeds. Notice the qualitative difference between the two methods: a study of the global form provided by Method 1 *versus* a local analysis obtained from Method 2.

2.3. *Smooth Seeds Are Represented by Curves with a Low Number of Harmonics*

Figures 1 and 2 show the original seed images and the EFT closed curves for the lateral and dorsal views of *S. colorata*, together with their corresponding curvature values along the curves. To obtain the EFT curves, 40 points were taken from the surface of the seed images, and the EFT images represented equations with 20 harmonics. In the lateral view, maximum curvature values correspond to the micropile region (Figure 1) and, in the dorsal view, to the two concave regions in the upper and lower sides of the seed image (Figure 2). The other peaks of curvature correspond to irregularities of the seed surface.

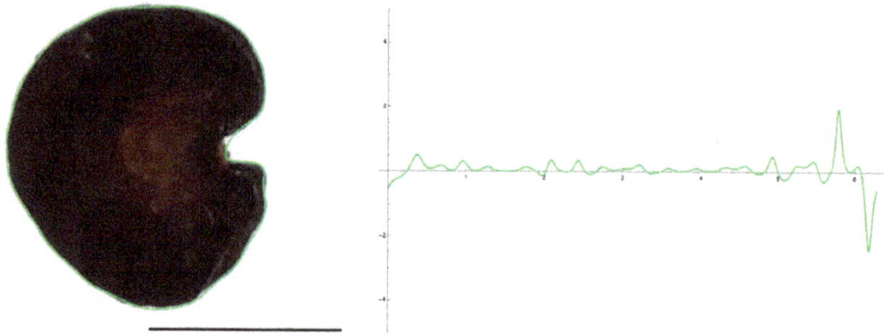

Figure 1. EFT and curvature analysis in a smooth seed. Lateral view of a seed of *S. colorata* with the EFT curve superimposed in green and the corresponding curvature values plotted. This is a smooth-type seed, lacking tubercles, and the maximum and minimum curvature values correspond to the micropilar region The EFT curve resulted from 43 points taken equidistantly along the seed silhouette and 20 harmonics, following the protocol published [20]. The program that provides the EFT curve contains the algorithm to plot the curvature values along it (see Supplementary Materials). Bar represents 1 mm.

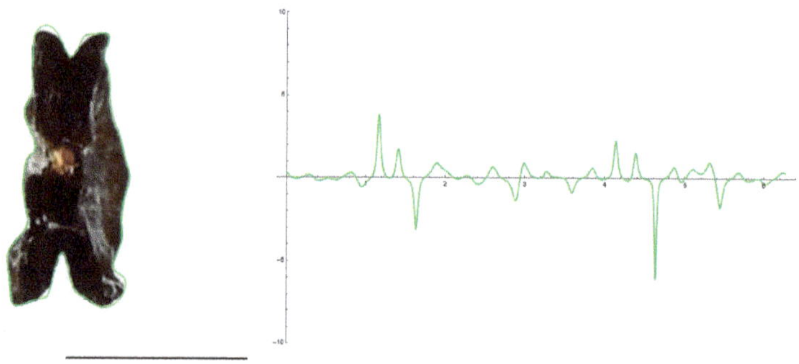

Figure 2. EFT and curvature analysis in a smooth seed. Dorsal view of a seed of *S. colorata* with the EFT curve superimposed in green and the corresponding curvature values plotted. This is a smooth-type seed, lacking tubercles and the maximum and minimum curvature values correspond to the upper and lower concavities. The EFT curve resulted from 38 points taken at regular distances along the silhouette and 20 harmonics, following the protocol published [20]. The program that provides the EFT curve contains the algorithm to obtain and represent the curvature values along it (see Supplementary Materials). Bar represents 1 mm.

Lateral and dorsal views of *Silene colorata* seeds were accurately represented by 40-point-derived curves with 20 to 30 harmonics. In contrast, for tuberculate seeds, models derived from 40-point curves resemble the general shape, without considering the tubercles. These models are easy to develop and can be useful to obtain visual information on the general seed shape (Figure 3) for characteristics such as solidity and roundness (or aspect ratio). However, adjusting the surface contour to include the tubercles requires models obtained with more points.

2.4. Seeds with Tubercles Require a Higher Number of Harmonics

The curves resulting from the application of the EFT with 100–250 points and 40–80 harmonics to the lateral and dorsal views of *S. chlorifolia*, *S. latifolia* and *S. perlmanii* are shown together with their corresponding curvature values (Figures 4–9).

In the lateral view of *S. chlorifolia* and *S. latifolia* seeds, there are less tubercles and with lower curvature values in the regions around the micropile. Curvature values in the tubercles comprise between −4 and 1 in *S. chlorifolia* and −8 and 5 in *S. latifolia*.

Nevertheless, the highest values are due to small irregularities in the process of generation of the curve, and real tubercle curvature comprises between −2 and 1 *S. chlorifolia* and between −5 and 4 for *S. latifolia*. Tubercle number and absolute curvature values are higher in *S. latifolia*. In the seeds analyzed, in both species, the tubercles are of regular size and shape.

In the dorsal view of seeds from both species, the tubercles were concentrated in the seed poles, being more pronounced and with higher curvature values than in the lateral sides (Figures 6 and 7).

Curves reproducing the silhouette of the lateral view of *S. perlmanii* were derived from the selection of 180–200 points in the application of EFT (Figure 8). The tubercles were larger than in *S. chlorifolia* and *S. latifolia* in relation to seed size. Estimated curvature values were between −3 and 1.

The curve representing the dorsal view of *S. perlmanii* reproduced 23 individual tubercles of curvature values between 1 and 35 (Figure 9).

Figure 3. EFT curves with a low number of harmonics. EFT curves representing the silhouettes corresponding to seeds of *Silene colorata*, *S. chlorifolia*, *S. latifolia* and *S. perlmanii* (left to right). Above: lateral views. Below: dorsal views. Fourier analysis [20] was performed taking 40 points from the seed surface, and the resulting EFT curves obtained from 20 harmonics.

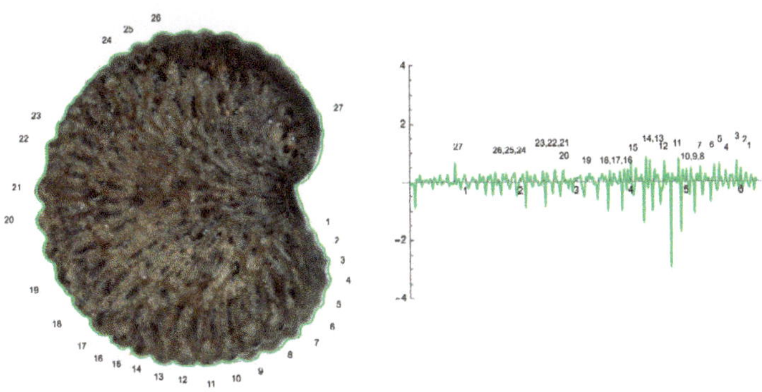

Figure 4. EFT and curvature analysis in a rugose seed. Lateral view of a seed of *S. colorata* with the EFT curve superimposed in green and the corresponding curvature values plotted. The EFT curve resulted from 193 points taken along the silhouette and 96 harmonics, following the protocol published [20]. The numbers indicate the correspondence between individual tubercles and their curvature values. The program that provides the EFT curve contains the algorithm to obtain and represent the curvature values along it (see Supplementary Materials). Rounded tubercles are disposed regularly and the maximum and minimum curvature values correspond to individual tubercles. Bar represents 1 mm.

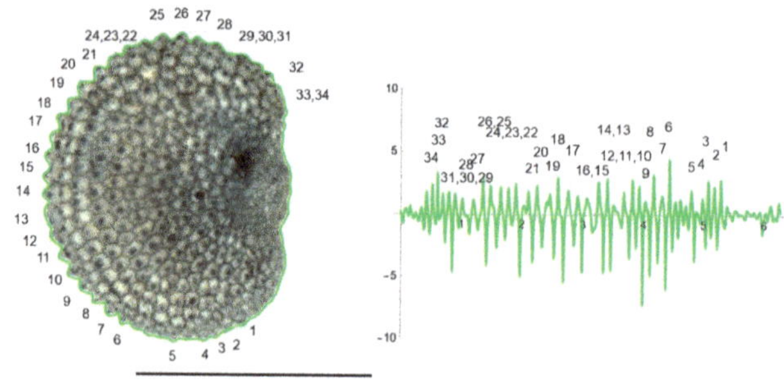

Figure 5. EFT and curvature analysis in an echinate-type seed. Lateral view of a seed of *S. latifolia* with the EFT curve superimposed in green and the corresponding curvature values plotted. The EFT curve resulted from 228 points taken along the silhouette and 74 harmonics, following the protocol published [20]. The numbers indicate the correspondence between individual tubercles and their curvature values. The program that provides the EFT curve contains the algorithm to obtain and represent the curvature values along it (see Supplementary Materials). This is an echinate-type seed, with acute tubercles disposed regularly and the maximum and minimum curvature values correspond to individual tubercles. Bar represents 1 mm.

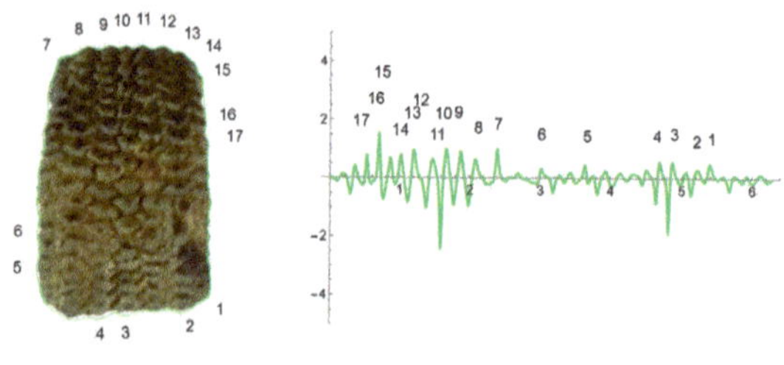

Figure 6. EFT and curvature analysis in a rugose seed. Dorsal view of a seed of with the EFT curve superimposed in green and the corresponding curvature values plotted. The EFT curve resulted from 102 points taken along the silhouette and 40 harmonics, following the protocol published [20]. The numbers indicate the correspondence between individual tubercles and their curvature values. The program that provides the EFT curve contains the algorithm to obtain and represent the curvature values along it (see Supplementary Materials). Maximum curvature values correspond to individual tubercles in the poles. Bar represents 1 mm.

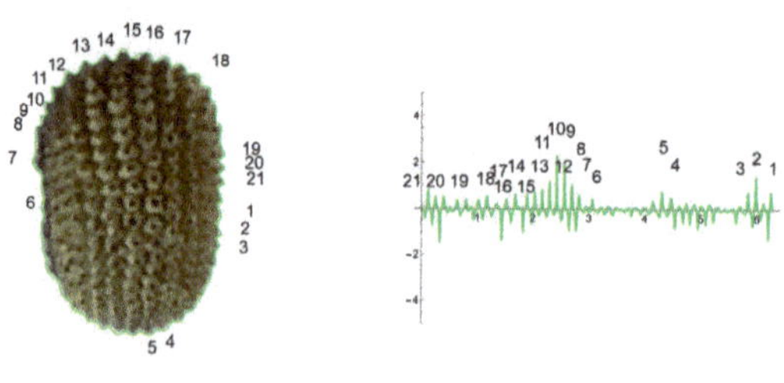

Figure 7. EFT and curvature analysis in an echinate-type seed. Dorsal view of a seed of *S. latifolia* with the EFT curve superimposed in green and the corresponding curvature values plotted. The EFT curve results from 98 points taken along the silhouette and 74 harmonics, following the protocol published [20]. The numbers indicate the correspondence between individual tubercles and their curvature values. The program that provides the EFT curve contains the algorithm to obtain and represent the curvature values along it (see Supplementary Materials). Maximum curvature values correspond to individual tubercles. Bar represents 1 mm.

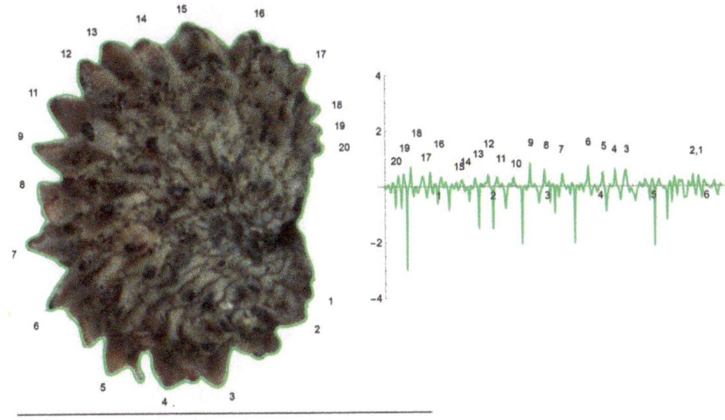

Figure 8. EFT and curvature analysis in a papillose-type seed. Lateral view of a seed of *S. perlmanni* with the EFT curve superimposed in green and the corresponding curvature values plotted. The EFT curve shown resulted from 189 points taken along the silhouette and 74 harmonics, following the protocol published [20]. The numbers indicate the correspondence between individual tubercles and their curvature values. The program that provides the EFT curve contains the algorithm to obtain and represent the curvature values along it (see Supplementary Materials). This is a papillose-type seed, with tubercles of varied shape disposed unevenly at the seed surface. The maximum and minimum curvature values correspond to particular tubercles. Bar represents 1 mm.

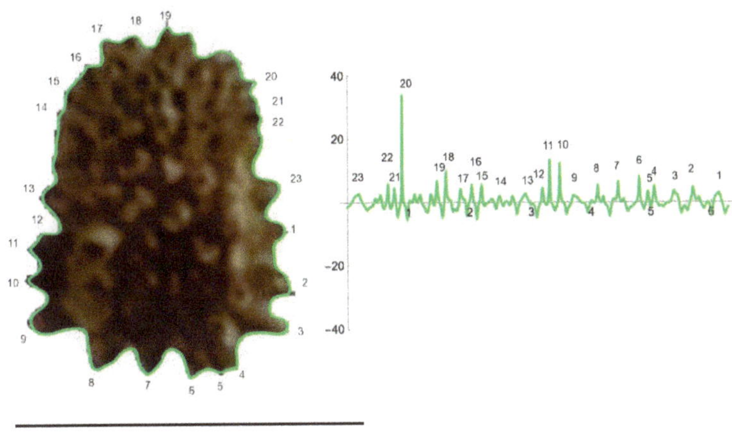

Figure 9. EFT and curvature analysis in a papillose-type seed. Dorsal view of a seed of *S. perlmanii* with the EFT curve superimposed in green and the corresponding curvature values plotted. The maximum and minimum curvature values correspond to individual tubercles. The EFT curve shown results from 184 points taken along the silhouette and 60 harmonics, following the protocol published [20]. The numbers indicate the correspondence between individual tubercles and their curvature values. Bar represents 1 mm. The program that provides the EFT curve contains the algorithm to obtain and represent the curvature values along it (see Supplementary Materials). Bar represents 1 mm.

2.5. Curvature Analysis on Individual Tubercles

Figure 10 shows the curvature analysis of individual tubercles, numbers 23, 1, 2 and 3, from the dorsal view of *S. perlmanii* represented in Figure 9. Curvature values are of 1.2 for tubercles 23, 1 and 2, and 2.1 for tubercle number 3. In contrast with *S. chlorifolia* and *S. latifolia*, the tubercles of *S. perlmanii* present great diversity in size and shape.

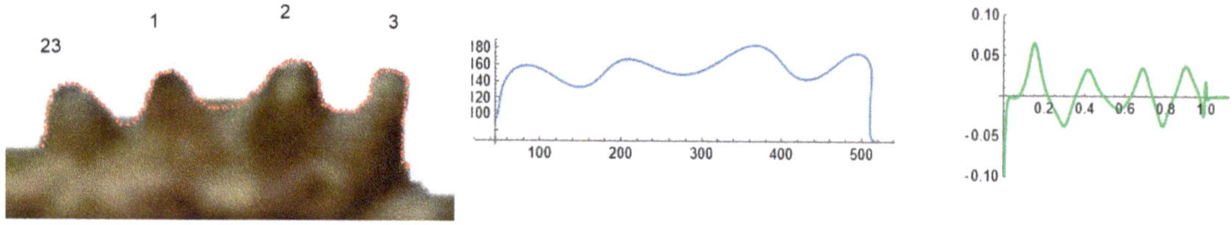

Figure 10. Curvature analysis of individual tubercles in *S. perlmanii.* Right: Four tubercles of *S. perlmanii*, numbers 23, 1, 2 and 3, from Figure 9 were selected to measure curvature individually. The corresponding Bézier curve (middle) and curvature plot (right) are shown in blue and green, respectively.

The results of Fourier analysis indicated differences in curvature values between species. To investigate this in more detail, curvature analysis was performed in *S. chlorifolia* and *S. perlmanii*, two species that have regular tubercles. Figure 11 shows representative samples of a tubercle of each species with their Bézier curve and corresponding curvature analysis. Table 2 presents the comparison of means (ANOVA) for maximum and average curvature values in six tubercles of each species. The differences were significant ($p < 0.05$) both for maximum and mean values in the comparison between *S. chlorifolia* and *S. perlmanii.*

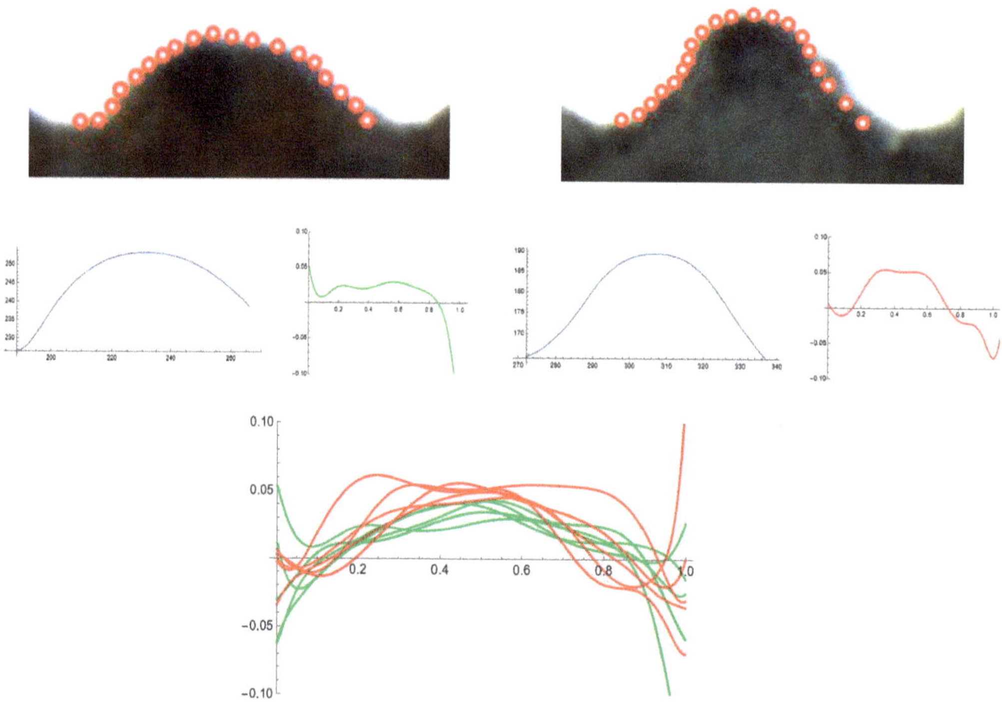

Figure 11. Curvature analysis for individual tubercles of *S. chlorifolia* and *S. latifolia.* Above: representative individual tubercles with points taken. Middle: curves and curvature plots. Below: plot of curvatures corresponding to five tubercles for each species. Green: *S. chlorifolia*; Red: *S. latifolia.*

Table 2. Summary of curvature results for the comparison between *S. chlorifolia* and *S. latifolia*. ANOVA for six individual tubercles of each species. Different letters in superscript indicate significant differences between species.

Species	*S. chlorifolia*	*S. latifolia*
Maximum curvature values in six tubercles (mean)	40.7 ^{a}	55.5 ^{b}
Mean curvature values in six tubercles (mean)	17.2 ^{a}	24.5 ^{b}

The curvature was also measured in six tubercles of each of three seeds (eighteen tubercles total) of *S. chlorifolia* and *S. latifolia* and the mean values were compared. The results are shown in Table 3.

Table 3. Summary of curvature results for the comparison between *S. chlorifolia* and *S. latifolia*. ANOVA for 18 individual tubercles corresponding to three seeds of each species. Different letters in superscript indicate significant differences between species. Different superscript letters indicate significant differences between files for the measurement indicated. The coefficients of variation are indicated between parentheses.

Species	*N*	*S. chlorifolia*	*S. latifolia*
Maximum curvature values	18	47.1 ^{a} (60.9)	97.8 ^{b} (42.7)
Mean curvature values	18	32.0 ^{a} (69.8)	61.3 ^{b} (38.5)

3. Discussion

The field of descriptive systematics aims at discovering the patterns in nature and how they vary between organisms, populations and species [27]. This requires the application of mathematical protocols to define, quantify and compare the shapes in the organisms [28].

The species of the genus *Silene* L. have a remarkable variation in geographical distribution, breeding systems and ecological relationships. Due to their short life-cycles, facility to breed and the growing availability of genetic resources, they can be considered as models for ecology and evolution [29]. In addition, *S. latifolia* has heteromorphic sex-determination with an evolving non-recombining y region rich in repetitive DNA that provides a unique system for the study of the origin and modification of sex chromosomes [30]. In addition, an interesting seed shape diversity makes *Silene* a useful model for studying variations in seed morphology [5–14].

In many species of *Silene*, as well as in other species of Caryophyllaceae, seeds have tubercles arranged along the seed surface. Typically, 20 to 60 tubercles are observed in the lateral view and a smaller number in the dorsal view. Size and shape of the tubercles, as well as the regularity of their distribution, vary among species and among populations of the same species, making it possible to search for associations between tubercle characteristics and genetic or environmental factors. Infraspecific variation concerning tubercle size and shape has been reported in other genus of the Caryophyllaceae, such as *Arenaria* L., *Acanthophyllum* L. as well as *Silene* [14,16,31], and the variations in tubercle shape have been attributed to geographical and ecological factors or, in contrast, to taxonomic differences [16]. The study of seed surface variation will benefit from new quantitative methods for the description of tubercle morphology and seed surface.

The observation by optical microscopy of seeds of 100 species of *Silene* allowed for their classification into four groups according to their silhouettes: smooth, rugose, echinate and papillose [10,11]. Accurate representation of the seed surface structure was obtained by the application of Fourier Transform to seed images [20]. Subsequently, in this article, we have investigated the differences between representative species of the four morphological types related to the representation of their seed silhouettes by EFT. The main geometric features of the silhouettes of smooth seeds are represented with EFT curves derived from 40 points selected in the seed surface and 20 harmonics. This result agrees with estimates of 10 harmonics for reproducing the shape of leaves [32] and makes EFT with low harmonic number an interesting tool for the representation of general aspects of seed shape. Nevertheless, Fourier analysis applied to the species of tuberculate seeds (rugose, echinate and papillose), required a higher number of harmonics to define well the individual protuberances. Thus, although Fourier analysis with a low number of harmonics can discriminate successfully between various seed morphotypes, only with a higher number of harmonics can the morphological properties, size, shape and distribution of the tubercles can be analyzed.

To obtain curves adjusting to the protuberances at least 100 points are required, and the accuracy increases with higher numbers up to 250 or even more. Equations of 60 to 80 harmonics are sufficient in most cases, but more may be necessary to have detailed representation of the tubercles.

In addition to curvature analysis on curves derived from EFT, the analysis was focused (involving higher precision) on individual tubercles. The comparison revealed lower curvature values in *S. chlorifolia* (rugose seeds) and higher in *S. latifolia* (echinate). The application of the method to diverse populations of these species is required to confirm that a range of curvature values is a property of each species. In addition, the application to different species of each of the groups (rugose, echinate and papillose) will tell whether curvature values may be associated with the general morphology of the tubercles. A constant curvature value observed in the seeds of *S. chlorifolia* is related to lower curvature values and the proximity between mean and maximum values.

The method presented here opens the way to the analysis of size, shape and distribution of tubercles along the seed surface. The reported results remark upon the difference among seed types based on cell surface, with smooth seeds being characterized by a profile represented by low number of points, and on the other side, papillose seeds with numerous large tubercles that can only be represented by EFT when a large number of points are considered. Both types are distinguished also by their extreme values of circularity (highest in smooth seeds, lowest in papillose seeds) [10]. In between these two types remain the other two groups, rugose and echinate, here represented by *S. chlorifolia* and *S. latifolia*, respectively. While the tubercles in both are distributed more regularly than in papillose seeds, the results show differences with increased curvature values in *S. latifolia*. The results with other species and populations will demonstrate whether this is a property of echinate seeds, in contrast with rugose seeds, or if these differences are due to the species or populations chosen.

4. Materials and Methods

4.1. *Silene* Seeds

The populations of seeds analyzed in this work are listed in Table 4.

Table 4. List of seed populations analyzed in this work. The populations JBUV 519, JBUV100 and JBUV 1444 were obtained from the carpoespermateca at the Botanical Garden of the University of Valencia and proceed from an exchange protocol between seed collections through the world.

Species	Source (Place of Origin)	Lifespan	Morphological Type [10,11]	Subgenus and Section [21]
S. perlmanii W.L.Wagner, D.R.Herbst and Sohmer	JBUV 519 * Botanischer Garten der Universität Zürich (U **)	Annual	Papillose	S. subg. *Silene* sect. Sclerophyllae (Chowdhuri) F.Jafari, Oxelman and Rabeler
S. colorata Poir. (AJ301)	Ana Juan (Villena, Alicante, Spain)	Annual	Smooth	S. subg. *Silene* sect. *Silene*
S. chlorifolia Sm.	JBUV100 BG der Martin-Luther-Univ. Halle-Wittenberg (U **)	Perennial	Rugose	S. subg. *Silene* sect. Sclerocalycinae (Boiss.) Schischk. in Komarov
S. latifolia Poir	JBUV 1444 * Humboldt University, Berlin (Germany, Brandenburg, Landkreis Märkisch-Oderland, Petershagen)	Perennial	Echinate	S. subg. *Behenantha* (Otth) Torr. and A.Gray sect. *Melandrium* (Röhl.) Rabeler

* JBUV = Jardín Botánico Universidad de Valencia. ** U stands for unknown.

4.2. Seed Images

For the analysis of individual tubercles, photographs were taken with a Nikon Stereomicroscope Model SMZ1500 (Nikon, Tokio, Japan) equipped with a 5.24 megapixel Nikon DS-Fi1 of camera (Nikon, Tokio, Japan); lateral and dorsal views used in FET analysis were taken with a Nikon Z6 camera (Nikon, Tokio, Japan), equipped with an objective AF-S Micro NIKKOR 60 mm f/2.8G ED (Nikon, Tokio, Japan).

4.3. Elliptic Fourier Transform (EFT)

The application of EFT to any closed plane figure results in a curve that mimics the silhouette of the original figure and is amenable to curvature analysis. For this, a series of points were selected at regular intervals on the seed silhouette (Figure 12). The function, whose graphic approximates the shape, is a combination of trigonometric expressions; then, its expression allows for calculating the curvature values along the curve [20]. The program to obtain an EFT curve from a series of points and the application to the four seed types was made available (see Supplementary Material). The points were taken starting from the right side of the seed image silhouette and moving clockwise. Different curves can result from the same image depending on the number and positions of the points taken, as well as the number of points selected in the curve construction process (number of harmonics). It is important to avoid the duplication of points and to take a similar number of points at equivalent distances in the different samples when a comparative analysis is sought.

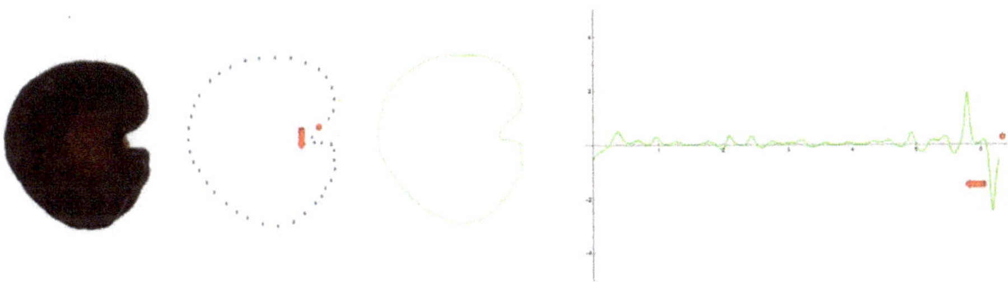

Figure 12. Example of the method (Method 1) to obtain EFT curve and curvature values from a seed image. Seed of *S. colorata* (lateral view), set of points selected, curve and results of curvature analysis. The red dot marks the initial point; the series of points follow clockwise, and the results of curvature analysis are shown counterclockwise, starting from the last point.

The process was divided in two consecutive methods. First (Method 1), 40 to 50 points were taken at regular intervals from the seed silhouettes of all four types and 20 or 30 points were selected for the Fourier curves. Method 1 resolved the silhouettes for lateral and dorsal views of *S. colorata* (smooth seeds), as well as the overall shape for all four species. For the tuberculate species (*S. chlorifolia*, *S. latifolia* and *S. perlmanii*), more points were taken at regular intervals from the seed silhouettes and Fourier curves performed with 60 harmonics or more were needed to fit the curve and the seed silhouette (Method 2). The program to obtain an EFT curve also contains the code for the calculation of curvature values along the curve and their corresponding plot (Figure 12).

4.4. Curvature Analysis

Curvature values were calculated either from the EFT curves (whole-seed images) or for individual tubercles (See Supplementary Materials). In the figures, the curvature values are represented in reverse sense to the direction of the curve (starting at the last point and moving counterclockwise; Figure 12). Curvature values below the horizontal axis belong to peaks pointing towards the center of the seed, while peaks with positive values correspond to protuberances. Curvature values were determined for individual tubercles of each species according to established procedures [23–26] (See Supplementary Materials). In the measurements of curvature for individual tubercles, the points were taken either manually, as represented in Figures 11 and 12, or automatically with the function Analyze line graph of Image J. In the first case (points taken manually; Figure 11 and Table 2), six tubercles were selected from the lateral views of representative seeds of *S. chlorifolia* and *S. latifolia*, and their maximum and mean curvature values determined. In the case of points taken automatically (Table 3), six tubercles of three representative seeds for each species were analyzed.

4.5. Statistical Analysis

ANOVA was used to show significant differences between populations for the measured variables. In the case of the comparison of morphological characters involving four populations, ANOVA was followed by Tukey test to provide specific information on which means were significantly different from one another. Statistical analyses (ANOVA) were carried out on IBM SPSS statistics v28 (SPSS 2021).

5. Conclusions

Fourier analysis has been applied to four representative morphotypes of *Silene* seeds based on seed surface structure: smooth, rugose, echinate and papillose. The method can successfully discriminate between various seed groups. The surface of smooth seeds, with no visible tubercles and higher circularity values, is represented by EFT equations with 40 harmonics. EFT opens the way to the analysis of seed surface protuberances. Curvature analysis applied to individual tubercles revealed differences between representative seeds of rugose and echinate groups, with lower values in rugose seeds and higher in tubercles of echinate seeds.

Supplementary Materials: The Mathematica® files with Fourier analysis, points for curvature analysis and curvature analysis of individual tubercles are available at https://www.mdpi.com/article/10.3390/plants12132439/s1. The file entitled "EFT for Silene seeds four types.nb" contains the EFT curves and curvature analysis shown in Figures 1, 2 and 4–9.

Author Contributions: Conceptualization, E.C., J.L.R.-L. and Á.T.; methodology, E.C., J.L.R.-L., J.J.M.-G. and Á.T.; software, E.C. and Á.T.; validation, E.C., J.L.R.-L. and Á.T.; formal analysis, E.C., J.L.R.-L., J.J.M.-G. and Á.T.; investigation, E.C., J.L.R.-L., J.J.M.-G. and Á.T.; resources, E.C., J.L.R.-L., J.J.M.-G. and Á.T.; data curation, E.C., J.L.R.-L., J.J.M.-G. and Á.T.; writing—original draft preparation, E.C.; writing—review and editing, E.C., J.L.R.-L., J.J.M.-G. and Á.T.; visualization, E.C., J.L.R.-L., J.J.M.-G. and Á.T.; supervision, E.C., J.L.R.-L. and Á.T.; project administration, E.C., J.L.R.-L. and Á.T.; funding acquisition, E.C., J.L.R.-L. and Á.T. All authors have read and agreed to the published version of the manuscript.

Funding: Project "CLU-2019-05-IRNASA/CSIC Unit of Excellence", funded by the Junta de Castilla y León and co-financed by the European Union (ERDF "Europe drives our growth").

Data Availability Statement: The data presented in this study are available in Supplementary Materials.

Acknowledgments: We thank Ana Juan of the University of Alicante and Bohuslav Janousek from IBP Brno for continuous collaboration and constructive criticism, and Ana Juan and Elena Estrelles and the Carpoespermateca of the Botanical Garden at the University of Valencia for providing the seeds used in this study.

Conflicts of Interest: The authors declare no conflict of interest.

References

1. Bittrich, V. Caryophyllaceae. In *Flowering Plants Dicotyledons: Magnoliid, Hamamelid, and Caryophyllid Families*; Kubitzki, K., Rohwer, J.G., Bittrich, V., Eds.; The Families and Genera of Vascular Plants; Springer: Berlin/Heidelberg, Germany, 1993; Volume 2, pp. 206–236.
2. Heywood, V.H.; Brummitt, R.K.; Culham, A.O.; Seberg, O. *Flowering Plant Families of the World*; Royal Botanic Gardens: Kew, UK, 2007; 424p.
3. Martin, A.C. The comparative internal morphology of seeds. *Am. Midl. Nat.* **1946**, *36*, 513–660. [CrossRef]
4. Johri, B.M.; Ambegaokar, K.B.; Srivastava, P.S. *Comparative Embryology of Angiosperms*; Springer: Berlin/Heidelberg, Germany, 1992; Volume 1.

5. Martín-Gómez, J.J.; Rewicz, A.; Rodriguez-Lorenzo, J.L.; Janousek, B.; Cervantes, E. Seed morphology in *Silene* based on geometric models. *Plants* **2020**, *9*, 1787. [CrossRef] [PubMed]
6. Juan, A.; Martín-Gómez, J.J.; Rodriguez-Lorenzo, J.L.; Janousek, B.; Cervantes, E. New techniques for seed shape description in *Silene* species. *Taxonomy* **2022**, *2*, 1–19. [CrossRef]
7. Martín-Gómez, J.J.; Porceddu, M.; Bacchetta, G.; Cervantes, E. Seed morphology in species from the *Silene mollissima* aggregate (Caryophyllaceae) by comparison with geometric models. *Plants* **2022**, *11*, 901. [CrossRef]
8. Rodríguez-Lorenzo, J.L.; Martín-Gómez, J.J.; Tocino, Á.; Juan, A.; Janoušek, B.; Cervantes, E. New geometric models for shape quantification of the dorsal view in seeds of *Silene* species. *Plants* **2022**, *11*, 958. [CrossRef]
9. Martín-Gómez, J.J.; Rodríguez-Lorenzo, J.L.; Janoušek, B.; Juan, A.; Cervantes, E. Comparison of seed images with geometric models, an approach to the morphology of *Silene* (Caryophyllaceae). *Taxonomy* **2023**, *3*, 109–132. [CrossRef]
10. Martín-Gómez, J.J.; Rodríguez-Lorenzo, J.L.; Tocino, Á.; Janousek, B.; Juan, A.; Cervantes, E. The outline of seed silhouettes: A morphological approach to *Silene* (Caryophyllaceae). *Plants* **2022**, *11*, 3383. [CrossRef]
11. Martín-Gómez, J.J.; Rodríguez-Lorenzo, J.L.; Juan, A.; Tocino, Á.; Janousek, B.; Cervantes, E. Seed morphological properties related to taxonomy in *Silene* L. species. *Taxonomy* **2022**, *2*, 298–323. [CrossRef]
12. Oxelman, B. A revision of the *Silene sedoides* group. *Willdenowia* **1995**, *25*, 143–169.
13. Yildiz, K.; Çirpici, A. Seed morphological studies in *Silene* L. from Turkey. *Pak. J. Bot.* **1998**, *30*, 173–178.
14. Tabaripour, R.; Koohdar, F.; Sheidai, M.; Gholipour, A. Intra-specific variations in *Silene*: Morphometry and micromorphometry analyses. *Afr. J. Biotech.* **2013**, *12*, 5208–5217.
15. Sadeghian, S.; Zarre, S.; Heubl, G. Systematic implication of seed micromorphology in *Arenaria* (Caryophyllaceae) and allied genera. *Flora* **2014**, *209*, 513–529.
16. Wofford, B.E. External seed morphology of *Arenaria* (Caryophyllaceae) of the southeastern United States. *Syst. Bot.* **1981**, *6*, 126–135. [CrossRef]
17. Wyatt, R. Intraspecific variation in seed morphology of *Arenaria uniflora* (Caryophyllaceae). *Syst. Bot.* **1984**, *9*, 423–431. [CrossRef]
18. Mostafavi, G.; Assadi, M.; Nejadsattari, T.; Sharifnia, F.; Mehregan, I. Seed micromorphological survey of the *Minuartia* species (Caryophyllaceae) in Iran. *Turk. J.Bot.* **2013**, *37*, 446–454. [CrossRef]
19. Minuto, L.; Roccotiello, E.; Casazza, G. New seed morphological features in *Moehringia* L. (Caryophyllaceae) and their taxonomic and ecological significance. *Plant Biosyst.* **2011**, *145*, 60–67. [CrossRef]
20. Cervantes, E.; Rodríguez-Lorenzo, J.L.; Gutiérrez del Pozo, D.; Martín-Gómez, J.J.; Janousek, B.; Tocino, Á.; Juan, A. Seed silhouettes as geometric objects: New applications of Elliptic Fourier Transform to seed morphology. *Horticulturae* **2022**, *8*, 974. [CrossRef]
21. McLellan, T.; Endler, J.A. The relative success of some methods for measuring and describing the shape of complex objects. *Syst. Biol.* **1998**, *47*, 264–281. [CrossRef]
22. Kuhl, F.P.; Giardina, C.R. Elliptic Fourier features of a closed contour. *Comput. Graph. Image Process.* **1982**, *18*, 236–258. [CrossRef]
23. Cervantes, E.; Tocino, A. Geometric analysis of Arabidopsis root apex reveals a new aspect of the ethylene signal transduction pathway in development. *J. Plant Physiol.* **2005**, *162*, 1038–1045. [CrossRef]
24. Noriega, A.; Tocino, A.; Cervantes, E. Hydrogen peroxide treatment results in reduced curvature values in the *Arabidopsis* root apex. *J. Plant Physiol.* **2009**, *166*, 554–558. [CrossRef]
25. Martín-Gómez, J.J.; Rewicz, A.; Goriewa-Duba, K.; Wiwart, M.; Tocino, Á.; Cervantes, E. Morphological description and classification of wheat kernels based on geometric models. *Agronomy* **2019**, *9*, 399. [CrossRef]
26. Cervantes, E.; Martín-Gómez, J.J.; Espinosa-Roldán, F.E.; Muñoz-Organero, G.; Tocino, Á.; Cabello Sáenz de Santamaría, F. Seed apex curvature in key Spanish grapevine cultivars. *Vitic. Data J.* **2021**, *3*, e66478. [CrossRef]
27. Landrum, L. What has happened to descriptive systematics? What would make it thrive? *Syst. Bot.* **2001**, *26*, 438–442.
28. Gielis, J.; Caratelli, D.; Fougerolle, Y.; Ricci, P.E.; Tavkelidze, I.; Gerats, T. Universal natural shapes: From unifying shape description to simple methods for shape analysis and boundary Value Problems. *PLoS ONE* **2012**, *7*, e29324. [CrossRef]
29. Bernasconi, G.; Antonovics, J.; Biere, A.; Charlesworth, D.; Delph, L.F.; Giraud, T.; Hood, M.E.; Marais, G.A.B.; Pannell, J.R.; Wolfe, L.M.; et al.. *Silene* as a model system in ecology and evolution. *Heredity* **2009**, *103*, 5–14. [CrossRef]
30. Filatov, D.A. Heterochiasmy and sex chromosome evolution in *Silene*. *Genes* **2023**, *14*, 543. [CrossRef]
31. Pirani, A.; Zarre, S.; Rabeler, R.; Assadi, M.; Joharchi, M.R.; Oxelman, B. Systematic significance of seed morphology in *Acanthophyllum* (Caryophyllaceae: Tribe Caryophylleae) in Iran. *Phytotaxa* **2019**, *387*, 105–118.
32. Hearn, D.J. Shape analysis for the automated identification of plants from images of leaves. *Taxon* **2009**, *58*, 934–954.

plants

MDPI

Article

Rapid Adaptation of *Chimonobambusa opienensis* Leaves to Crown–Thinning in Giant Panda Ecological Corridor, Niba Mountain

Di Fang [1], Junren Xian [1,*], Guopeng Chen [2], Yuanbin Zhang [3], Hantang Qin [1,4], Xin Fu [1], Liyang Lin [1], Yuxuan Ai [1], Zhanbiao Yang [1], Xiaoxun Xu [1], Yuanxiang Yang [1] and Zhang Cheng [1]

1 College of Environmental Sciences, Sichuan Agricultural University, Chengdu 611130, China; fangdi3434_cici@163.com (D.F.); qin-hantang@outlook.com (H.Q.); fu__xin@163.com (X.F.); 2394559793@foxmail.com (L.L.); 250516811@foxmail.com (Y.A.); yzb195@126.com (Z.Y.); xuxiaoxu2013@163.com (X.X.); 529877087@foxmail.com (Y.Y.); 29442926@foxmail.com (Z.C.)
2 College of Forestry, Gansu Agricultural University, Lanzhou 730070, China; chgp1986@gmail.com
3 Institute of Mountain Hazards and Environment, Chinese Academy of Sciences, Chengdu 610041, China; zhangyb@imde.ac.cn
4 School of Ecological and Environmental Sciences, East China Normal University, Shanghai 201100, China
* Correspondence: xianjr@126.com

Abstract: Leaf traits reflect the ecological strategy in heterogeneous contexts and are widely used to explore the adaption of plant species to environmental change. However, the knowledge of short-term effect of canopy management on understorey plant leaf traits is still limited. Here, we studied the short-term effect of crown–thinning on the leaf morphological traits of bamboo (*Chimonobambusa opienensis*), an important understorey plant and staple food for the giant panda (*Ailuropoda melanoleuca*) of Niba Mountain. Our treatments were two crown–thinnings (spruce plantation, CS, and deciduous broad-leaved forest, CB) and two controls (broad-leaved forest canopy, FC, and the bamboo grove of clearcutting, BC). The results showed that: the CS enhanced the annual leaf length, width, area, and thickness, CB decreased almost all annual leaf traits, and perennial leaf traits in CS and CB were the opposite. The log-transformed allometric relationships of length vs. width, biomass vs. area were significantly positive while those of specific leaf area vs. thickness were significantly negative, which varied largely in treatments and age. The leaf traits and allometric relationships suggested that the CS created a more suitable habitat for bamboo growth. This study highlighted that the understorey bamboo leaf traits could adapt the improved light environment induced by crown–thinning rapidly.

Keywords: *Chimonobambusa opienensis*; leaf traits; trade-off; short-term effect; crown–thinning

Citation: Fang, D.; Xian, J.; Chen, G.; Zhang, Y.; Qin, H.; Fu, X.; Lin, L.; Ai, Y.; Yang, Z.; Xu, X.; et al. Rapid Adaptation of *Chimonobambusa opienensis* Leaves to Crown–Thinning in Giant Panda Ecological Corridor, Niba Mountain. *Plants* **2023**, *12*, 2109. https://doi.org/10.3390/plants12112109

Academic Editor: Yasutomo Hoshika

Received: 20 April 2023
Revised: 19 May 2023
Accepted: 24 May 2023
Published: 26 May 2023

1. Introduction

The leaf is a sensitive organ of plant adaptation to environmental changes. The leaves have been widely explored the adaptative strategies at altitude [1], latitude [2] and success [3] gradients. Plants often improve the performance via adjusting the functional traits (e.g., leaf) to acquire resources [4] for growth and reproduction. This suggests that leaf traits correlate and covary with the other traits [2,5,6], which are often explored by $y = \beta x^{\alpha}$ (linearized under the form lg (y) = lg $(\beta) + \alpha$ lg (x), x and y being the determines whether the relationship is isometric ($\alpha = 1.0$) or allometric ($\alpha \neq 1$)) [5,7–9]. This trait-based approaches are now widely explored from organs (e.g., leaf, stems and roots) [10,11] to whole-plant [5] and ecosytem [12] at the scale from local [13] to global [14], and used to study resource acquisition [5], adaptation to environment change and disturbance [15], community assembly [16] and ecosystem function [15,17]. These prior studies also shows that leaf traits vary largely across lineages, life forms [5], ontogenetic stages, size [18], functional groups [19] and environments [16,17,20,21]. Thinning is an essential silvicultural approach to improve light intensity for understorey [22,23], and have potential effects on forest structure (e.g., species richness and composition) and function (e.g., recovery, regeneration and productivity) [24]. However, the trade-offs of understorey leaf morphological traits (leaf traits, hereafter) induced by thinning have not yet been studied.

The Giant Panda Ecological Corridor in the Niba Mountain (GPECN), has been designed and constructed to connect the giant panda populations in the Daxiangling Mountains and the Qionglai Mountain [25]. *Chimonobambusa opienensis* (bamboo, hereafter) is one of the key staple food for the giant panda (*Ailuropoda melanoleuca*) in the GPECN, naturally distributes in the Liangshan Mountain, Daxiangling Mountain and Qionglai Mountain, ranging from 950 m to 2200 m [26]. The delicious new bamboo shoot is the main forest well-being for the local farmers. After decades of recovery, more dense-canopy (coverage > 90%) secondary deciduous broad-leaved forests (DBF) and spruce (*Picea asperata*) plantations (SP), and some dense bamboo groves of clearcutting (BC) have been well developed in the Niba Mountain. However, the closed canopy of DBF and SP limit bamboo to grow, and dense bamboo groves also limit giant pandas to spend time [27]. This suggests that suitable canopy is beneficial to bamboo growth for giant pandas. Hence, in the Autumn of 2018, the crown–thinning of deciduous broad-leaved forest (CB) and spruce plantations (CS) had been implemented to improve giant panda habitat via promoting the understorey bamboo to grow and develop [28]. This provided a good platform for studying the short-term effect of crown–thinning on bamboo leaf trade-off. In the late August of 2020, after in situ survey, the annual and perennial bamboo leaves in the plots of crown–thinning (CB and CS) and controls (BC and deciduous broad-leaved forest canopy, FC) had been sampled, and the leaf traits had been measured. Our objectives were to: (1) determine the short-term effect of crown–thinning on the different-age bamboo leaf traits; and (2) clarify short-term effect of crown–thinning on the trade-off between core leaf traits.

2. Results

2.1. Variations of Leaf Traits in Age Categories and Treatments

In the GPECN, the bamboo leaf length, width, thickness, LSI, area, biomass and SLA were 125.05 ± 21.46 mm, 20.37 ± 3.71 mm, 0.14 ± 0.03 mm, 6.23 ± 0.99, 19.34 ± 6.59 cm^2, 0.101 ± 0.040 g and 201.71 ± 49.73 $cm^2\ g^{-1}$, respectively. The GLM analyses (Table 1) showed that the age had significant effects on all traits (all $p < 0.01$); Treatments had minor effect on thickness ($p = 0.\ 078$), area ($p = 0.115$) and biomass ($p = 0.088$), and significant effects on the other traits (all $p < 0.01$); Treatments × age had significant effect on length ($p = 0.018$), area ($p = 0.023$) and the other traits (all $p < 0.01$), minor effect on LSI ($p = 0.312$), respectively. The age had significant effect on most leaf traits, but had minor effects on the thickness of BC, the SLA of CB and CS, and the LSI of CS (Figure 1).

Table 1. The effects (*F*- and *p*-value) of treatments and age categories on leaf traits.

Variance Sources	Treatments		Age		Treatments × Age	
	F	*P*	*F*	*P*	*F*	*P*
Length	28.504	<0.001	15.933	<0.001	46.012	<0.001
Width	12.201	<0.001	34.696	<0.001	17.197	<0.001
Area	18.653	<0.001	39.326	<0.001	43.575	<0.001
Thickness	2.394	0.069	30.578	<0.001	26.949	<0.001
Biomass	12.174	<0.001	0.611	0.435	64.173	<0.001
SLA	9.275	<0.001	23.41	<0.001	28.657	<0.001
LSI	20.03	<0.001	10.95	0.001	4.2	0.006

Note: LSI was leaf shape index.

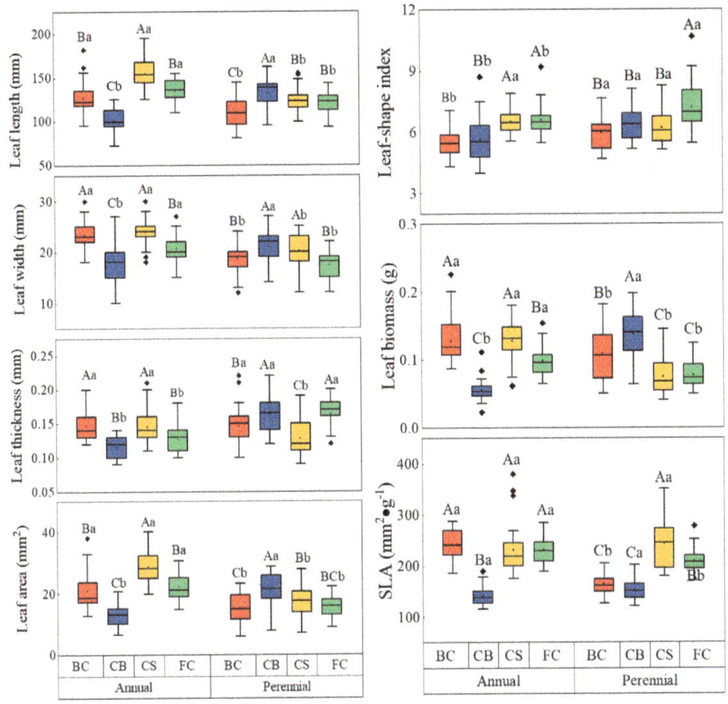

Figure 1. Variations of leaf traits in age categories and treatments. Notes: CS: Crown–thinning of spruce plantation; CB: crown–thinning of deciduous broad-leaved forest; FC: broad-leaved forest canopy; BC: bamboo grove of clearcutting. The same as below. The same capital and lowercase letter were not significantly different ($p < 0.05$) between the treatments and ages.

The boxes of annual leaf traits were showed in Figure 1. The length (154.43 ± 17.26 mm) and area (28.64 ± 5.11 cm^2) of CS were significant than those of FC (134.40 ± 12.93 mm and 22.34 ± 4.4 cm^2, respectively) and BC (126.47 ± 19.01 mm and 20.92 ± 5.91 cm^2). The width (23.90 ± 2.83 mm), thickness (0.145 ± 0.027 mm) and biomass (0.128 ± 0.03 g) of CS were significantly higher than those of FC (20.60 ± 2.43 mm, 0.127 ± 0.017 mm and 0.098 ± 0.022 g, respectively). The LSI of CS (6.49 ± 0.59) was significantly higher than that of BC (5.47 ± 0.69). The CB decreased almost all annual leaf traits significantly. The width, thickness and biomass of BC were significantly higher than those of FC, while the LSI of BC was the opposite. The other leaf traits between different treatments did not have significant difference.

The boxes of perennial leaf traits were showed in Figure 1. The length (131.13 ± 15.70 mm), thickness (0.163 ± 0.027 mm), area (20.94 ± 5.63 cm^2), and biomass (0.136 ± 0.036 g) of CB were significantly higher than the other treatments, while the SLA (152.06 ± 20.67 cm^2 g^{-1}) was the opposite. The SLA (244.41 ± 45.14 cm^2 g^{-1}) of CS were significantly higher than the other treatments and the thickness (0.127 ± 0.025 mm) was the opposite. The length (122.97 ± 13.78 mm) and area (17.56 ± 4.52 cm^2) of CS were significantly higher than those of BC (108.90 ± 16.23 mm and 15.15 ± 4.84 cm^2). The LSI of FC (7.17 ± 1.23) was significantly higher than the other treatments. The length (121.43 ± 12.30 mm), thickness (0.163 ± 0.02 mm) and SLA (208.82 ± 22.58 cm^2 g^{-1}) of FC were significantly higher than those of BC (108.90 ± 16.23mm, 0.146 ± 0.028 mm and 163.56 ± 17.27 cm^2 g^{-1}, respectively), while the biomass (0.077 ± 0.02 g) of FC was significantly lower than that of BC (0.108 ± 0.038 g). The other leaf traits between different treatments did not have significant difference.

2.2. Allometric Relationships between Core Leaf Traits

2.2.1. Length vs. Width

The relationships of length vs. width were allometric except for the perennial leaf in FC (p = 0.150, Figure 2A,B). Most relationships of length vs. width followed the "diminishing returns" except for the annual leaf in BC, and there existed significant differences between slopes and 1.0 except for annual leaf in BC, and perennial leaf in CS and FC (Table 2). The SCSs and ISEs between treatments were significant except for annual leaf ISE of CS-FC and CB-BC, annual leaf SCS of FC-CB, perennial leaf ISEs of CB-CS, BC-CS and FC-BC, and perennial leaf SCS of CB-CS and BC-FC (Table 2), respectively. The inter-annual allometric relationships varied only largely in FC, and the allometric relationship of annual leaves in FC was significant (p = 0.002, Table 3) while that of perennial leaves was slightly (p = 0.150). The ISEs and SCSs were significant except for ISEs in FC (p = 0.302) and BC (p = 0.064).

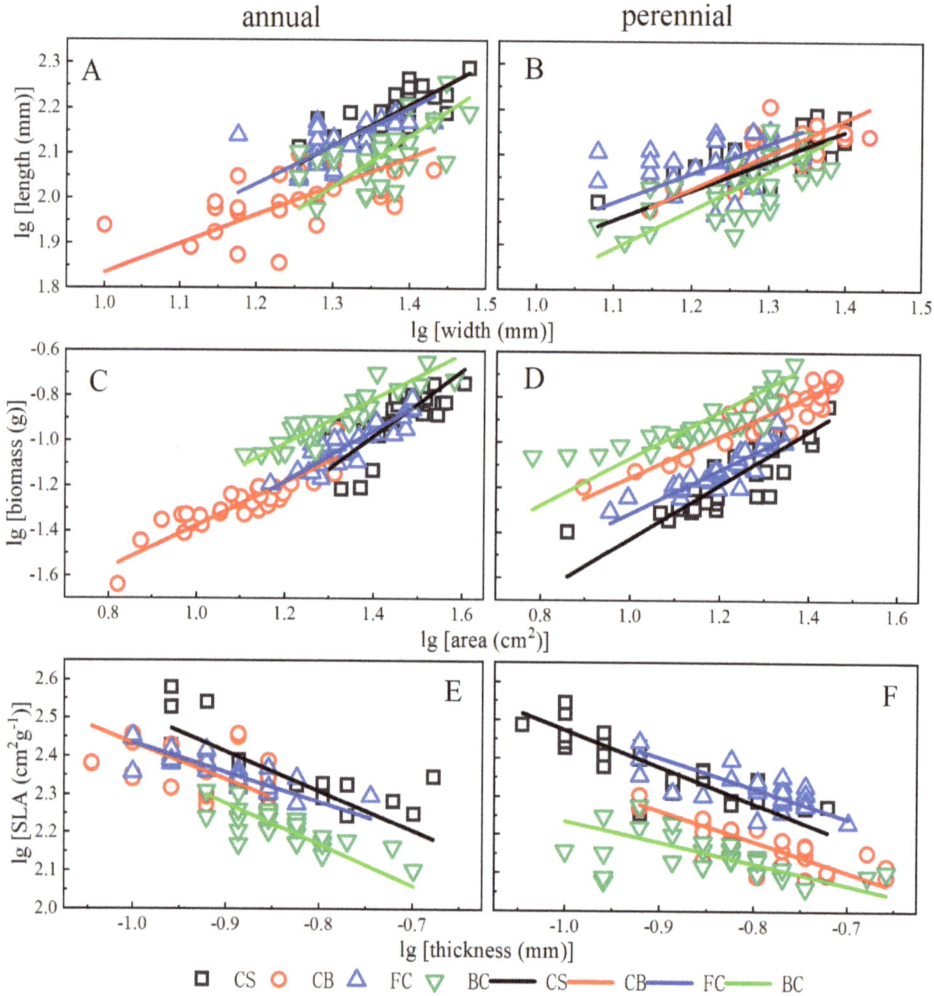

Figure 2. Allometric relationships of length vs. width, biomass vs. area and SLA vs. thickness in age categories and treatments.

Table 2. Effects of crown–thinning on the allometric relationships of length vs. width, biomass vs. area and SLA vs. thickness.

Trade-Off	Year	Treatments	R^2	Slope (95% CI)	Intercept	Common Slope (95% CI)	Intercepts Shift in Elevation			Shifts along the Common Slope		
							CS	CB	FC	CS	CB	FC
Length vs. width	annual	CS	0.494 **	0.910 ^{ns} (0.694, 1.195)	0.934 (0.589, 1.279)	0.874 (0.747, 1.021)	1			1		
		CB	0.266 **	0.642 ** (0.463, 0.889)	1.194 (0.926, 1.462)		**	1		**	1	
		FC	0.291 **	0.844 ^{ns} (0.612, 1.162)	1.020 (0.660, 1.381)		ns	**	1	**	**	1
		BC	0.342 **	1.125 ^{ns} (0.826, 1.533)	0.564 (0.082, 1.046)		**	ns	**	**	**	ns
	perennial	CS	0.392 **	0.660 ** (0.490, 0.888)	1.230 (0.971, 1.490)	0.743 (0.640, 0.863)	1			1		
		CB	0.501 **	0.773 ^{ns} (0.590, 1.013)	1.099 (0.820, 1.377)		ns	1		ns	1	
		FC	0.073 ^{ns}	0.657 *(0.456, 0.946)	1.273 (0.971, 1.575)		**	ns	1	*	**	1
		BC	0.434 **	0.858 ^{ns} (0.644, 1.143)	0.95 (0.634, 1.266)		*	**	**	**	**	ns
Biomass vs. area	annual	CS	0.541 **	1.462 ** (1.128, 1.896) ^{a}	−3.026 (−3.584, −2.469)							
		CB	0.805 **	0.946 ^{ns} (0.798, 1.122) ^{b}	−2.319 (−2.499, −2.140)							
		FC	0.764 **	1.125 ^{ns} (0.933, 1.356) ^{ab}	−2.528 (−2.812, −2.243)							
		BC	0.843 **	1.002 ^{ns} (0.860, 1.168) ^{b}	−2.214 (−2.416, −2.013)							
	perennial	CS	0.691 **	1.198 ^{ns} (0.967, 1.483)	−2.624 (−2.943, −2.306)	1.014 (0.934, 1.100)	1			1		
		CB	0.824 **	0.915 ^{ns} (0.778, 1.076)	−2.067 (−2.262, −1.873)		**	1		**	1	
		FC	0.786 **	0.973 ^{ns} (0.814, 1.163)	−2.285 (−2.494, −2.077)		**	**	1	ns	**	1
		BC	0.893 **	1.041 ^{ns} (0.917, 1.181)	−2.197 (−2.35, −2.043)		**	ns	**	ns	**	ns
SLA vs. thickness	annual	CS	0.547 **	−1.035 ^{ns} (−1.339, −0.800)	1.481 (1.252, 1.71)	−0.918 (−0.794, −1.064)	1			1		
		CB	0.043 ^{ns}	−0.943 ^{ns} (−1.365, −0.651)	1.265 (0.929, 1.601)		**	1		**	1	
		FC	0.66 **	−0.777 *(−0.972, −0.621)	1.66 (1.502, 1.819)		**	**	1	ns	**	1
		BC	0.04 ^{ns}	−1.084 ^{ns} (−1.571, −0.748)	1.47 (1.125, 1.816)		ns	**	**	ns	**	ns
	perennial	CS	0.713 **	−0.958 ^{ns} (−1.176, −0.779)	1.516 (1.337, 1.696)	−0.821 (−0.720, −0.932)	1			1		
		CB	0.653 **	−0.801 ^{ns} (−1.004, −0.638)	1.544 (1.398, 1.689)		**	1		**	1	
		FC	0.466 **	−0.798 ^{ns} (−1.055, −0.604)	1.685 (1.506, 1.865)		*	**	1	**	**	1
		BC	0.045 ^{ns}	−0.558 ** (−0.808, −0.386)	1.74 (1.561, 1.92)		**	ns	**	**	**	**

Note: BC was bamboo grove of clearcutting. * $p < 0.05$; ** $p < 0.01$; ns, $p > 0.05$. 95% CI were 95% confidence intervals. The superscript labels after R^2 meant the significant-level of allometry relationship, the superscript label after slope meant the significant-level between slope and slope = 1.0, the different superscript lowercase letters after (95% CI) meant that there existed the significant difference ($p < 0.05$) between the slopes, respectively. The same as below.

Table 3. Effects of age on the allometric relationships of length vs. width, biomass vs. area and SLA vs. thickness.

Trade-Off	Treatments	Common Slope (95% CI)	Intercepts Shift in Elevation	Shifts along the Common Slope
Length vs. width	CS	0.7866 (0.639, 0.966)	**	**
	CB	0.717 (0.582, 0.882)	**	**
	FC	0.7568 (0.593, 0.963)	ns	**
	BC	0.9725 (0.78691, 1.205)	ns	**
Biomass vs. area	CS	1.298 (1.099, 1.536)	ns	**
	CB	0.9297 (0.829, 1.044)	**	**
	FC	1.0422 (0.915, 1.188)	**	**
	BC	1.025 (0.931, 1.128)	**	**
SLA vs. thickness	CS	−0.9868 (−0.842, −1.15862)	**	**
	CB	−0.836 (−0.691, −1.016)	**	**
	FC	−0.7853 (−0.661, −0.9337)	**	**

Note: The SLA vs. thickness in BC had no common slope ($p = 0.012$), and the slope of perennial leaves was significantly higher than that of annual leaves ($p = 0.01$).

2.2.2. Biomass vs. Area

All biomass vs. area were significantly allometric relationship (all $p < 0.001$, Figure 2C,D), and only the difference between the slope and 1.0 of annual leaves in CS was significant. The perennial leaves in CB and FC followed the "diminishing returns", while the others followed the "increasing returns". The biomass vs. area of annual leaves had no common slope ($p = 0.044$), and the slope in CS was significantly higher than that in CB-BC (Table 3). Based on the common slope of perennial leaves, ISEs (except for CB-BC, and CS-FC, BC-CS) and SCSs (except for FC-BC) were significant. The inter-annual allometric relationships did exist in all treatments significantly, and all ISEs and SCSs were significant except for ISE in CS ($p = 0.205$, Table 3).

2.2.3. SLA vs. Thickness

The relationships of SLA vs. thickness (Figure 2E,F) were significantly allometric except for that of annual leaves in CB and BC, and perennial leaves in BC, respectively. Only the difference between slope and 1.0 of annual leaves in CS and perennial leaves in BC were significant. The SCSs (except for annual leaves of BC-CS, FC-BC and FC-CS) and ISCs (except for annual leaves of BC-CS, and perennial leaves of BC-CB) were significant (Table 2).

The inter-annual allometric relationships (Table 3) varied only largely in CB, and the SLA vs. thickness of annual leaves was not allometric ($p = 0.272$) while that of perennial leaves was significantly allometric ($p < 0.01$). SLA vs. thickness of each treatment had common slopes except for BC ($p = 0.012$), and the slop of annual leaves in BC was significantly higher than that of perennial leaves ($p < 0.01$). The SCSs and ISCs were significantly in CS, CB and FC (all $p < 0.001$), respectively.

3. Materials and Methods

3.1. Site

The study was carried out in the core of GPECN, which is located in the Sichuan Daxiangling Provincial Nature Reserve (DNR; 102°29′36″–102°52′24″ E, 29°28′33″–29°43′54″ N, 690–3666 m a.s.l.). As the Largest Reintroduction Base for Giant Panda and southern of Giant Panda National Park, the DNR locates in the southwest of Yingjing County of Sichuan Province, and the area is about 29,000 hm^2. The major soil is yellow-brown soil, which is classified as Alfisols in the soil taxonomy system of China [29]. The climate is humid, and the rainfall is hefty, with up to 200 rainy days per year [30]. The DNR vegetations vary from the evergreen broad-leaved forest (below 1500 m), coniferous forest (1500–2500 m), and coniferous and broad-leaved mixed forest (2100–2600 m) to the coniferous forest (above 2500 m) [30], respectively.

3.2. Experimental Design and Survey

In the Autumn of 2018, near to Management and Protection Station of the DNR (29°40′15.960″ N, 102°36′38.592″ E, 2096 m a.s.l.), the crown–thinning experiment had been carried out in the DBF and SP by removing the one-tree crown of every tree, which was equivalent to basal-area removal by about 50% [27] and decreased the canopy coverage to 40–50%. In the later Autumn of 2020, 8 plots (2 m × 2 m) of four treatments (BC, SC, FC and BC), total 32 bamboo plots bad been set and surveyed. The plots were distributed randomly, and the distance between the two plots was more than 20 m for all treatments. The stand and bamboo characteristics of all treatments were listed in the Table 4. After two-year, the canopy coverage of CF and CB increased in some distance (Table 4), and the bamboo growth varied largely in four treatments. Because the sudden outbreak of COVID-19 at the beginning of 2020, the people around DNR tended to pick and cut new shoots rather than went out to work. So, we did not investigate the current-year bamboo in all plots.

Table 4. Tree-layer characteristics of study plots.

Treatments	Canopy Tree	Average DBH (cm)	Canopy Coverage	Average Height (m)
CB	*Betula* and Litsea	10.21 ± 1.73 a	About 55%	15.44 ± 4.74 a
CS	*Picea asperata*	9.96 ± 1.88 a	About 50%	14.78 ± 4.17 a
FC	*Betula* and Litsea	9.83 ± 1.72 a	About 95%	14.48 ± 4.38 a

Notes: CS: Crown–thinning of spruce plantation; CB: crown–thinning of deciduous broad-leaved forest; FC: broad-leaved forest canopy. Columns (mean ± S.D.) followed by the same letter (s) were not significantly different ($p < 0.05$).

3.3. Leaf Sampling and Measurement

We defined that the annual leaf was the leaves (not the current year) from annual bamboos and perennial leaf was the leaves (>2 years) from perennial bamboos (>2 years) [31]. Leaf sample collection and traits determination had been carried out: 3 mean leaves from the middle to the top part of mean annual and perennial bamboo were harvested to measure leaf traits (area, length, thickness and width, $n = 4$ treatments $\times$ 2 ages $\times$ 30 leaves = 240). The individual leaf area was measured using CI-203 Portable Laser Area Meter (1%, CID, Washington, DC, USA), biomass (killed at 105 °C for 15 min and then dried at 85 °C for 48 h; Taisite, WHL45B, Tianjin, China) were measured with an electronic balance (0.0001 g, Zhuojing Experimental Equipment Co., Ltd., BMS, Shanghai, China) [32], and length, thickness and width were measured using a digital vernier caliper (0.01 mm, Deli, model DL91300, Deli Group Co., Ltd., Ningbo, Zhejiang, China), based on which the specific leaf area (SLA, $cm^2\ g^{-1}$) and leaf shape index (LSI = length/width) were calculated.

3.4. Statistical Analysis

Before performing analysis, we examined the data for normality and homogeneity of variances, and log-transformed to correct deviations from these assumptions when needed. We examined the differences in leaf traits among treatments, age, and their interactions by General Linear Model (GLM). We compared the traits by Fisher's least significant difference (LSD) procedure. The statistical analyses were performed using IBM SPSS 20.0 (IBM Corp., Armonk, NY, USA). We created the figures by OriginPro 2019b (OriginLab Corp., Northampton, MA, USA). The tests were considered significant at $p < 0.05$ level. The data were mean $\pm$ S.D.

Based on the lg-transformed data, we analyzed trade-offs between core leaf traits of leaf quality vs. photosynthetic-area (biomass vs. area) [8], light acquisition [4] and photosynthetic potential [33] vs. cold resistance (SLA vs. thickness) [34] and leaf shape (length vs. width) [35], respectively. The allometric relationships between leaf traits were analyzed using the standardized major axis method (SMA) with the SMATR Version 2.0 (http://bio.mq.edu.au/research/groups/comparative/SMATR/download.html, accessed on 19 April 2023) [7]. According to Warton et al. (2006) [7] and Chmura et al. (2017) [9], we tested slope differences among compared treatments. The slope values were compared with Sidak correction (when more than two treatments were compared) when significant differences were found in test 1.0. In those cases where the slopes did not differ statistically, we tested for significance of intercept shift in elevation (ISE) and shift along the common slope (SCS).

4. Discussion

Our crown–thinning enhanced the annual leaf size (length, width, thickness, biomass and area) in CS largely [22], and had significant negative effect on those in CB. The reasons might be that: the canopy-openess promoted the relative higher understorey plants to grow [36], the stump sprouting increased shading in sometimes [37], and the height of annual bamboos in CB (148 $\pm$ 13 cm) were significant shorter than that of the perennial (200 $\pm$ 7 cm, $p < 0.01$, unpublished), and the interaction shading of tree and perennial neighbor bamboo affected the leaf heavily [38], which decreased the bamboo photosynthesis [18]. On the contrary, the height of annual bamboo in FC (143 $\pm$ 6 cm) were slightly higher than that of the perennial (137 $\pm$ 16 cm, $p = 0.721$, unpublished), and could use the scattered light transformed through the canopy [39], which leaded the leaf to be better than that of CB. Furthermore, another complex response might be that bamboo was a typical clone plant, which daughter ramets (i.e., the annual bamboo in our study) could get resources by itself and share from mother ramets [40] (i.e., the perennial bamboo in our study).

In the present study, the SLA change showed that the bamboo leaf economic spectrum shifted from "slow investment-return" of higher light intensity (i.e., BC) and lower light intensity (i.e., FC and annual leaf of CB) to "fast investment-return" of middle light intensity (i.e., CS) [14]. Combined with other different-age leaf traits, we found that moderate canopy (i.e., CS) rather than sunny (i.e., BC) and dense canopy (i.e., FC and annual leaf of CB) would improve bamboo to grow and develop [41]. This is consistent with the moderate disturbance theory in ecological management [42]. Therefore, bamboo leaves have shown obvious adaptation to the light intensity shift, which meant that our canopy management via crown–thinning could improve bamboo growth. In fact, our treatments were a canopy-shading gradient for bamboo (i.e., BC < CB < CS < FC, Table 4). Unlike more experiments [43], bamboo leaf traits of our study showed a unimodal curve with the gradient rather than a linear relationship, which is consistent with the leaf anatomical traits at levels of plant functional group from eastern China [44], and vascular plant leaf area vs. leaf biomass within the Bailongjiang from northwestern China [45].

In the present study, the relationships of lg-transformed length vs. width and biomass vs. area were positively, while that of the SLA vs. thickness was the converse. Furthermore, the trade-off between SLA and leaf thickness suggested that future management shold pay more attention on cold. Furthermore, the trade-offs of annual leaf in CB were shifted largely, and the shifts did not decrease the leaf quality. For instance, the leaf length, thickness, width and area increased within the crown–thinning conditions (esp. CB). The enhanced light improved the habitat for annual bamboo in CB [4,8,33–35]. In other words, the bamboo leaf traits were unimodal curves following our crown–thinning light gradients, which was inconsistent with other studies, their functional traits usually were liner-relationship with latitude [46], elevation [43] and enhancing-light [47,48].

In the persent study, the CB stumps are easy to sprout while the CS stumps are difficult to sprout, and the CB has higher understorey plant-species pool than CS. This leaded the CB shading increased largely than that of CS, which hindered the annual bamboo in CB [36,37]. This is why the similar crown–thinning lead to different results, i.e., the successful crown–thinning was CS. Hence, our future management should pay more attention on the contexts of target stand rather than simply canopy-removal. The result was only the early effects of crown–thinning on bamboo leaf traits. And the comprehensive knowledge of crown–thinning on bamboo need more and longer systematic explorations.

5. Conclusions

The crown–thinning had obviously influenced bamboo leaf traits, which were high trait- and age-specific. The bamboo leaf could adapt to changing environments quickly, and covaried in traits did not reduce leaf quality. The CS increased light intensity and improved the growth environment for bamboo, and the canopy-removal intensity might appropriately increase aimed to fast-growing and density canopy forests.

Author Contributions: Conceptualization, J.X., G.C., D.F. and Y.Z.; methodology, D.F., G.C. and J.X.; software, validation, formal analysis, investigation, resources and data curation, D.F., X.F., L.L., H.Q., Y.A. and J.X.; writing—original draft preparation, D.F., J.X., G.C., Z.Y., X.X., Y.Y. and Z.C.; writing—review and editing, D.F., J.X. and G.C.; visualization, D.F. and J.X.; supervision, J.X., G.C. and Y.Z.; project administration, D.F., J.X. and Y.Z.; funding acquisition, Y.Z., G.C., H.Q. and D.F. All authors have read and agreed to the published version of the manuscript.

Funding: This research was funded by the Key Research Project for Science and Technology Program of Sichuan Province (2020YFS0023), National Natural Science Foundation of China (31800352) and Provincial College Students' Innovative Entrepreneurial Training Plan Program of Sichuan (202010626066).

Data Availability Statement: The data presented in this study are available on request from the corresponding author.

Conflicts of Interest: The authors declare no conflict of interest.

References

1. Zhu, G.; Niklas, K.J.; Li, M.; Sun, J.; Lyu, M.; Chen, X.; Wang, M.; Zhong, Q.; Cheng, D. "Diminishing Returns" in the scaling between leaf area and twig size in three forest communities along an elevation gradient of Wuyi Mountain, China. *Forests* **2019**, *10*, 1138. [CrossRef]
2. Li, Y.; Liu, C.; Sack, L.; Xu, L.; Li, M.; Zhang, J.; He, N. Leaf trait network architecture shifts with species-richness and climate across forests at continental scale. *Ecol. Lett.* **2022**, *25*, 1442–1457. [CrossRef] [PubMed]
3. Chen, G.; Yang, K.; Zhang, J.; Wang, M.; Wang, L.; Xian, J. Effects of succession stages and altitudinal gradient on leaf surface area and biomass allocation of typical plants in the subalpine of Eastern Tibetan Plateau. *Glob. Ecol. Conserv.* **2021**, *27*, e01590. [CrossRef]
4. Freschet, G.T.; Violle, C.; Bourget, M.Y.; Scherer-Lorenzen, M.; Fort, F. Allocation, morphology, physiology, architecture: The multiple facets of plant above- and below-ground responses to resource stress. *New Phytol.* **2018**, *219*, 1338–1352. [CrossRef]
5. Li, J.; Chen, X.; Niklas, K.J.; Sun, J.; Wang, Z.; Zhong, Q.; Hu, D.; Cheng, D. A whole-plant economics spectrum including bark functional traits for 59 subtropical woody plant species. *J. Ecol.* **2022**, *110*, 248–261. [CrossRef]
6. Liu, Y.; Li, G.; Wu, X.; Niklas, K.J.; Yang, Z.; Sun, S. Linkage between species traits and plant phenology in an alpine meadow. *Oecologia* **2021**, *195*, 409–419. [CrossRef]
7. Warton, D.I.; Wright, I.J.; Falster, D.S.; Westoby, M. Bivariate line-fitting methods for allometry. *Biol. Rev.* **2006**, *81*, 259–291. [CrossRef]
8. Ashby, B.; Watkins, E.; Lourenço, J.; Gupta, S.; Foster, K.R. Competing species leave many potential niches unfilled. *Nat. Ecol. Evol.* **2017**, *1*, 1495–1501. [CrossRef]
9. Chmura, D.J.; Guzicka, M.; Rożkowski, R.; Chałupka, W. Allometry varies among related families of Norway spruce. *Ann. For. Sci.* **2017**, *74*, 36. [CrossRef]
10. Poorter, L.; Bongers, F. Leaf traits are good predictors of plant performance across 53 rain forest species. *Ecology* **2006**, *87*, 1733–1743. [CrossRef]
11. Reich, P.B. The world-wide 'fast–slow' plant economics spectrum: A traits manifesto. *J. Ecol.* **2014**, *102*, 275–301. [CrossRef]
12. He, N.; Liu, C.; Piao, S.; Sack, L.; Xu, L.; Luo, Y.; He, J.; Han, X.; Zhou, G.; Zhou, X.; et al. Ecosystem traits linking functional traits to macroecology. *Trends Ecol. Evol.* **2019**, *34*, 200–210. [CrossRef]
13. Yang, K.; Chen, G.; Xian, J.; Yu, X.; Wang, L. Scaling relationship between leaf mass and leaf area: A case study using six alpine Rhododendron species in the Eastern Tibetan Plateau. *Glob. Ecol. Conserv.* **2021**, *30*, e01754. [CrossRef]
14. Wright, I.J.; Reich, P.B.; Westoby, M.; Ackerly, D.D.; Baruch, Z.; Bongers, F.; Cavender-Bares, J.; Chapin, T.; Cornelissen, J.H.C.; Diemer, M.; et al. The worldwide leaf economics spectrum. *Nature* **2004**, *428*, 821–827. [CrossRef]
15. Wang, H.; Wang, R.; Harrison, S.P.; Prentice, I.C. Leaf morphological traits as adaptations to multiple climate gradients. *J. Ecol.* **2022**, *110*, 1344–1355. [CrossRef]
16. Lasky, J.R.; Uriarte, M.; Boukili, V.K.; Chazdon, R.L. Trait-mediated assembly processes predict successional changes in community diversity of tropical forests. *Proc. Natl. Acad. Sci. USA* **2014**, *111*, 5616–5621. [CrossRef]
17. Poorter, L.; Craven, D.; Jakovac, C.C.; Sande, M.T.v.d.; Amissah, L.; Bongers, F.; Chazdon, R.L.; Farrior, C.E.; Kambach, S.; Meave, J.A.; et al. Multidimensional tropical forest recovery. *Science* **2021**, *374*, 1370–1376. [CrossRef]
18. Bin, Y.; Li, Y.; Russo, S.E.; Cao, H.; Ni, Y.; Ye, W.; Lian, J. Leaf trait expression varies with tree size and ecological strategy in a subtropical forest. *Funct. Ecol.* **2022**, *36*, 1010–1022. [CrossRef]
19. Osnas, J.L.D.; Katabuchi, M.; Kitajima, K.; Wright, S.J.; Reich, P.B.; Bael, S.A.V.; Kraft, N.J.B.; Samaniego, M.J.; Pacala, S.W.; Lichstein, J.W. Divergent drivers of leaf trait variation within species, among species, and among functional groups. *Proc. Natl. Acad. Sci. USA* **2018**, *115*, 5480–5485. [CrossRef] [PubMed]
20. Wei, N.; Cronn, R.; Liston, A.; Ashman, T.-L. Functional trait divergence and trait plasticity confer polyploid advantage in heterogeneous environments. *New Phytol.* **2019**, *221*, 2286–2297. [CrossRef] [PubMed]
21. He, N.; Li, Y.; Liu, C.; Xu, L.; Li, M.; Zhang, J.; He, J.; Tang, Z.; Han, X.; Ye, Q.; et al. Plant trait networks: Improved resolution of the dimensionality of adaptation. *Trends Ecol. Evol.* **2020**, *35*, 908–918. [CrossRef] [PubMed]
22. Denslow, J.S.; Schultz, J.C.; Vitousek, P.M.; Strain, B.R. Growth responses of tropical shrubs to treefall gap environments. *Ecology* **1990**, *71*, 165–179. [CrossRef]
23. Kishore, K.; Rupa, T.R.; Samant, D. Influence of shade intensity on growth, biomass allocation, yield and quality of pineapple in mango-based intercropping system. *Sci. Hortic.* **2021**, *278*, 109868. [CrossRef]
24. Ren, S.; Ali, A.; Liu, H.; Yuan, Z.; Yang, Q.; Shen, G.; Zhou, S.; Wang, X. Response of community diversity and productivity to canopy gap disturbance in subtropical forests. *For. Ecol. Manag.* **2021**, *502*, 119740. [CrossRef]
25. Wei, F.; Swaisgood, R.; Hu, Y.; Nie, Y.; Yan, L.; Zhang, Z.; Qi, D.; Zhu, L. Progress in the ecology and conservation of giant pandas. *Conserv. Biol.* **2015**, *29*, 1497–1507. [CrossRef]
26. Shi, J.Y.; Zhang, Y.X.; Zhou, D.Q.; Ma, L.S.; Yao, J. Qiongzhuea Hsueh et TP Yi. In *Illustrated Flora of Bambusoideae in China*; Shi, J., Zhang, Y., Zhou, D., Ma, L., Yao, J., Eds.; Springer: Singapore, 2020; pp. 555–572.
27. Taylor, A.H.; Qin, Z. Structure and composition of selectively cut and uncut Abies-Tsuga forest in Wolong Natural Reserve and implications for Panda conservation in China. *Biol. Conserv.* **1989**, *47*, 83–108. [CrossRef]

28. Negishi, Y.; Eto, Y.; Hishita, M.; Negishi, S.; Suzuki, M.; Masaka, K.; Seiwa, K. Role of thinning intensity in creating mixed hardwood and conifer forests within a Cryptomeria japonica conifer plantation: A 14-year study. *For. Ecol. Manag.* **2020**, *468*, 118184. [CrossRef]
29. Gong, Z. *Chinese Soil Taxonomy: Theory, Method and Practice*; Science Press: Beijing, China, 1999. (In Chinese)
30. Fu, M.; Pan, H.; Song, X.; Dai, Q.; Qi, D.; Ran, J.; Hou, R.; Yang, X.; Gu, X.; Yang, B.; et al. Back-and-forth shifts in habitat selection by giant pandas over the past two decades in the Daxiangling Mountains, southwestern China. *J. Nat. Conserv.* **2022**, *66*, 126129. [CrossRef]
31. Shi, J.Y.; Zhang, Y.X.; Zhou, D.Q.; Ma, L.S.; Yao, J. Morphology of Bamboos. In *Illustrated Flora of Bambusoideae in China*; Shi, J., Zhang, Y., Zhou, D., Ma, L., Yao, J., Eds.; Springer: Singapore, 2020; pp. 1–18.
32. Guo, J.; Zhao, C.; Zhang, L.; Han, Y.; Cao, R.; Liu, Y.; Sun, S. Water table decline alters arthropod community structure by shifting plant communities and leaf nutrients in a Tibetan peatland. *Sci. Total Environ.* **2022**, *814*, 151944. [CrossRef]
33. Liu, M.; Wang, Z.; Li, S.; Lü, X.; Wang, X.; Han, X. Changes in specific leaf area of dominant plants in temperate grasslands along a 2500-km transect in northern China. *Sci. Rep.* **2017**, *7*, 10780. [CrossRef]
34. Gorsuch, P.A.; Pandey, S.; Atkin, O.K. Temporal heterogeneity of cold acclimation phenotypes in Arabidopsis leaves. *Plant Cell Environ.* **2010**, *33*, 244–258. [CrossRef]
35. Li, Y.; Zou, D.; Shrestha, N.; Xu, X.; Wang, Q.; Jia, W.; Wang, Z. Spatiotemporal variation in leaf size and shape in response to climate. *J. Plant Ecol.* **2019**, *13*, 87–96. [CrossRef]
36. Carter, T.A.; Fornwalt, P.J.; Dwire, K.A.; Laughlin, D.C. Understory plant community responses to widespread spruce mortality in a subalpine forest. *J. Veg. Sci.* **2022**, *33*, e13109. [CrossRef]
37. Sample, R.D.; Boggess, C.M.; Brooke, J.M.; Jenkins, M.A. Stump sprouting performance of common tree species in the midstory of hardwood forests in Indiana. *For. Sci.* **2022**, *68*, 440–446. [CrossRef]
38. Niinemets, Ü. A review of light interception in plant stands from leaf to canopy in different plant functional types and in species with varying shade tolerance. *Ecol. Res.* **2010**, *25*, 693–714. [CrossRef]
39. Chen, K.; Hu, K.; Xi, F.; Wang, H.; Kohnen, M.V.; Gao, P.; Liao, J.; Wei, W.; Liu, X.; Zhang, H.; et al. High-efficient and transient transformation of Moso bamboo (*Phyllostachys edulis*) and Ma bamboo (*Dendrocalamus latiflorus* Munro). *J. Plant Biol.* **2023**, *66*, 75–86. [CrossRef]
40. Zhai, W.; Wang, Y.; Luan, J.; Liu, S. Effects of nitrogen addition on clonal integration between mother and daughter ramets of Moso bamboo: A ^{13}C-CO_2 pulse labeling study. *J. Plant Ecol.* **2022**, *15*, 756–770. [CrossRef]
41. Hao, X.; Jia, J.; Mi, J.; Yang, S.; Khattak, A.M.; Zheng, L.; Gao, W.; Wang, M. An optimization model of light intensity and nitrogen concentration coupled with yield and quality. *Plant Growth Regul.* **2020**, *92*, 319–331. [CrossRef]
42. Baard, P. Managing climate change: A view from deep Ecology. *Ethics Environ.* **2015**, *20*, 23–44. [CrossRef]
43. Fadrique, B.; Baraloto, C.; Bravo-Avila, C.H.; Feeley, K.J. Bamboo climatic tolerances are decoupled from leaf functional traits across an Andean elevation gradient. *Oikos* **2022**, *22*, e09229. [CrossRef]
44. He, N.; Liu, C.; Tian, M.; Li, M.; Yang, H.; Yu, G.; Guo, D.; Smith, M.D.; Yu, Q.; Hou, J. Variation in leaf anatomical traits from tropical to cold-temperate forests and linkage to ecosystem functions. *Funct. Ecol.* **2018**, *32*, 10–19. [CrossRef]
45. Yang, K.; Chen, G.; Xian, J.; Chen, W. Varying relationship between vascular plant leaf area and leaf biomass along an elevational gradient on the eastern Qinghai-Tibet Plateau. *Front. Plant Sci.* **2022**, *13*, 824461. [CrossRef] [PubMed]
46. Wang, W.; Jiang, Y.; Li, B.; Xi, N.; Chen, Y.; He, D.; Feng, J.; Chu, C. Species abundance is jointly determined by functional traits and negative density dependence in a subtropical forest in southern China. *J. Plant Ecol.* **2021**, *14*, 491–503. [CrossRef]
47. Blondeel, H.; Perring, M.P.; Depauw, L.; De Lombaerde, E.; Landuyt, D.; De Frenne, P.; Verheyen, K. Light and warming drive forest understorey community development in different environments. *Glob. Chang. Biol.* **2020**, *26*, 1681–1696. [CrossRef]
48. De Pauw, K.; Sanczuk, P.; Meeussen, C.; Depauw, L.; De Lombaerde, E.; Govaert, S.; Vanneste, T.; Brunet, J.; Cousins, S.A.O.; Gasperini, C.; et al. Forest understorey communities respond strongly to light in interaction with forest structure, but not to microclimate warming. *New Phytol.* **2022**, *233*, 219–235. [CrossRef]

plants

MDPI

Article

Plasticity in the Morphology of Growing Bamboo: A Bayesian Analysis of Exogenous Treatment Effects on Plant Height, Internode Length, and Internode Numbers

Chongyang Wu [1], Yucong Bai [1], Zhihua Cao [2], Junlei Xu [1], Yali Xie [1], Huifang Zheng [1], Jutang Jiang [1], Changhong Mu [1], Wenlong Cheng [1], Hui Fang [1] and Jian Gao [1,*]

[1] Key Laboratory of National Forestry and Grassland Administration, Beijing for Bamboo & Rattan Science and Technology/International Center for Bamboo and Rattan, Beijing 100102, China; wcy@icbr.ac.cn (C.W.); bai.yucong@icbr.ac.cn (Y.B.); xjl@icbr.ac.cn (J.X.); xieyali@icbr.ac.cn (Y.X.); jiangjutang@icbr.ac.cn (J.J.); muchanghong@icbr.ac.cn (C.M.); chengwenlong@icbr.ac.cn (W.C.); fanghui@icbr.ac.cn (H.F.)

[2] Anhui Academy of Forestry, Hefei 230036, China

[*] Correspondence: gaojian@icbr.ac.cn

Abstract: Sucrose (Suc) and gibberellin (GA) can promote the elongation of certain internodes in bamboo. However, there is a lack of field studies to support these findings and no evidence concerning how Suc and GA promote the plant height of bamboo by regulating the internode elongation and number. We investigated the plant height, the length of each internode, and the total number of internodes of Moso bamboo (*Phyllostachys edulis*) under exogenous Suc, GA, and control group (CTRL) treatments in the field and analyzed how Suc and GA affected the height of Moso bamboo by promoting the internode length and number. The lengths of the 10th–50th internodes were significantly increased under the exogenous Suc and GA treatments, and the number of internodes was significantly increased by the exogenous Suc treatment. The increased effect of Suc and GA exogenous treatment on the proportion of longer internodes showed a weakening trend near the plant height of 15–16 m compared with the CTRL, suggesting that these exogenous treatments may be more effective in regions where bamboo growth is suboptimal. This study demonstrated that both the exogenous Suc and GA treatments could promote internode elongation of Moso bamboo in the field. The exogenous GA treatment had a stronger effect on internode elongation, and the exogenous Suc treatment had a stronger effect on increasing the internode numbers. The increase in plant height by the exogenous Suc and GA treatments was promoted by the co-elongation of most internodes or the increase in the proportion of longer internodes.

Keywords: bamboo; plant morphology; sucrose; gibberellin; Bayesian analysis; plasticity

Citation: Wu, C.; Bai, Y.; Cao, Z.; Xu, J.; Xie, Y.; Zheng, H.; Jiang, J.; Mu, C.; Cheng, W.; Fang, H.; et al. Plasticity in the Morphology of Growing Bamboo: A Bayesian Analysis of Exogenous Treatment Effects on Plant Height, Internode Length, and Internode Numbers. *Plants* **2023**, *12*, 1713. https://doi.org/10.3390/plants12081713

Academic Editors: Jie Gao, Weiwei Huang, Johan Gielis and Peijian Shi

Received: 15 March 2023
Revised: 15 April 2023
Accepted: 17 April 2023
Published: 20 April 2023

1. Introduction

Bamboos (Poaceae, Bambusoideae) can be divided into woody and herbaceous species based on their culm and flowering characteristics and are one of the most important forest resources in the world [1,2]. Woody bamboo has a higher economic value than herbaceous bamboo since bamboo materials and shoots are more valuable. The economic value of bamboo is directly related to its culm height, and thus, culm height growth is an aspect of bamboo research. The woody bamboos are distributed in tropical and subtropical regions. Owing to their high ecological and economic value [3–7], woody bamboos are widely cultivated in many countries outside their natural range [8,9].

China may be the origin region of bamboo. At present, China has the world's most diverse bamboo resources [10]; there are 534 bamboo species in 34 genera [11]. The bamboo forest area covers 64,116 km^2 [12]. China is the world's largest producer, consumer, and exporter of bamboo products. In 2020, the total import and export trade of bamboo products in China was USD 2.21 billion [13]. Moso bamboo (*Phyllostachys edulis*) is a perennial woody plant that covers an area of 46,778 km^2, accounting for 72.96% of the total bamboo forest area [12]. This species is the most crucial dual-purpose plant, producing bamboo shoots and material.

Bamboo material has excellent rigidity, hardness, flexural strength, and ductility, and it is widely used as a material for engineering and in daily applications [7,14,15]. Bamboo-flattening technology is considered one of the most advanced and mainstream technologies. This technique uses intact internodes for pressure flattening or planing flattening; as a result, longer internodes can add higher value to bamboo products, such as flooring, cutting boards, and furniture [16–19]. Meanwhile, from an ecological perspective, longer internodes mean increased biomass, leading to higher bamboo height, thicker culms, and more branches and leaves, which are features that are directly related to the carbon sinks of bamboo forests [20,21]. Although longer internodes are important to bamboo processing and ecology, how to increase the internodes at a low cost is an unsolved problem.

Sugar (e.g., sucrose, glucose, fructose) is the energy source within plants; sucrose (Suc) is not only the main photosynthetic product but is also the main form of carbohydrate storage in many higher plants [22–24]. The Suc is transported through the phloem to tissues and organs that do not perform photosynthesis, and it plays a vital role in various stages of plant growth [24–26]. In a study of tomato and sorghum, it was found that Suc was closely related to the growth of axillary buds, and Suc affects the length of lateral branches by regulating some development-related genes [27,28]. In other studies, exogenous Suc treatment was found to promote the growth of cucumber seedlings [29] and the root growth of rice seedlings [30]. Different concentrations of Suc can induce stomatal opening or closing in *Arabidopsis thaliana* [31]. Although the external application of Suc can promote plant growth, excessive Suc concentration can cause changes in osmotic pressure during growth, resulting in plant death [32,33]. Sugar also plays a vital role in the rapid growth of bamboo. The catabolism of starch and soluble sugars was significantly increased in the spring shoots of *P. edulis* [34]. The culm sheath uses water as a medium to transport organic acids, sugars, and other compounds to the internodes for the growth of *P. edulis* shoots [35]. Wang et al. [36] found that sugar content was high in the fast elongation and mature internodes of *Fargesia yunnanensis*, implying that the elongation of bamboo internodes requires a large amount of sugar. Several other studies further supported this conjecture. Wei et al. [37] applied 5 g L^{-1} Suc to *Bambusa multiplex* and found positive correlations between Suc content and the elongation rate of internodes. A study of *F. yunnanensis* showed that the starch content in shoot buds was inadequate for the consumption of bamboo shoots during the rapid growth period, and more carbohydrates needed to be obtained from the parent plant [36]. Suc could not be detected until the end of the rapid growth period of *Fargesia yunnanensis*, and the Suc transported to the internodes was entirely hydrolyzed, suggesting that the energy in the bamboo shoots may not be sufficient to fully support the rapid elongation of the internodes [36]. All of the above results show that sugar is highly correlated with the rapid growth of bamboo. Is it possible to increase the internode length by supplementing plants with sugar during the rapid growth period?

Gibberellin (GA) was first discovered from the pathogenic fungus *Gibberella fujikuroi* of *Oryza sativa*, and this discovery triggered the first Green Revolution [38]. GA has many roles in the developmental processes of plants, including seed germination [39] and flower and fruit formation and development [23,39–41]. It is noteworthy that the most apparent effect of GA is to regulate stem length and internode elongation during the growth period by inducing cell division and elongation [42,43]. This was demonstrated in model plants and crops, such as *Arabidopsis* [44,45], rice [46], maize [23,47], Napier grass [48], and *Medicago truncatula* [49]. GA is primarily used for dwarfing in agriculture, and there have been many reports on this in crops, such as wheat [50–53], rice [54,55], maize [56,57], and watermelon [58,59]. Crops need to be dwarfed by GA to increase yield. In contrast, bamboo requires greater plant height to increase economic and ecological value. When the shoot height of *P. edulis* exceeds 1.5 m, the GA content in the middle of the shoot is higher than in the tip and bottom [60]. GA signaling may affect the growth of

P. edulis by regulating gibberellic-acid-stimulated transcript genes or general regulatory factor proteins [61,62]. However, the effects of GA on bamboo internodes have been less well studied. Wang et al. [63] reported that *P. edulis* f. *tubaeformis* (a dwarf variant of Moso bamboo) has shorter internodes and plant heights than wild Moso bamboo, and the GA content in the internodes of the variant is also lower than that of wild Moso bamboo. Similar results were found in *P. edulis* seedlings; the seedlings under GA treatment had longer internodes and higher plant height [64], faster growth rates, and longer internode cells than the controls [65].

To date, studies suggest that Suc and GA may promote plant height and internode elongation in bamboos [37,63–65]. However, most studies focused on specific internodes or seedlings of bamboo, and there is no field investigation evidence concerning the changes in plant height or internode length under exogenous Suc and GA treatments. Because plant height and internode length are very important to the economic and ecological value of bamboo, focusing on the effects of Suc and GA on plant height and internode length of *P. edulis*, we posed the following questions:

(1) Does exogenous Suc or GA treatment promote the plant height of *P. edulis* in the wild?
(2) Is it a large elongation of a few internodes or a small elongation of all the internodes that promotes an increase in the plant height of *P. edulis*?
(3) Does having more internodes promote the increase in plant height of *P. edulis*?
(4) Does the promotion of internode length under exogenous Suc and GA treatments diminish with distance from the injection point?

In this study, we subjected *P. edulis* in a wild bamboo forest to exogenous Suc and GA treatments and measured the plant height, internode length, and internode number under different treatments. Using Bayesian modeling to analyze the effects of exogenous Suc and GA on plant height, internode length, and internode number, we discuss how Suc and GA promote the plant growth of *P. edulis* in view of the above four questions.

2. Results

2.1. Effects of Exogenous Treatments on the Plant Height, Internode Numbers, and Internode Lengths

Exogenous Suc and GA treatments resulted in significantly higher plant heights than in the CTRL (Figure 1a). The observed values of plant height under exogenous Suc treatments were more concentrated than others, and no *P. edulis* was smaller than 10 m in all 79 samples. There were more internodes under exogenous the Suc treatments than in the CTRL, and there were no significant differences between exogenous GA treatments and the CTRL (Figure 1b). The internode length was significantly higher under exogenous Suc and GA treatments than in the CTRL (Figure 1c–f). Details of the statistical results are shown in Tables S1 and S2.

2.2. Data Characteristics and Correlations between Variables

The multi-density plots in Figure 2 show the distribution of each variable under the CTRL, exogenous Suc, and GA treatments. The distributions of plant heights in different treatments were similar, but the data in the exogenous Suc treatment were concentrated around 16 m. There were more plants higher than 18 m in the exogenous GA treatment. The distribution of internode numbers coincided in the CTRL and exogenous GA treatments, and the distribution was shifted to the right in the exogenous Suc treatment. For the lengths of the 10th–50th internodes in the exogenous Suc and GA treatments, the distribution was shifted to the right and was more concentrated around 0.30–0.35 m compared with the CTRL. The numbers in Figure 2 show the Pearson correlation analysis. There were strong correlations between the plant height and the lengths of the 30th–40th internodes, plant height and the lengths of the 40th–50th internodes, the lengths of the 20th–30th internodes and the lengths of the 10th–20th internodes, and the lengths of the 20th–30th internodes and the lengths of the 30th–40th internodes ($r > 0.8$, $p < 0.001$). However, in comparing all variables, the correlations in the CTRL treatment were higher than those of the exogenous Suc and GA treatments in almost all cases (Figure 2).

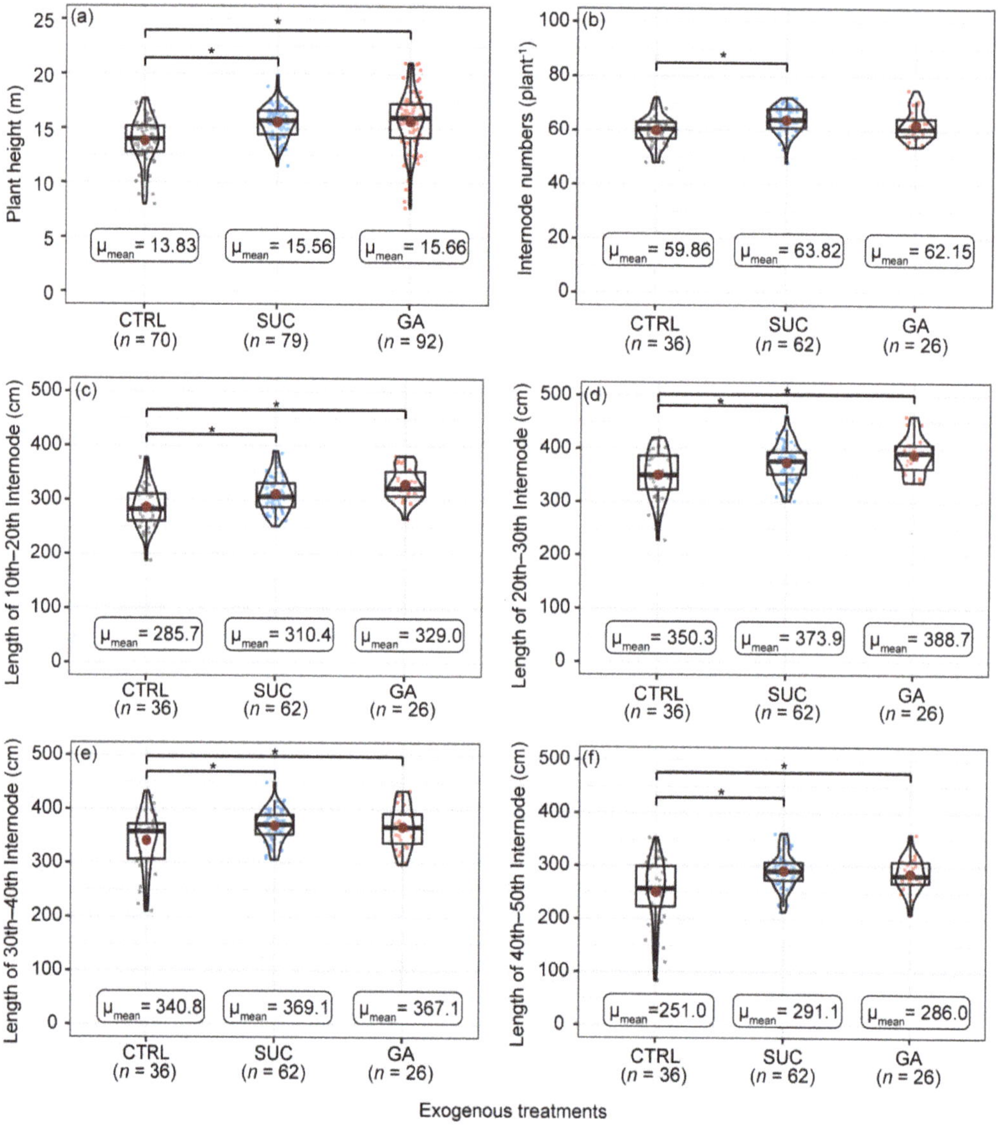

Figure 1. Comparisons of the plant height (**a**), internode number (**b**), lengths of the 10th–20th internodes (**c**), lengths of the 20th–30th internodes (**d**), lengths of the 30th–40th internodes (**e**), and lengths of the 40th–50th internodes (**f**) under different exogenous treatments. Box plots and violin plots show the data distribution of each variable. Points represent the measured values of each variable. Red points indicate mean values. The numbers in brackets represent sampling numbers. An asterisk means a significant difference according to the Bayesian GLMM.

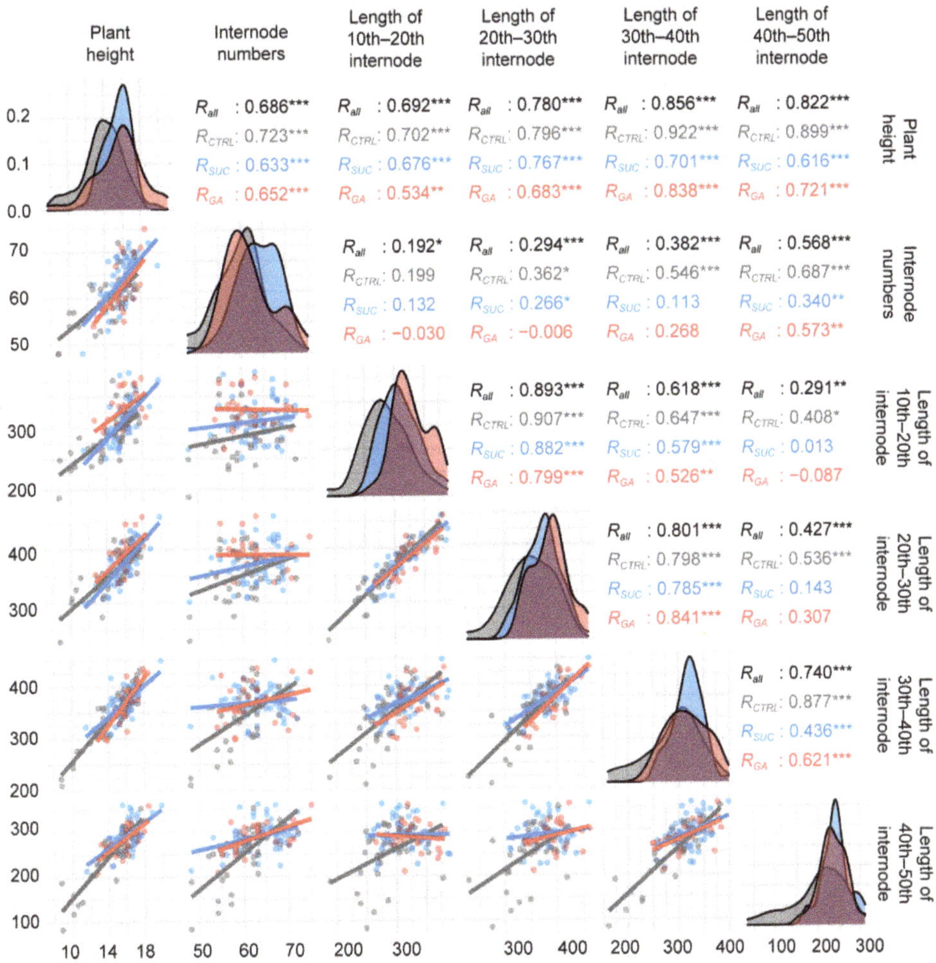

Figure 2. Correlations between the plant height; internode numbers; and lengths of the 10th–20th internodes, 20th–30th internodes, 30th–40th internodes, and 40th–50th internodes under different exogenous treatments. The different colors show each exogenous treatment. Numbers indicate Pearson correlation coefficients (* $p < 0.05$, ** $p < 0.01$, *** $p < 0.001$). Multidensity plots and scatterplots show the distributions of the variables.

2.3. *Effects of Exogenous Treatments on the Internode Lengths*

Figure 3 shows the differences in the internode length under the CTRL, exogenous Suc, and GA treatments. The internode length increased from the base, reached a peak near the 25th–30th internode, and then gradually decreased until the tip of the tree. Exogenous Suc and GA treatments resulted in significantly longer total lengths of the 10th–20th, 20th–30th, 30th–40th, and 40th–50th internodes than the CTRL. The length increase in the useful internodes (>20 cm) was approximately 1.2 m in the exogenous Suc treatment and approximately 1.4 m in the exogenous GA treatment. The details of these statistical results are shown in Table S1.

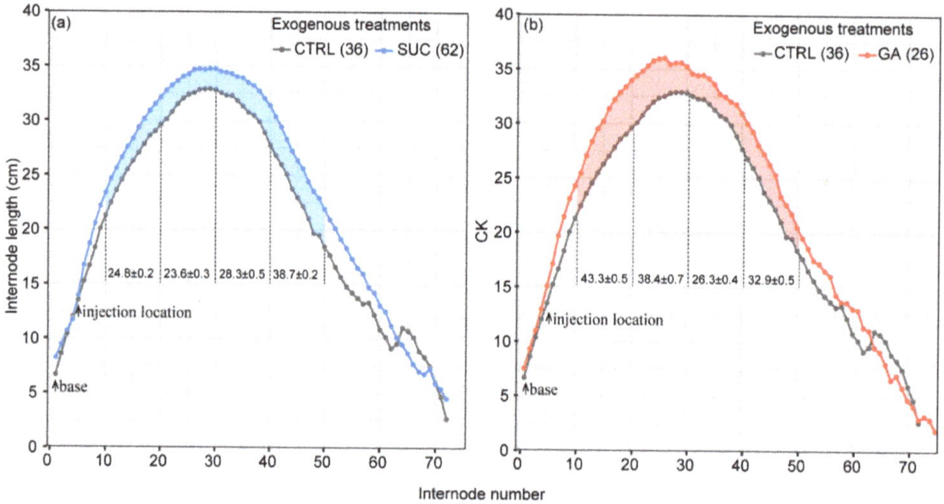

Figure 3. The internode lengths at different internodes under Suc (**a**) and GA (**b**) treatments compared with the CTRL. The points indicate the mean value of the internode length. The colored shapes indicate a significant difference in the length of every ten internodes between the Suc or GA treatment and the CTRL. The mean values and standard error of 10 internodes with a significant difference are shown between the lines. The numbers in brackets represent the sample sizes.

2.4. Variation in Promotion of Internode Elongation under Exogenous Treatments with Distance

To know whether the promoting effect of internode length under different exogenous treatments varied with distance, we examined the interaction effect between exogenous treatments and distance from the injection point. The interaction effect between different exogenous (CTRL–Suc) treatments and the internode number was not significant (Figure 4, Table S3). This result means that the effect of exogenous Suc treatment on internode length did not enhance or attenuate with increasing distance from the injection point. However, the interaction effect between different exogenous (CTRL–GA) treatments and internode number was significantly positive in the 10th–20th and significantly negative in the 20th–30th internodes and was not significant in the 30th–40th or 40th–50th internodes (Figure 4, Table S3).

2.5. Effects of Exogenous Treatments on the Percentage of Internodes with Different Lengths

The proportion of 30–40 cm internodes was significantly higher under exogenous Suc treatment compared with the CTRL, but there was no significant difference in other lengths of internodes (Figure 5, Table S4). The results indicate that the increasing internode numbers under exogenous Suc treatment were largely from the 30–40 cm internodes. The proportions of 10–20 cm and 20–30 cm internodes were significantly lower, and those of the 30–40 cm and 40–50 cm internodes were significantly higher in the exogenous GA treatment compared with the CTRL (Figure 5, Table S4). The decreasing numbers with short internodes counterbalanced the increasing numbers with long internodes. Under exogenous Suc and GA treatments, increases were seen in the long internodes; the longer the internodes, the greater their utility for applications.

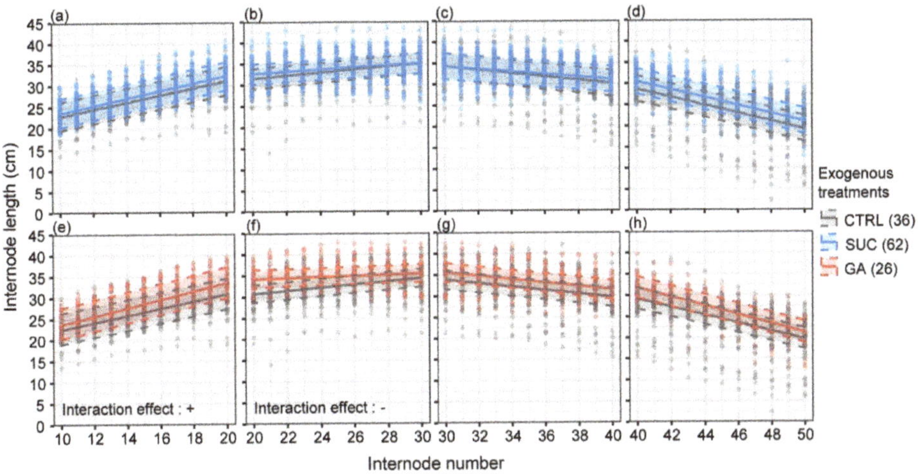

Figure 4. The effect of the interaction between exogenous treatments and internode number on the internode length. The points represent the measured values of the internode length. The shaded intervals within the dashed lines show the 95% Bayesian credible intervals. The numbers in brackets represent the sample sizes (**a–h**). The symbols + or − mean a significant positive interaction or negative interaction, respectively.

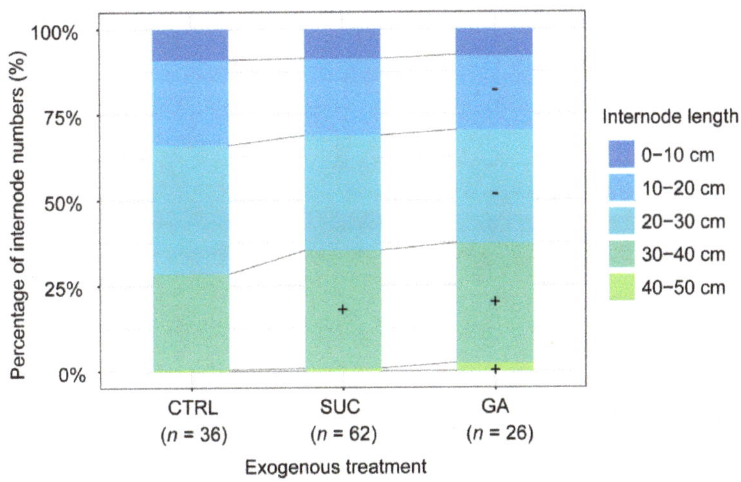

Figure 5. The percentages of internodes with different lengths under different exogenous treatments. The numbers in brackets represent the sample sizes. The symbols + or − mean a significant increase or significant decrease in the proportion of internodes compared with the CTRL, respectively.

2.6. *Effects of Exogenous Treatments on the Percentages of Internodes at Different Plant Heights*

Figure 6 and Table S5 show the interaction effects between different exogenous treatments and plant height to further confirm whether plant height affected the exogenous hormonal effects on the proportions of internode lengths. We found similar trends between the exogenous Suc and GA treatments. The proportion of 0–10 cm internodes was significantly higher, and the proportions of 20–30 cm, 30–40 cm, and 40–50 cm internodes were significantly lower than in the CTRL with increasing plant height. Judging from the intersection of the 95% Bayesian credible intervals, the increased effect of the exogenous treatments on the proportion of longer internodes (>20 cm) showed a weakening trend near the plant height of 15–16 m compared with the CTRL.

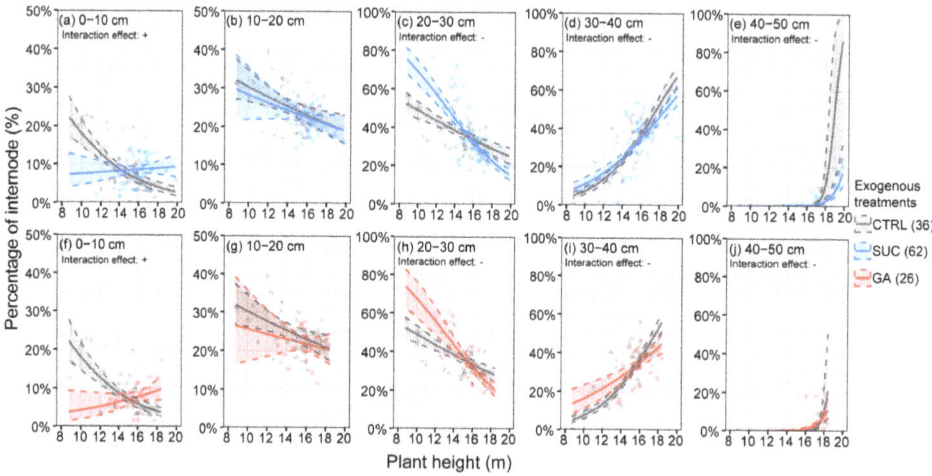

Figure 6. The effects of the interactions between the exogenous treatments and plant height on the percentage of internodes with different lengths ((**a**,**f**) 0−10 cm, (**b**,**g**) 10−20 cm, (**c**,**h**) 20−30 cm, (**d**,**i**) 30−40 cm, and (**e**,**j**) 40−50 cm). The points represent the percentage of internodes of each length at different plant heights. The shaded intervals within the dashed line show the 95% Bayesian credible intervals. The numbers in brackets represent the sample sizes. The symbols + or − mean a significant positive or negative interaction, respectively.

2.7. Effects of Exogenous Treatments on Ecological Benefits

To evaluate the effect of exogenous treatments on the ecological benefits of *P. edulis*, we estimated biomass and carbon storage per plant for different exogenous treatments. The biomass and carbon storage were significantly higher under the exogenous Suc and GA treatments than in the CTRL, and under the exogenous Suc treatment, the increases were approximately 1.2-fold; under the exogenous GA treatment, the increases were approximately 1.4-fold higher than under the CTRL (Figure 7, Table S6).

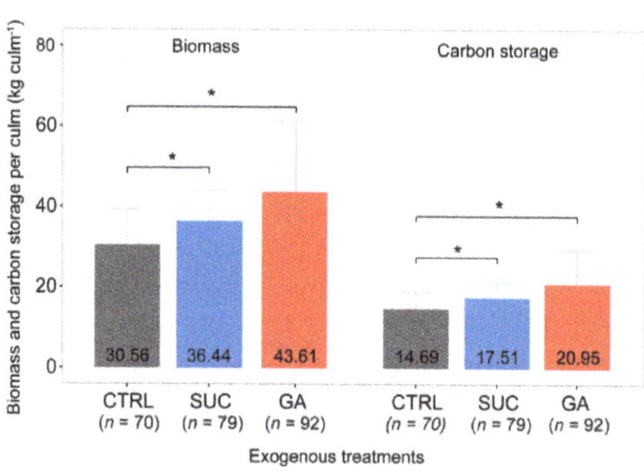

Figure 7. Comparison of the estimated biomass and carbon storage under different exogenous treatments. The numbers at the bottom of the bar graph represent the average values. The error bars show the 95% Bayesian credible intervals. The asterisk means a significant difference according to the Bayesian GLMM. The numbers in brackets represent the sample sizes.

3. Discussion

In the present study, we measured the plant height, internode length, and internode numbers of *P. edulis* under exogenous Suc and GA treatments. Owing to the field experiment of large bamboos requiring significant resources and manpower, this is the first published report demonstrating the effects of exogenous Suc and GA treatments on the plant height, internode length, and internode number of *P. edulis*. Our results show that both Suc and GA had a significant positive effect on the plant height; Suc had a stronger effect on internode number compared with GA, while GA had a stronger effect on internode length than Suc. In addition, we did not find the greater number of internodes (75) or higher plant height (21.0 m) of *P. edulis* that were published previously [63,65–74]. Here, the characteristics of the culm structure, the promoting effects of the exogenous Suc and GA on the plant height, internode length, and internode number, the variation in the effect of exogenous treatment with distance, and the ecological benefits brought by increased plant height are discussed. These also cover the four questions raised when presenting the objectives of this study.

3.1. Exogenous Treatments Did Not Affect the Characteristics of the Culm Morphological and Internode Length Distributions

Our results showed that the distribution of internode length in *P. edulis* had distinct features: gradually increasing from the bottom, peaking near the 25th–30th internode, then gradually getting shorter again, showing a shape similar to a Gaussian distribution (Figure 3). These results are in accord with those of recent studies [65], which indicated that the diameter of the internodes gradually decreases from the bottom to the top, becoming extremely thin at the tip; the culm shape is similar to a cone [68,74,75]. *P. edulis* is generally taller than 10 m, and some individuals reach about 20 m [70–73,76]. This height is rapidly achieved within a few months during the rapid growth period, and there is no doubt that the internodes at the bottom are subjected to greater mechanical stress compared with the tip. This mechanical stress may be one of the reasons for the shortening of the internode length at the bottom, where lignification has not yet been completed [37]. The mechanical stress effect in plant development is a general phenomenon [37,77–80].

Additionally, we considered that the short internodes at the tip may be related to the lack of water and nutrients. The height of *P. edulis* increases rapidly, but the emergence of branches and leaves does not appear during the rapid growth period [81,82]; without leaf transpiration, it may be difficult to transport water to internodes at the tip simply using root pressure [83–85]. It is well known that water can be used as a vehicle to transfer carbohydrates and nutrients [81,86,87]. Accordingly, we can infer that the internodes closer to the tip may lack water and nutrients, causing the internodes to be thin and short during the growing stage. Therefore, the mechanical stress on the internodes of the bottom and the nutritional deficiency of internodes at the tip may be why the internode length presented a Gaussian distribution and the culm was in the shape of a cone. However, this hypothesis lacks empirical support. For example, we can control the pressure by truncating the upper internodes and then measuring the internode length under different pressures during the rapid growth period. Alternatively, water transport and nutrient content of different internodes (bottom, middle, and tip) can be measured. Our results indicated that the exogenous Suc and GA treatments did not significantly alter the characteristics of culm morphology or node length distribution.

3.2. Effects of the Exogenous Treatments on the Internode Lengths

Our analysis showed that the GA treatment significantly promoted internode elongation (Figures 1c–f and 3b). These results were in accord with recent studies indicating that GA affects internode length by stimulating cell division and elongation in rice, which is a plant that is also in the Poaceae family [42,88–90]. Other studies also found that GA can promote internode elongation by increasing the cell length and cell number of *P. edulis* seedlings during the growth period. Thus, it is reasonable to consider that the promoting effect of exogenous GA on internode length is also achieved by increasing the cell length and number during the rapid growth period.

As an energy source and one of the main storage forms of carbohydrates in plants [25,26], sugar affects plant growth in many ways. Sugar can promote the growth of lateral branches and roots, improve the photosynthetic capacity of leaves, and control the opening and closing of stomata [91,92]. The exogenous Suc treatment significantly promoted internode elongation (Figures 1c–f and 3a). This finding is consistent with those of Wei et al. [37], who found that increasing soluble sugar content inhibited *BmSnf1*, a growth-inhibiting gene, in *Bambusa multiplex* (Lour.) Raeusch. ex Schult, and helped internodes enter into rapid growth because *BmSnf1* is known to be active under low-energy conditions and acts as a growth inhibitor [93]. The sugar content increased exponentially in the division and elongation zones in the internodes during the rapid growth period, indicating that the growth of internodes requires a large amount of sugar [37]. Meanwhile, in a study of *F. yunnanensis*, it was also demonstrated that the starch storage in bamboo shoots could not meet the elongation of the internodes during the rapid growth period, and an additional supply was required from the parental bamboo through rhizomes [36,87,94]. Therefore, we believe that exogenous Suc treatment supplemented the lack of starch and sugar during the rapid growth of *P. edulis* and promoted internode elongation.

We modeled the interaction effects on internode length between exogenous treatments and internode number, analyzing whether the promotion effects of exogenous Suc and GA on internode elongation varied with distance from the injection point. In our results, the effect of the exogenous Suc treatment did not diminish with distance from the injection point, while the effect of the exogenous GA treatment became stronger between the 10th and 20th internodes and became weaker between the 20th and 30th internodes with distance (Figure 4). The same trend can be seen in Figure 3; starting from the injection point, the promoting effect of exogenous GA on the internode length becomes stronger and reaches a peak at around the 25th internode. The GA_3 used in this study was shown to promote internode elongation in various crops, such as soybeans [95], oats [96], and corn [97]. The promotion of internode elongation by GA_3 varied with the distance; we believe that because the injected dose was not saturated relative to the demands of the rapid growth period, this led to large consumption of the exogenous GA_3 near the injection point. These findings support the hypothesis of Chen et al. [65], who barely detected GA_3 in *P. edulis* shoots during the rapid growth period (18th internode), consistent with our unpublished results. These results may mean that the endogenous GA_3 in *P. edulis* shoots during the rapid growth period could not meet the demand and was completely depleted. We may be able to use larger doses of GA_3 to promote internode elongation in the future.

3.3. *Effects of the Exogenous Treatments on the Internode Numbers*

A total of 7728 internodes of 124 individuals of *P. edulis* were measured in the present study. Our results showed that the internode numbers of *P. edulis* were not significantly increased under exogenous GA treatment compared with the CTRL, consistent with Honi et al. [98], who found no significant change in the internode numbers of jute under GA treatment. There are few studies concerning internode numbers, and almost all such studies investigated crops. For example, high temperatures can increase the internode numbers in soybeans [99–101], while drought can decrease the internode numbers in sugarcane crops [102]. We only found one report on the internode numbers in bamboo species describing how the internode length was significantly positively correlated with the internode number [67].

Nodes are one of the important characteristics of bamboo culms because nodes will strengthen the culms and prevent the deformation and cracking of internodes [103,104]. The internode numbers under the exogenous Suc treatment were significantly higher than those of the CTRL in the present study, and the average internode number increased by about four (Figure 1b). This result has not previously been reported. The bamboo shoots usually go dormant in the soil and wait for germination [105–107]. Bamboo shoots start to differentiate nodes and internodes in the S-4 developmental stages, and internodes are completely differentiated in the S-6 developmental stages [105]. It is generally believed that the internode numbers will not change after the shoots emerge from the soil. Nevertheless, even for the shoots in the ground, the shoot tips are not completely lignified. Is it possible that the supply of Suc promotes the differentiation of the undifferentiated cells in apical meristem? These results must be interpreted with caution because so many factors can affect the development of bamboo in the wild. Our results cannot confirm this conjecture, and we need to sample the bamboo shoots with a finer period division and observe the changes in the apical meristem.

The internode numbers were significantly increased under exogenous Suc treatment, but we also wanted to know whether the added internodes were short or long, as this is important for post-harvest bamboo material processing [16–19]. Figure 5 depicts the results; compared with the CTRL, the proportion of 30–40 cm internode–s in the exogenous Suc treatment increased significantly and accounted for more than 30% of the total. Although the proportions of 10–20 cm and 20–30 cm internodes were lower under exogenous GA treatment, the proportions of 30–40 cm and 40–50 cm internodes increased significantly. This result suggests that the increase in the internode numbers under Suc treatment could not only help to increase the stability of the bamboo culm but it can also have economic implications in bamboo material processing. In Figure 6, we illustrate the variation in the proportion of internode numbers of different lengths with plant height. The proportion of the shortest internodes (0–10 cm) in the CTRL decreased with the increase in plant height, while there was a slight increase in the exogenous Suc and GA treatments (Figure 6a,f) owing to the internodes being short and thin at the bamboo tip. There were more tall bamboos and more short and thin internodes under the exogenous Suc and GA treatments, and this may have been related to the difficulty in supplying water and nutrients to the higher places we mentioned earlier. The proportion of shorter internodes (10–20 cm and 20–30 cm) decreased with the increase in plant height, and the proportion of longer internodes (30–40 cm and 40–50 cm) increased with the increase in plant height. This trend was stronger under exogenous Suc and GA treatments than in the CTRL (Figure 6b–e,g–j). This coincided with the results shown in Figure 5; taller culms increased the proportion of long internodes and decreased the proportion of short internodes under the exogenous Suc and GA treatments.

An interesting finding is that the intersection of the interaction effects (Figure 6b–e,g–j) was at around 16 m in plant height. This result indicates that once the plant height of *P. edulis* exceeds 16 m, the regulation effects of exogenous Suc and GA on the ratios of long or short internodes begin to weaken, and the rate of return will start to decrease. There are physical limits to the heights of plants; these limits depend on their geographical location, precipitation, physical structure, and water transport capacity [108–112]. For taller bamboo plants (close to the physical limit), even a small increase in height can have a huge potential cost. In contrast, using exogenous Suc and GA on bamboo shoots that should be stunted may obtain a higher rate of return. The exogenous Suc and GA treatments may be more effective in areas where bamboo is not growing well.

3.4. *Effects of Exogenous Treatments on the Plant Height and Ecological Benefits*

The exogenous Suc and GA treatments significantly increased the plant height of *P. edulis* compared with the CTRL (Figure 1a). The plant height of bamboo comprises two parts: the internode length and internode number. The exogenous GA treatment significantly increased the internode length, while the exogenous Suc treatment increased the internode length and internode number (Figure 1b–f). Combined with the visualization results shown in Figure 3, we concluded that the promotion of internode length under exogenous Suc and GA treatments acted on most internodes. In other words, most of the internode elongation by the exogenous treatments resulted in an increase in plant height rather than only on some specific internodes. The internode elongation under the exogenous Suc treatment was weaker than that under the exogenous GA treatment, but the increase in internode number was greater. The internodes have low strength along the grain [103], and the nodes have an important mechanical function on the culm, preventing bending, buckling, and axial cracking [103,104]. This means that the increased nodes under the exogenous Suc treatment may be better from the perspective of culm stability, especially in some mountainous areas with harsh terrain. However, considering the utilization of bamboo material, the increased internode length under exogenous GA treatment has greater economic value; these two methods should be used differently depending on the occasion.

Although the exogenous Suc and GA treatments provided positive results in this study, considering the physiological integration ability of bamboos [113–115], we still believe that the promotion effect was underestimated. This is because when the exogenous treatment supplies the starch or hormones that the bamboo needs during the rapid growth period, the parent plant may redistribute its extra resources to another bamboo through the rhizomes (which may have included the control in the experiment). The results shown in Figure 1a also demonstrate the following: there were fewer bamboos with lower plant height under the exogenous Suc and GA treatments, and the data points were more concentrated compared with the control, suggesting that the exogenous treatments helped some *P. edulis* shoots that had limited growth potential to grow taller. If these samples were taken out of the data, the effects of the exogenous Suc and GA on the plant height and internode length might have been greater than what is shown in the results. In future studies, it may be necessary to set up an enclosed area or perform rhizome truncation treatments [113,114,116], and reevaluate the effect of exogenous Suc and GA treatments.

China plants a large amount of bamboo yearly, and bamboo can accumulate 76% of the biomass over the next five years during the rapid growth period [70,117]. The average carbon stock of *P. edulis* forests is 2.39 times higher than that of China fir (*Cunninghamia lanceolata*) forests, and thus, has a huge carbon sequestration potential [70,117–120]. Our results demonstrated that the biomass and carbon storage per plant of *P. edulis* increased significantly under the exogenous Suc and GA treatments, with average increases of 5.9 kg and 13.0 kg in biomass and average increases of 2.8 kg and 6.3 kg as a carbon sink (Figure 7). Although the plant height and diameter of bamboo cannot increase after the rapid growth period [117], a higher culm can support more branches and leaves. Considering the ecological benefits, the effects of exogenous Suc and GA treatments on plant height are meaningful until bamboo felling. The felled bamboo is processed into building materials [7,121–123], which is a process that ensures that this part of the sequestered carbon will not return to nature within a short time. The global forest area has continued to decline over the past 30 years, but bamboo forests have grown at an average annual rate of 3% [124]. *P. edulis* is China's most widely planted bamboo species [12]. Using low-cost Suc or GA to enhance *P. edulis* can increase the output value and help bamboo forests play an important role in mitigating climate change.

In the context of bamboo processing, the increase in internode length evidently contributes to the enhancement of economic benefits, as longer internodes allow for the production of a greater number of bamboo products. For bamboo handicrafts such as fans, walking sticks, and furniture, an increase in the length of individual internodes can significantly elevate the added value of the products. However, to quantify the impact of increased internode length on the economic benefits of these products, it will be necessary to gather more relevant data in future work, such as the price variations of walking sticks of different lengths.

The present study is the first to verify the effects of exogenous Suc and GA treatments on the plant height of *P. edulis* in the field. Because many bamboo species have similar physiological and ecological characteristics [125,126], we have reason to believe that the promoting effects of Suc and GA on the internode length and internode numbers may not be limited to *P. edulis*, but may also be effective for other bamboo species. This low-cost method of increasing the plant height and internode length may be highly profitable in some areas where bamboo is not growing well, both economically and ecologically. In future research, we will try to remove the effect of physiological integration and increase the number of bamboo species and hormone types.

4. Materials and Methods

4.1. Study Site and Experiment Design

In the present study, all experimental subjects were selected from pure stands of *P. edulis* located in Guangde City of Anhui Province (119.384 E, 30.799 N), Yanling County of Hunan Province (114.033 E, 26.548 N), and Xianning City of Hubei Province (114.313 E, 29.815 N) in the south of China (Figure 8). The soils were predominantly yellow loam; the temperatures and precipitation (Table S7) were suitable for *P. edulis* growth as a major distribution area in China (annual average temperature: 15.6–19.5 °C, annual total precipitation: 1436.6–1876.2 mm) [127]. We randomly labeled 160 bamboo shoots at a height of close to 50 cm with a similar ground diameter and recorded the ground diameter of all shoots in each study site at the end of March 2021. We injected 50 mL of 20 μM Suc (Shanghai Yuanye Bio-Technology Co., Ltd., Shanghai, China; S11055) and 20 μM GA_3 (Shanghai Yuanye Bio-Technology Co., Ltd., Shanghai, China; S28506) solutions into 60 shoots, and as a control, we injected 50 mL of pure water into another 60 shoots in each study site. We did not perform the exogenous treatment via spraying because the solution is usually sprayed on the leaves; however, bamboo grows very quickly during the rapid growth period and can reach heights that cannot be sprayed in a short time. The injection location was the fifth internode from the bottom. The injection was carried out every three days, five times in total in each study site from 28 March 2021 to 16 April 2021. Parenthetically, we found the suitable dose of Suc and GA to increase the plant height of bamboo shoots via pre-experiments from 2019 to 2020 in Guangde City, Anhui Province (119.384 E, 30.799 N). Virtually all bamboo shoots stopped their rapid growth by September. Hence, we measured the plant heights of the bamboo (n = 241) using the criterion RD 1000 (Laser Technology, USA) at the end of August 2021, except for other degraded bamboo shoots. Meanwhile, we felled 124 bamboo shoots (CTRL: 36; Suc: 62; GA: 26) and measured the internode length after labeling every internode. Owing to local laws and manpower constraints, we were not able to fell more bamboo plants.

4.2. Statistical Analysis

We analyzed the relationships between variables using a generalized linear mixed model (GLMM) with the Bayesian Markov chain Monte Carlo (MCMC) method. There are many advantages of Bayesian modeling approaches, including having the flexibility to set the distribution family of the models according to the characteristics of the data and fitting the data well even with small sample sizes. Therefore, the Bayesian approach is used in a wide range of applications [128–130]. In the present study, the internodes were divided into 10th–20th, 20th–30th, 30th–40th, and 40th–50th ranges for analysis because in the culm morphology of *P. edulis*, the closer to the tip, the thinner the culm. The 1st–10th internodes near the base and after the 50th internodes are shorter in length and have fewer applications. We set the plant height, internode number, internode length, and percentage of internodes with different lengths, estimated biomass, and carbon storage per culm as the dependent variables, with exogenous treatment, distance from the injection point, and plant height as the independent variables for modeling. The details of all model structures are shown in Table 1. For each model, we set 4 independent chains with 1 interval and obtained 5000 posterior samples from each chain after a burn-in of 5000 iterations. Prior distributions of all models used default settings in the brms package 2.17.0 [131]. We considered that zero not being within the 95% credible interval of variables indicated a statistically significant result. We calculated the R-hat values of all parameters and checked that weather was less than 1.01 [132].

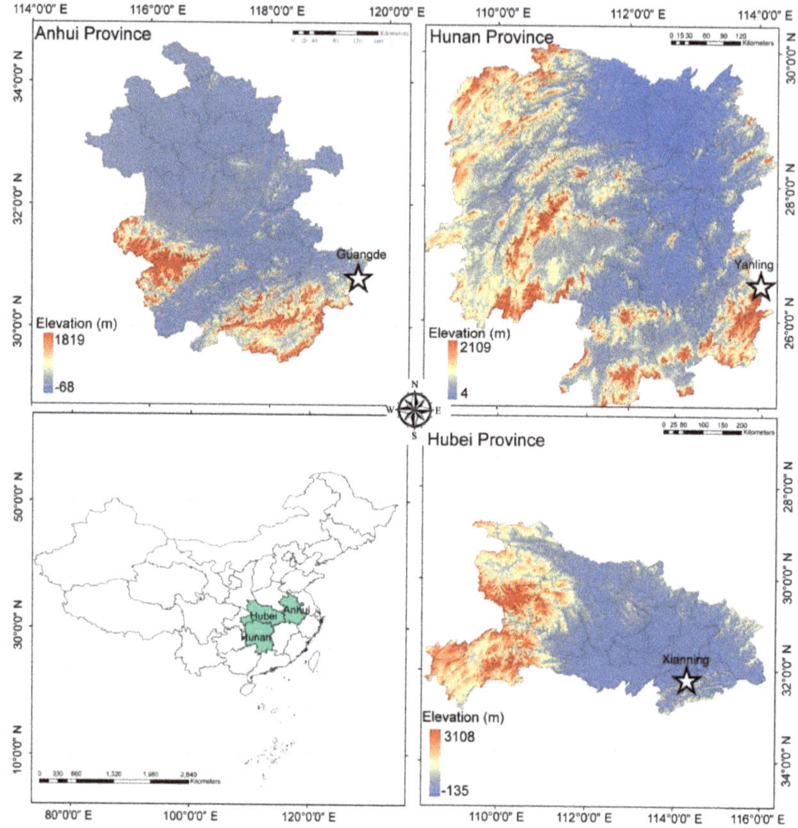

Figure 8. Study area.

Table 1. The structure of the Bayesian GLMM.

Model Structure	*i*	*j*	*k*	*l*
Ph_i or Eb_i or $Ec_i \sim Normal\ (\mu_i, \sigma^2{}_i)$ $\mu_i = \beta_0 + \beta_1 S_j + \beta_2 G_j + r_{1i} + r_{2l}$	1, 2, . . . , 241	0 or 1		1, 2, 3
$In_i \sim Poisson\ (\lambda_i)$ $log(\lambda_i) = \beta_0 + \beta_1 S_j + \beta_2 G_j + r_{1i} + r_{2l}$	1, 2, . . . , 124	0 or 1		1, 2, 3
$Il^{10\text{–}20th\ or\ 20\text{–}30th\ or\ 30\text{–}40th\ or\ 40\text{–}50th}{}_i \sim Normal\ (\mu_i, \sigma^2{}_i)$ $\mu_i = \beta_0 + \beta_1 S_j + \beta_2 G_j + r_{1i} + r_{2l}$	1, 2, . . . , 124	0 or 1		1, 2, 3
$Il^{10\text{–}20th\ or\ 20\text{–}30th\ or\ 30\text{–}40th\ or\ 40\text{–}50th}{}_i \sim Normal\ (\mu_i, \sigma^2{}_i)$ $\mu_i = \beta_0 + \beta_1 S_j + \beta_2 G_j + \beta_3 D_k + \beta_4 S_j D_k + \beta_5 G_j D_k + r_{1i} + r_{2l}$	1, 2, . . . , 124	0 or 1	1, 2, . . . , 75	1, 2, 3
$In^{0\text{–}10\ cm\ or\ 10\text{–}20\ cm\ or\ 20\text{–}30\ cm\ or\ 30\text{–}40\ cm\ or\ 40\text{–}50\ cm}{}_i \sim Binomial\ (In_i, logistic\ (p_i))$ $p_i = \beta_0 + \beta_1 S_j + \beta_2 G_j + r_{1i} + r_{2l}$	1, 2, . . . , 124	0 or 1		1, 2, 3
$In^{0\text{–}10\ cm\ or\ 10\text{–}20\ cm\ or\ 20\text{–}30\ cm\ or\ 30\text{–}40\ cm\ or\ 40\text{–}50\ cm}{}_i \sim Binomial\ (In_i, logistic\ (p_i))$ $\mu_i = \beta_0 + \beta_1 S_j + \beta_2 G_j + \beta_3 Ph_i + \beta_4 S_j Ph_i + \beta_5 G_j Ph_i + r_{1i} + r_{2l}$	1, 2, . . . , 124	0 or 1		1, 2, 3

Ph_i: plant height of culm *i*; Eb_i: estimated biomass of each culm *i*; Ec_i: estimated carbon storage of each culm *i*; In_i: internode numbers of each culm *i*; $Il^{10th\text{–}20th}{}_i$: 10th–20th internoder length of each culm *i*; $Il^{20th\text{–}30th}{}_i$: 20th–30th internode length of each culm *i*; $Il^{30th\text{–}40th}{}_i$: 30th–40th internode length of each culm *i*; $Il^{40th\text{–}50th}{}_i$: 40th–50th internode length of each culm *i*; $In^{0\text{–}10\ cm}{}_i$: 0–10 cm internode numbers of each culm *i*; $In^{10\text{–}20\ cm}{}_i$: 10–20 cm internode numbers of each culm *i*; $In^{20\text{–}30\ cm}{}_i$: 20–30 cm internode numbers of each culm *i*; $In^{30\text{–}40\ cm}{}_i$: 30–40 cm internode numbers of each culm *i*; $In^{40\text{–}50cm}{}_i$: 40–50 cm internode numbers of each culm *i*; β: intercept or regression coefficient for each variable; r_{1i}: random effect of ground diameter in each culm *i*; r_{2l}: random effect of each study site *l*; *S*: exogenous SUC treatment for each culm *i*; G: exogenous GA treatment for each culm *i*; D: distance from the jection point of each culm *i*.

The above-ground biomass and carbon storage per culm were calculated using the formulas below [133]:

$$B_{culm} = \frac{2t + \frac{R}{t} + 8}{100t + 20\frac{R}{t} - 920} \times H_{culm} D^{[150825 - 564.8t + 98.6\frac{R}{t}]10^{-5}} \quad (1)$$

$$C_{culm} = B_{culm} \times cc_{culm} \times 100\% \quad (2)$$

where B_{culm} is the above-ground biomass per culm, t is the annual average temperature, R is the annual average precipitation, H_{culm} is the culm height, D is the ground diameter, C_{culm} is the carbon storage per culm, and cc_{culm} is the carbon content of the *P. edulis* culm [134].

Analyses and visualizations were done using R v4.2.0 and ArcMap v10.2.1.

5. Conclusions

To the best of our knowledge, this study demonstrated for the first time that the plant height of *P. edulis* was promoted by regulating the co-elongation of most internodes and internode numbers during the rapid growth period. To address the questions in the introduction, we drew four main conclusions: (1) exogenous GA affected the internode length, while exogenous Suc affected the internode length and internode numbers; (2) most of the internode elongation resulted in the increase of plant height rather than only on some specific internodes; (3) the promotion of internode elongation using exogenous GA decreased with distance; and (4) the promotion of plant height using exogenous Suc and GA was better for bamboos below 16 m. Because many bamboo species have similar physiological and ecological characteristics, we have reason to believe that these insights may not be limited to *P. edulis*, but may also be effective for other bamboo species. These conclusions are helpful to understand the physiological and morphological characteristics of bamboo culms during the rapid growth period and provide a low-cost way to improve the economic and ecological benefits of Moso bamboo forests.

Supplementary Materials: The following supporting information can be downloaded from https://www.mdpi.com/article/10.3390/plants12081713/s1, Table S1: The posterior distribution for each parameter of plant height, internode number, and lengths of the 10th–20th, 20th–30th, 30th–40th, 40th–50th internodes. Table S2: Mean value, standard error, maximum value, and minimum value of plant height, internode number, and lengths of the 10th–20th, 20th–30th, 30th–40th, and 40th–50th internodes. Table S3: The posterior distribution for each parameter of internode length. Table S4: The posterior distribution for each parameter of the percentage of internodes with different internode lengths (0–10 cm, 10–20 cm, 20–30 cm, 30–40 cm, 40–50 cm). Table S5: The posterior distribution for each parameter of the percentage of internodes with different lengths (0–10 cm, 10–20 cm, 20–30 cm, 30–40 cm, 40–50 cm). Table S6: The posterior distribution for each parameter of estimated biomass and carbon storage. Table S7: Monthly temperature and precipitation of each study area in 2021.

Author Contributions: J.G. and C.W. conceived the study, designed the experiments, discussed the results, and finalized the manuscript. Y.B., Z.C., J.X., Y.X., H.Z., J.J., C.M., W.C. and H.F. collected the samples and discussed the results. All authors have read and agreed to the published version of the manuscript.

Funding: The Project was Supported by The National Key Research and Development Program of China (2021YFD2200505), the Fundamental Research Funds of ICBR (1632020005), and the National Key Research and Development Program of China (2018YFD0600101).

Data Availability Statement: Not applicable.

Acknowledgments: Special thanks to Zaijun Gu, Minggang Fang, Miao Gong, Biao Leng, and Xuexiang Liu for their contributions to this work. We thank Zhanchao Cheng, Juan Li, and Shaohua Mu of the International Center for Bamboo and Rattan for their suggestions on the experimental design. We also thank the anonymous reviewers and editors for their comments and editing.

Conflicts of Interest: The authors declare no conflict of interest.

References

1. Sharma, Y. *Bamboos in the Asia Pacific Region*; Lessard, G., Chouinard, A., Eds.; International Development Research Centre: Ottawa, ON, Canada, 1980.
2. Scurlock, J.M.; Dayton, D.C.; Hames, B. Bamboo: An overlooked biomass resource? *Biomass Bioenergy* **2000**, *19*, 229–244. [CrossRef]
3. Zhou, B.Z.; Fu, M.Y.; Xie, J.Z.; Yang, X.S.; Li, Z.C. Ecological functions of bamboo forest: Research and Application. *J. For. Res.* **2005**, *16*, 143–147. [CrossRef]
4. Song, X.Z.; Zhou, G.M.; Jiang, H.; Yu, S.Q.; Fu, J.H.; Li, W.Z.; Wang, W.F.; Ma, Z.H.; Peng, C.H. Carbon sequestration by Chinese bamboo forests and their ecological benefits: Assessment of potential, problems, and future challenges. *Environ. Rev.* **2011**, *19*, 418–428. [CrossRef]
5. Choudhury, D.; Sahu, J.K.; Sharma, G.D. Value addition to bamboo shoots: A review. *J. Food Sci. Technol.* **2012**, *49*, 407–414. [CrossRef]
6. Sawarkar, A.D.; Shrimankar, D.D.; Kumar, A.; Kumar, A.; Singh, E.; Singh, L.; Kumar, S.; Kumar, R. Commercial clustering of sustainable bamboo species in India. *Ind. Crops Prod.* **2020**, *154*, 112693. [CrossRef]
7. Gao, J. *The Moso Bamboo Genome*; Springer International Publishing: Cham, Switzerland, 2021.
8. Buziquia, S.T.; Lopes, P.V.F.; Almeida, A.K.; de Almeida, I.K. Impacts of bamboo spreading: A review. *Biodivers. Conserv.* **2019**, *28*, 3695–3711. [CrossRef]
9. Xu, Q.F.; Liang, C.F.; Chen, J.H.; Li, Y.C.; Qin, H.; Fuhrmann, J.J. Rapid bamboo invasion (expansion) and its effects on biodiversity and soil processes +. *Glob. Ecol. Conserv.* **2020**, *21*, e00787. [CrossRef]
10. Ahmad, Z.; Upadhyay, A.; Ding, Y.L.; Emamverdian, A.; Shahzad, A. Bamboo: Origin, habitat, distributions and global prospective. In *Biotechnological Advances in Bamboo*; Ahmad, Z., Ding, Y., Shahzad, A., Eds.; Springer: Singapore, 2021; pp. 1–31.
11. Chen, S.L.; Li, D.Z.; Zhu, G.H.; Wu, Z.L.; Lu, S.L.; Liu, L.; Wang, Z.P.; Wang, C.P.; Sun, B.X.; Sun, B.S.; et al. *Flora of China*; Science Press: Beijing, China, 2018; Volume 22.
12. State Forestry and Grassland Administration of China. *Report of Forest Resources in China (2014–2018)*; China Forestry Publishing House: Beijing, China, 2019.
13. INBAR. *China Bamboo and Rattan Commodity International Trade Report*; Wu, J.Q., Wang, X.D., Eds.; International Bamboo and Rattan Organisation: Beijing, China, 2020.
14. Dixon, P.G.; Gibson, L.J. The structure and mechanics of moso bamboo material. *J. R. Soc. Interface* **2014**, *11*, 20140321. [CrossRef]
15. Chen, M.L.; Ye, L.; Li, H.; Wang, G.; Chen, Q.; Fang, C.H.; Dai, C.P.; Fei, B.H. Flexural strength and ductility of moso bamboo. *Constr. Build. Mater.* **2020**, *246*, 118418. [CrossRef]
16. Fang, C.H.; Jiang, Z.H.; Sun, Z.J.; Liu, H.R.; Zhang, X.B.; Zhang, R.; Fei, B.H. An overview on bamboo culm flattening. *Constr. Build. Mater.* **2018**, *171*, 65–74. [CrossRef]
17. Li, Y.J.; Lou, Z.C.; Jiang, Y.J.; Wang, X.Z.; Yuan, T.C.; Yang, M.F. Flattening technique without nicked in curved bamboo strips. *For. Mach Woodwork Equip.* **2020**, *48*, 28–30+34. [CrossRef]
18. Li, Z.; Chen, C.; Mi, R.; Gan, W.; Dai, J.; Jiao, M.; Xie, H.; Yao, Y.; Xiao, S.; Hu, L. A strong, tough, and scalable structural material from fast-growing bamboo. *Adv. Mater.* **2020**, *32*, e1906308. [CrossRef] [PubMed]
19. Lou, Z.C.; Wang, Q.Y.; Sun, W.; Zhao, Y.H.; Wang, X.Z.; Liu, X.R.; Li, Y.J. Bamboo flattening technique: A literature and patent review. *Eur. J. Wood Wood Prod.* **2021**, *79*, 1035–1048. [CrossRef]
20. Muktar, M.S.; Habte, E.; Teshome, A.; Assefa, Y.; Negawo, A.T.; Lee, K.W.; Zhang, J.; Jones, C.S. Insights into the genetic architecture of complex traits in napier grass (*Cenchrus purpureus*) and QTL regions governing forage biomass yield, water use efficiency and feed quality traits. *Front. Plant Sci.* **2021**, *12*, 678862. [CrossRef] [PubMed]
21. Yan, Q.; Li, J.; Lu, L.Y.; Gao, L.J.; Lai, D.W.; Yao, N.; Yi, X.F.; Wu, Z.Y.; Lai, Z.Q.; Zhang, J.Y. Integrated analyses of phenotype, phytohormone, and transcriptome to elucidate the mechanism governing internode elongation in two contrasting elephant grass (*Cenchrus purpureus*) cultivars. *Ind. Crops Prod.* **2021**, *170*, 113693. [CrossRef]
22. Wind, J.; Smeekens, S.; Hanson, J. Sucrose: Metabolite and signaling molecule. *Phytochemistry* **2010**, *71*, 1610–1614. [CrossRef]
23. Taiz, L.; Zeiger, E.; Møller, I.M.; Murphy, A. *Plant Physiology and Development*; Sinauer Associates Incorporated: Sunderland, MA, USA, 2015.
24. Aluko, O.O.; Li, C.; Wang, Q.; Liu, H. Sucrose utilization for improved crop yields: A review article. *Int. J. Mol. Sci.* **2021**, *22*, 4704. [CrossRef] [PubMed]
25. Doehlert, D.C.; Kuo, T.M.; Felker, F.C. Enzymes of sucrose and hexose metabolism in developing kernels of two inbreds of maize. *Plant Physiol.* **1988**, *86*, 1013–1019. [CrossRef]
26. Patrick, J.W. Does Don Fisher's high-pressure manifold model account for phloem transport and resource partitioning? *Front. Plant Sci.* **2013**, *4*, 184. [CrossRef]
27. Kebrom, T.H.; Mullet, J.E. Photosynthetic leaf area modulates tiller bud outgrowth in sorghum. *Plant Cell Environ.* **2015**, *38*, 1471–1478. [CrossRef]
28. Xia, X.; Dong, H.; Yin, Y.; Song, X.; Gu, X.; Sang, K.; Zhou, J.; Shi, K.; Zhou, Y.; Foyer, C.H.; et al. Brassinosteroid signaling integrates multiple pathways to release apical dominance in tomato. *Proc. Natl. Acad. Sci. USA* **2021**, *118*, e2004384118. [CrossRef] [PubMed]
29. Cao, Y.Y.; Yang, M.T.; Chen, S.Y.; Zhou, Z.Q.; Li, X.; Wang, X.J.; Bai, J.G. Exogenous sucrose influences antioxidant enzyme activities and reduces lipid peroxidation in water-stressed cucumber leaves. *Biol. Plant.* **2015**, *59*, 147–153. [CrossRef]

30. Zhou, Z.P.; Yuan, Y.Z.; Zhou, W.; Zhang, C.F. Effects of exogenously supplied sucrose on *OsSUTs* and *OsSPSs* transcript abundances and rice root ammonium assimilation. *Acta Physiol. Plant.* **2016**, *38*, 274. [CrossRef]
31. Medeiros, D.B.; Perez Souza, L.; Antunes, W.C.; Araujo, W.L.; Daloso, D.M.; Fernie, A.R. Sucrose breakdown within guard cells provides substrates for glycolysis and glutamine biosynthesis during light-induced stomatal opening. *Plant J.* **2018**, *94*, 583–594. [CrossRef]
32. Ishioka, N.; Tanimoto, S.; Harada, H. Roles of nitrogen and carbohydrate in floral-bud formation in *Pharbitis* Apex Cultures. *J. Plant Physiol.* **1991**, *138*, 573–576. [CrossRef]
33. Donnison, I.S.; Francis, D. Experimental control of floral reversion in isolated shoot apices of the long-day plant *Silene coeli-rosa*. *Physiol. Plant.* **1994**, *92*, 329–335. [CrossRef]
34. Wang, X.; Geng, X.; Yang, L.; Chen, Y.; Zhao, Z.; Shi, W.; Kang, L.; Wu, R.; Lu, C.; Gao, J. Total and mitochondrial transcriptomic and proteomic insights into regulation of bioenergetic processes for shoot fast-growth initiation in moso bamboo. *Cells* **2022**, *11*, 1240. [CrossRef]
35. Zheng, H.F.; Cai, M.M.; Bai, Y.C.; Xu, J.L.; Xie, Y.L.; Song, H.J.; Li, J.; Gao, J. The effect of guttation on the growth of bamboo shoots. *Forests* **2021**, *13*, 31. [CrossRef]
36. Wang, S.; Pei, J.; Li, J.; Tang, G.; Zhao, J.; Peng, X.; Nie, S.; Ding, Y.; Wang, C. Sucrose and starch metabolism during *Fargesia yunnanensis* shoot growth. *Physiol. Plant.* **2020**, *168*, 188–204. [CrossRef]
37. Wei, Q.; Guo, L.; Jiao, C.; Fei, Z.; Chen, M.; Cao, J.; Ding, Y.; Yuan, Q. Characterization of the developmental dynamics of the elongation of a bamboo internode during the fast growth stage. *Tree Physiol.* **2019**, *39*, 1201–1214. [CrossRef]
38. Hori, S. Some observations on 'Bakanae'disease of the rice plant. *Mem. Agric. Res. Sta.* **1898**, *12*, 110–119.
39. Randoux, M.; Jeauffre, J.; Thouroude, T.; Vasseur, F.; Hamama, L.; Juchaux, M.; Sakr, S.; Foucher, F. Gibberellins regulate the transcription of the continuous flowering regulator, *RoKSN*, a rose *TFL1* homologue. *J. Exp. Bot.* **2012**, *63*, 6543–6554. [CrossRef] [PubMed]
40. Hedden, P.; Sponsel, V. A century of gibberellin research. *J. Plant Growth Regul.* **2015**, *34*, 740–760. [CrossRef] [PubMed]
41. Zhang, S.; Zhang, D.; Fan, S.; Du, L.; Shen, Y.; Xing, L.; Li, Y.; Ma, J.; Han, M. Effect of exogenous GA_3 and its inhibitor paclobutrazol on floral formation, endogenous hormones, and flowering-associated genes in 'Fuji' apple (*Malus domestica* Borkh.). *Plant Physiol. Biochem.* **2016**, *107*, 178–186. [CrossRef] [PubMed]
42. Stowe, B.B.; Yamaki, T. The history and physiological action of the gibberellins. *Annu. Rev. Plant Physiol.* **1957**, *8*, 181–216. [CrossRef]
43. Yan, Q.; Li, J.; Lu, L.Y.; Yi, X.F.; Yao, N.; Lai, Z.Q.; Zhang, J.Y. Comparative transcriptome study of the elongating internode in elephant grass (*Cenchrus purpureus*) seedlings in response to exogenous gibberellin applications. *Ind. Crops Prod.* **2022**, *178*, 114653. [CrossRef]
44. Koornneef, M.; van der Veen, J.H. Induction and analysis of gibberellin sensitive mutants in *Arabidopsis thaliana* (L.) heynh. *Theor. Appl. Genet.* **1980**, *58*, 257–263. [CrossRef]
45. Fu, X.; Harberd, N.P. Auxin promotes *Arabidopsis* root growth by modulating gibberellin response. *Nature* **2003**, *421*, 740–743. [CrossRef]
46. Liao, Z.; Yu, H.; Duan, J.; Yuan, K.; Yu, C.; Meng, X.; Kou, L.; Chen, M.; Jing, Y.; Liu, G.; et al. SLR1 inhibits MOC1 degradation to coordinate tiller number and plant height in rice. *Nat. Commun.* **2019**, *10*, 2738. [CrossRef]
47. Chen, Y.; Hou, M.; Liu, L.; Wu, S.; Shen, Y.; Ishiyama, K.; Kobayashi, M.; McCarty, D.R.; Tan, B.C. The maize *DWARF1* encodes a gibberellin 3-oxidase and is dual localized to the nucleus and cytosol. *Plant Physiol.* **2014**, *166*, 2028–2039. [CrossRef]
48. Ishii, Y.; Yamano, A.; Idota, S. Effects of short-day and gibberellic acid treatments on summer vegetative propagation of napier grass (*Pennisetum purpureum* Schumach). *Int. J. Agron.* **2016**, *2016*, 9606914. [CrossRef]
49. Zhang, X.; He, L.; Zhao, B.; Zhou, S.; Li, Y.; He, H.; Bai, Q.; Zhao, W.; Guo, S.; Liu, Y.; et al. *Dwarf and Increased Branching 1* controls plant height and axillary bud outgrowth in *Medicago truncatula*. *J. Exp. Bot.* **2020**, *71*, 6355–6365. [CrossRef] [PubMed]
50. Rebetzke, G.J.; Richards, R.A. Gibberellic acid-sensitive dwarfing genes reduce plant height to increase kernel number and grain yield of wheat. *Aust. J. Agric. Res.* **2000**, *51*, 235–246. [CrossRef]
51. Zhang, Y.; Ni, Z.; Yao, Y.; Nie, X.; Sun, Q. Gibberellins and heterosis of plant height in wheat (*Triticum aestivum* L.). *BMC Genet* **2007**, *8*, 40. [CrossRef] [PubMed]
52. Wang, Y.S.; Du, Y.Y.; Yang, Z.Y.; Chen, L.; Condon, A.G.; Hu, Y.G. Comparing the effects of GA-responsive dwarfing genes *Rht13* and *Rht8* on plant height and some agronomic traits in common wheat. *Field Crops Res.* **2015**, *179*, 35–43. [CrossRef]
53. Tian, X.; Xia, X.; Xu, D.; Liu, Y.; Xie, L.; Hassan, M.A.; Song, J.; Li, F.; Wang, D.; Zhang, Y.; et al. *Rht24b*, an ancient variation of *TaGA2ox-A9*, reduces plant height without yield penalty in wheat. *New Phytol.* **2022**, *233*, 738–750. [CrossRef]
54. Mitsunaga, S.; Tashiro, T.; Yamaguchi, J. Identification and characterization of gibberellin-insensitive mutants selected from among dwarf mutants of rice. *Theor. Appl. Genet.* **1994**, *87*, 705–712. [CrossRef]
55. Qiao, F.; Yang, Q.; Wang, C.L.; Fan, Y.L.; Wu, X.F.; Zhao, K.J. Modification of plant height via RNAi suppression of *OsGA20ox2* gene in rice. *Euphytica* **2007**, *158*, 35–45. [CrossRef]
56. Voorend, W.; Nelissen, H.; Vanholme, R.; De Vliegher, A.; Van Breusegem, F.; Boerjan, W.; Roldan, R.I.; Muylle, H.; Inze, D. Overexpression of *GA20-OXIDASE1* impacts plant height, biomass allocation and saccharification efficiency in maize. *Plant Biotechnol. J.* **2016**, *14*, 997–1007. [CrossRef]

57. Hu, S.; Wang, C.; Sanchez, D.L.; Lipka, A.E.; Liu, P.; Yin, Y.; Blanco, M.; Lubberstedt, T. Gibberellins promote brassinosteroids action and both increase heterosis for plant height in maize (*Zea mays* L.). *Front. Plant Sci.* **2017**, *8*, 1039. [CrossRef]
58. Gebremeskel, H.; Dou, J.; Li, B.; Zhao, S.; Muhammad, U.; Lu, X.; He, N.; Liu, W. Molecular mapping and candidate gene analysis for GA_3 responsive short internode in watermelon (*Citrullus lanatus*). *Int. J. Mol. Sci.* **2019**, *21*, 290. [CrossRef] [PubMed]
59. Dong, W.; Wu, D.; Wang, C.; Liu, Y.; Wu, D. Characterization of the molecular mechanism underlying the dwarfism of dsh mutant watermelon plants. *Plant Sci. Lett.* **2021**, *313*, 111074. [CrossRef] [PubMed]
60. Li, L.; Cheng, Z.; Ma, Y.; Bai, Q.; Li, X.; Cao, Z.; Wu, Z.; Gao, J. The association of hormone signalling genes, transcription and changes in shoot anatomy during moso bamboo growth. *Plant Biotechnol. J.* **2018**, *16*, 72–85. [CrossRef]
61. Hou, D.; Bai, Q.S.; Li, J.; Xie, L.H.; Li, X.Y.; Cheng, Z.C.; Gao, J. The gibberellic acid-stimulated transcript gene family in moso bamboo: A genome-wide survey and expression profiling during development and abiotic stresses. *J. Plant Growth Regul.* **2018**, *37*, 1135–1147. [CrossRef]
62. Cai, M.; Cheng, W.; Bai, Y.; Mu, C.; Zheng, H.; Cheng, Z.; Gao, J. *PheGRF4e* initiated auxin signaling during moso bamboo shoot development. *Mol. Biol. Rep.* **2022**, *49*, 8815–8825. [CrossRef] [PubMed]
63. Wang, T.; Liu, L.; Wang, X.; Liang, L.; Yue, J.; Li, L. Comparative analyses of anatomical structure, phytohormone levels, and gene expression profiles reveal potential dwarfing mechanisms in shengyin bamboo (*Phyllostachys edulis* f. *tubaeformis*). *Int. J. Mol. Sci.* **2018**, *19*, 1697. [CrossRef] [PubMed]
64. Zhang, H.; Wang, H.; Zhu, Q.; Gao, Y.; Wang, H.; Zhao, L.; Wang, Y.; Xi, F.; Wang, W.; Yang, Y.; et al. Transcriptome characterization of moso bamboo (*Phyllostachys edulis*) seedlings in response to exogenous gibberellin applications. *BMC Plant Biol.* **2018**, *18*, 125. [CrossRef]
65. Chen, M.; Guo, L.; Ramakrishnan, M.; Fei, Z.; Vinod, K.K.; Ding, Y.; Jiao, C.; Gao, Z.; Zha, R.; Wang, C.; et al. Rapid growth of moso bamboo (*Phyllostachys edulis*): Cellular roadmaps, transcriptome dynamics, and environmental factors. *Plant Cell* **2022**, *34*, 3577–3610. [CrossRef]
66. Isagi, Y.; Torii, A. Range expansion and its mechanisms in a naturalized bamboo species, *Phyllostachys pubescens*, in Japan. *J. Sustain. For.* **1997**, *6*, 127–141. [CrossRef]
67. Inoue, A.; Kuraoka, K.; Kitahara, F. Mathematical expression for the relationship between internode number and internode length for bamboo, *Phyllostachys pubescens*. *J. For. Res.* **2012**, *23*, 435–439. [CrossRef]
68. Inoue, A. Culm form analysis for bamboo, *Phyllostachys pubescens*. *J. For. Res.* **2013**, *24*, 525–530. [CrossRef]
69. Gao, X.; Li, Z.D.; Yu, H.M.; Jiang, Z.H.; Wang, C.; Zhang, Y.; Qi, L.H.; Shi, L. Modeling of the height–diameter relationship using an allometric equation model: A case study of stands of *Phyllostachys edulis*. *J. For. Res.* **2015**, *27*, 339–347. [CrossRef]
70. Yen, T.M. Culm height development, biomass accumulation and carbon storage in an initial growth stage for a fast-growing moso bamboo (*Phyllostachy pubescens*). *Bot. Stud.* **2016**, *57*, 10. [CrossRef] [PubMed]
71. Inoue, A.; Suga, H. Relationships of culm surface area to other culm dimensions for bamboo, *Phyllostachys pubescens*. *J. For. Res.* **2017**, *14*, 236–239. [CrossRef]
72. Suga, H.; Inoue, A.; Kitahara, F. Derivation of two-way volume equation for bamboo, *Phyllostachys pubescens*. *J. For. Res.* **2017**, *16*, 261–267. [CrossRef]
73. Zhao, X.; Zhao, P.; Zhang, Z.; Zhu, L.; Hu, Y.; Ouyang, L.; Ni, G.; Ye, Q. Culm age and rhizome affects night-time water recharge in the bamboo *Phyllostachys pubescens*. *Front. Plant Sci.* **2017**, *8*, 1928. [CrossRef]
74. Inoue, A.; Sato, M.; Shima, H. A new taper index based on form-factor: Application to three bamboo species (*Phyllostachys* spp.). *Eur. J. For. Res.* **2021**, *140*, 1533–1542. [CrossRef]
75. Banik, R.L. Morphology and growth. In *Bamboo*; Springer: Berlin/Heidelberg, Germany, 2015; pp. 43–89.
76. Zhou, Z.S.; Tang, Y.; Xu, H.X.; Wang, J.Z.; Hu, L.L.; Xu, X.J. Dynamic changes in leaf biomass and the modeling of individual moso bamboo (*Phyllostachys edulis* (Carrière) J. Houz) under intensive management. *Forests* **2022**, *13*, 693. [CrossRef]
77. Galaud, J.-P.; Gaspar, T.; Boyer, N. Inhibition of internode growth due to mechanical stress in *Bryonia dioica*: Relationship between changes in DNA methylation and ethylene metabolism. *Physiol. Plant.* **1993**, *87*, 25–30. [CrossRef]
78. Valinger, E.; Lundqvist, L.; Sundberg, B. Mechanical stress during dormancy stimulates stem growth of Scots pine seedlings. *For. Ecol. Manag.* **1994**, *67*, 299–303. [CrossRef]
79. Takano, M.; Takahashi, H.; Suge, H. Mechanical stress and gibberellin: Regulation of hollowing induction in the stem of a bean plant, *Phaseolus vulgaris* L. *Plant Cell Physiol.* **1995**, *36*, 101–108. [CrossRef]
80. Huber, H.; de Brouwer, J.; von Wettberg, E.J.; During, H.J.; Anten, N.P.R. More cells, bigger cells or simply reorganization? Alternative mechanisms leading to changed internode architecture under contrasting stress regimes. *New Phytol.* **2014**, *201*, 193–204. [CrossRef] [PubMed]
81. Li, R.; Werger, M.; During, H.; Zhong, Z.C. Carbon and nutrient dynamics in relation to growth rhythm in the giant bamboo *Phyllostachys pubescens*. *Plant Soil* **1998**, *201*, 113–123. [CrossRef]
82. Zhou, G.; Jiang, P.; Xu, Q. *Carbon Fixing and Transition in the Ecosystem of Bamboo Stands*; Science Press: Beijing, China, 2010.
83. O'Toole, J.C.; Cruz, R.T.; Singh, T.N. Leaf rolling and transpiration. *Plant Sci. Lett.* **1979**, *16*, 111–114. [CrossRef]
84. Wheeler, T.D.; Stroock, A.D. The transpiration of water at negative pressures in a synthetic tree. *Nature* **2008**, *455*, 208–212. [CrossRef]
85. Vadez, V.; Kholova, J.; Medina, S.; Kakkera, A.; Anderberg, H. Transpiration efficiency: New insights into an old story. *J. Exp. Bot.* **2014**, *65*, 6141–6153. [CrossRef]

86. Peel, A.J. *Transport of Nutrients in Plants*; Elsevier: London, UK, 2013.
87. Song, X.; Peng, C.; Zhou, G.; Gu, H.; Li, Q.; Zhang, C. Dynamic allocation and transfer of non-structural carbohydrates, a possible mechanism for the explosive growth of moso bamboo (*Phyllostachys heterocycla*). *Sci. Rep.* **2016**, *6*, 25908. [CrossRef]
88. Kende, H.; van der Knaap, E.; Cho, H.T. Deepwater rice: A model plant to study stem elongation. *Plant Physiol.* **1998**, *118*, 1105–1110. [CrossRef]
89. Kaneko, M.; Itoh, H.; Inukai, Y.; Sakamoto, T.; Ueguchi, M.; Ashikari, M.; Matsuoka, M. Where do gibberellin biosynthesis and gibberellin signaling occur in rice plants? *Plant J.* **2003**, *35*, 104–115. [CrossRef]
90. Nagai, K.; Mori, Y.; Ishikawa, S.; Furuta, T.; Gamuyao, R.; Niimi, Y.; Hobo, T.; Fukuda, M.; Kojima, M.; Takebayashi, Y.; et al. Antagonistic regulation of the gibberellic acid response during stem growth in rice. *Nature* **2020**, *584*, 109–114. [CrossRef]
91. Qazi, H.A.; Paranjpe, S.; Bhargava, S. Stem sugar accumulation in sweet sorghum—Activity and expression of sucrose metabolizing enzymes and sucrose transporters. *J. Plant Physiol.* **2012**, *169*, 605–613. [CrossRef] [PubMed]
92. Qin, G.; Zhu, Z.; Wang, W.; Cai, J.; Chen, Y.; Li, L.; Tian, S. A tomato vacuolar invertase inhibitor mediates sucrose metabolism and influences fruit ripening. *Plant Physiol.* **2016**, *172*, 1596–1611. [CrossRef] [PubMed]
93. Robaglia, C.; Thomas, M.; Meyer, C. Sensing nutrient and energy status by SnRK1 and TOR kinases. *Curr. Opin. Plant Biol.* **2012**, *15*, 301–307. [CrossRef]
94. Fang, D.; Mei, T.; Roll, A.; Holscher, D. Water transfer between bamboo culms in the period of sprouting. *Front. Plant Sci.* **2019**, *10*, 786. [CrossRef] [PubMed]
95. Shan, F.; Zhang, R.; Zhang, J.; Wang, C.; Lyu, X.; Xin, T.; Yan, C.; Dong, S.; Ma, C.; Gong, Z. Study on the regulatory effects of GA_3 on soybean internode elongation. *Plants* **2021**, *10*, 1737. [CrossRef]
96. Mendu, N.; Silflow, C.D. Elevated levels of tubulin transcripts accompany the GA_3-induced elongation of Oat internode segments. *Plant Cell Physiol.* **1993**, *34*, 973–983. [CrossRef]
97. Xu, N.F.; York, K.; Miller, P.; Cheikh, N. Co-regulation of ear growth and internode elongation in corn. *Plant Growth Regul.* **2004**, *44*, 231–241. [CrossRef]
98. Honi, U.; Amin, M.R.; Kabir, S.M.T.; Bashar, K.K.; Moniruzzaman, M.; Jahan, R.; Jahan, S.; Haque, M.S.; Islam, S. Genome-wide identification, characterization and expression profiling of gibberellin metabolism genes in jute. *BMC Plant Biol.* **2020**, *20*, 306. [CrossRef]
99. Schaik, P.H.; Probst, A.H. Effects of some environmental factors on flower production and reproductive efficiency in soybeans. *Agron. J.* **1958**, *50*, 192–197. [CrossRef]
100. Thomas, J.F.; Raper, C.D. Morphological response of soybeans as governed by photoperiod, temperature, and age at treatment. *Bot. Gaz.* **1977**, *138*, 321–328. [CrossRef]
101. Allen, L.H.; Zhang, L.X.; Boote, K.J.; Hauser, B.A. Elevated temperature intensity, timing, and duration of exposure affect soybean internode elongation, mainstem node number, and pod number per plant. *Crop J.* **2018**, *6*, 148–161. [CrossRef]
102. Misra, V.; Solomon, S.; Mall, A.K.; Prajapati, C.P.; Hashem, A.; Abd Allah, E.F.; Ansari, M.I. Morphological assessment of water stressed sugarcane: A comparison of waterlogged and drought affected crop. *Saudi J. Biol. Sci.* **2020**, *27*, 1228–1236. [CrossRef] [PubMed]
103. Wang, F.L.; Shao, Z.P.; Wu, Y.J.; Wu, D. The toughness contribution of bamboo node to the Mode I interlaminar fracture toughness of bamboo. *Wood Sci. Technol.* **2013**, *48*, 1257–1268. [CrossRef]
104. Shima, H.; Sato, M.; Inoue, A. Self-adaptive formation of uneven node spacings in wild bamboo. *Phys. Rev. E* **2016**, *93*, 022406. [CrossRef] [PubMed]
105. Wei, Q.; Jiao, C.; Guo, L.; Ding, Y.; Cao, J.; Feng, J.; Dong, X.; Mao, L.; Sun, H.; Yu, F.; et al. Exploring key cellular processes and candidate genes regulating the primary thickening growth of moso underground shoots. *New Phytol.* **2017**, *214*, 81–96. [CrossRef] [PubMed]
106. Shen, Z.; Zhang, Y.H.; Zhang, L.; Li, Y.; Sun, Y.D.; Li, Z.Y. Changes in the distribution of endogenous hormones in *Phyllostachys edulis* 'Pachyloen' during bamboo shooting. *PLoS ONE* **2020**, *15*, e0241806. [CrossRef]
107. Zhang, J.; Ma, R.; Ding, X.; Huang, M.; Shen, K.; Zhao, S.; Xiao, Z.; Xiu, C. Association among starch storage, metabolism, related genes and growth of moso bamboo (*Phyllostachys heterocycla*) shoots. *BMC Plant Biol.* **2021**, *21*, 477. [CrossRef]
108. Falster, D.S.; Westoby, M. Plant height and evolutionary games. *Trends Ecol. Evol.* **2003**, *18*, 337–343. [CrossRef]
109. Woodruff, D.R.; Bond, B.J.; Meinzer, F.C. Does turgor limit growth in tall trees? *Plant Cell Environ.* **2004**, *27*, 229–236. [CrossRef]
110. Ryan, M.G.; Phillips, N.; Bond, B.J. The hydraulic limitation hypothesis revisited. *Plant Cell Environ.* **2006**, *29*, 367–381. [CrossRef]
111. Niklas, K.J. Maximum plant height and the biophysical factors that limit it. *Tree Physiol.* **2007**, *27*, 433–440. [CrossRef] [PubMed]
112. Moles, A.T.; Warton, D.I.; Warman, L.; Swenson, N.G.; Laffan, S.W.; Zanne, A.E.; Pitman, A.; Hemmings, F.A.; Leishman, M.R. Global patterns in plant height. *J. Ecol.* **2009**, *97*, 923–932. [CrossRef]
113. Li, R.; Werger, M.J.A.; Kroon, H.; During, H.J.; Zhong, Z.C. Interactions between shoot age structure, nutrient availability and physiological integration in the giant bamboo *Phyllostachys pubescens*. *Plant Biol.* **2000**, *2*, 437–446. [CrossRef]
114. Saitoh, T.; Seiwa, K.; Nishiwaki, A. Importance of physiological integration of dwarf bamboo to persistence in forest understorey: A field experiment. *J. Ecol.* **2002**, *90*, 78–85. [CrossRef]
115. Zhai, W.L.; Wang, Y.; Luan, J.W.; Liu, S.R. Effects of nitrogen addition on clonal integration between mother and daughter ramets of moso bamboo: A ^{13}C-CO_2 pulse labeling study. *J. Plant Ecol.* **2021**, *15*, 756–770. [CrossRef]

116. Lu, Z.; Franklin, S.B. Clonal integration and regeneration in bamboo *Bashania fargesii*. *For. Ecol. Manag.* **2022**, *523*, 120504. [CrossRef]
117. Yen, T.M.; Lee, J.S. Comparing aboveground carbon sequestration between moso bamboo (*Phyllostachys heterocycla*) and China fir (*Cunninghamia lanceolata*) forests based on the allometric model. *For. Ecol. Manag.* **2011**, *261*, 995–1002. [CrossRef]
118. Chen, X.G.; Zhang, X.Q.; Zhang, Y.P.; Booth, T.; He, X.H. Changes of carbon stocks in bamboo stands in China during 100 years. *For. Ecol. Manag.* **2009**, *258*, 1489–1496. [CrossRef]
119. Jyoti Nath, A.; Das, G.; Das, A.K. Above ground standing biomass and carbon storage in village bamboos in North East India. *Biomass Bioenergy* **2009**, *33*, 1188–1196. [CrossRef]
120. Wu, W.G.; Liu, Q.; Zhu, Z.; Shen, Y.Q. Managing bamboo for carbon sequestration, bamboo stem and bamboo shoots. *Small-Scale For.* **2014**, *14*, 233–243. [CrossRef]
121. Gupta, A.; Kumar, A. Potential of bamboo in sustainable development. *Asia Pac. Bus. Rev.* **2008**, *4*, 100–107. [CrossRef]
122. Chele, E.S.; Ricardo, M.C.; Ana, P.M.; Teresa, M.R. Bamboo, from traditional crafts to contemporary design and architecture. *Procedia Soc. Behav. Sci.* **2012**, *51*, 777–781. [CrossRef]
123. Manandhar, R.; Kim, J.H.; Kim, J.T. Environmental, social and economic sustainability of bamboo and bamboo-based construction materials in buildings. *J. Asian Archit. Build. Eng.* **2019**, *18*, 49–59. [CrossRef]
124. FAO. *Global Forest Resources Assessment 2010: Main Report. FAO Forestry Paper 163*; Food and Agriculture Organization of the United Nations: Rome, Italy, 2010.
125. Liese, W. Research on bamboo. *Wood Sci. Technol.* **1987**, *21*, 189–209. [CrossRef]
126. Liese, W.; Köhl, M. *Bamboo: The Plant and Its Uses*; Springer International Publishing: Berlin/Heidelberg, Germany, 2015.
127. McNally, A. *GES DISC Dataset: FLDAS Noah Land Surface Model L4 Global Monthly 0.1 × 0.1 Degree (MERRA-2 and CHIRPS) (FLDAS_NOAH01_C_GL_M 001)*; NASA [Data Set]; NASA: Washington, DC, USA, 2018. [CrossRef]
128. Elmariah, S.; Mauri, L.; Doros, G.; Galper, B.Z.; O'Neill, K.E.; Steg, P.G.; Kereiakes, D.J.; Yeh, R.W. Extended duration dual antiplatelet therapy and mortality: A systematic review and meta-analysis. *Lancet* **2015**, *385*, 792–798. [CrossRef]
129. Feehan, D.M.; Mahmud, A.S. Quantifying population contact patterns in the United States during the COVID-19 pandemic. *Nat. Commun.* **2021**, *12*, 893. [CrossRef]
130. Harrington, S.M.; Wishingrad, V.; Thomson, R.C. Properties of markov chain monte carlo performance across many empirical alignments. *Mol. Biol. Evol.* **2021**, *38*, 1627–1640. [CrossRef]
131. Bürkner, P.C. Brms: An R package for Bayesian multilevel models using stan. *J. Stat. Softw.* **2017**, *80*, 1–28. [CrossRef]
132. Gelman, A.; Hill, J. *Data Analysis Using Regression and Hierarchical/Multilevel Models*; Cambridge University Press: Cambridge, UK, 2007.
133. Zhou, F.C. *Bamboo Forest Cultivation*; China Forestry Publishing House: Beijing, China, 1998.
134. Zhang, H.X.; Zhuang, S.Y.; Sun, B.; Ji, H.B.; Li, C.M.; Zhou, S. Estimation of biomass and carbon storage of moso bamboo (*Phyllostachys pubescens* Mazel ex Houz.) in southern China using a diameter–age bivariate distribution model. *Forestry* **2014**, *87*, 674–682. [CrossRef]

plants

MDPI

Communication

Time Effects of Global Change on Forest Productivity in China from 2001 to 2017

Jiangfeng Wang [1], Yanhong Li [1,*] and Jie Gao [1,2,*]

1 College of Life Sciences, Xinjiang Normal University, Urumqi 830054, China
2 Key Laboratory of Earth Surface Processes of Ministry of Education, College of Urban and Environmental Sciences, Peking University, Beijing 100871, China
* Correspondence: liyh1330824@163.com (Y.L.); jiegao72@gmail.com (J.G.)

Abstract: With global warming, the concentrations of fine particulate matter ($PM_{2.5}$) and greenhouse gases, such as CO_2, are increasing. However, it is still unknown whether these increases will affect vegetation productivity. Exploring the impacts of global warming on net primary productivity (NPP) will help us understand how ecosystem function responds to climate change in China. Using the Carnegie-Ames-Stanford Approach (CASA) ecosystem model based on remote sensing, we investigated the spatiotemporal changes in NPP across 1137 sites in China from 2001 to 2017. Our results revealed that: (1) Mean Annual Temperature (MAT) and Mean Annual Precipitation (MAP) were significantly positively correlated with NPP ($p < 0.01$), while $PM_{2.5}$ concentration and CO_2 emissions were significantly negatively correlated with NPP ($p < 0.01$). (2) The positive correlation between temperature, rainfall and NPP gradually weakened over time, while the negative correlation between $PM_{2.5}$ concentration, CO_2 emissions and NPP gradually strengthened over time. (3) High levels of $PM_{2.5}$ concentration and CO_2 emissions had negative effects on NPP, while high levels of MAT and MAP had positive effects on NPP.

Keywords: global warming; greenhouse effect; productivity; spatialtemporal change

Citation: Wang, J.; Li, Y.; Gao, J. Time Effects of Global Change on Forest Productivity in China from 2001 to 2017. *Plants* **2023**, *12*, 1404. https://doi.org/10.3390/plants12061404

Academic Editor: Yasutomo Hoshika

Received: 22 February 2023
Revised: 15 March 2023
Accepted: 21 March 2023
Published: 22 March 2023

1. Introduction

Net primary productivity (NPP) refers to the organic carbon fixed by plants through photosynthesis [1,2]. Ecosystems with high productivity not only maintain high biodiversity but also play a critical role in conserving water and soil nutrients [3]. However, it is still unclear whether global warming will significantly affect ecosystem productivity, despite the changing distribution patterns and dominant species of vegetation in recent years [4]. Additionally, the greenhouse effect resulting from global warming has an obvious spatial–temporal impact. Thus, exploring the spatiotemporal changes in the relationship between vegetation productivity and global warming can provide a theoretical basis for predicting the impact of global change on forest function [5–7].

In recent years, China's rapid economic development and urbanization have resulted in severe air pollution problems, particularly with regard to haze (atmospheric particulate: PM). Numerous studies have shown that $PM_{2.5}$, particulate matter with a diameter of 2.5 microns or less in the atmosphere, is the main pollutant in many Chinese cities [8]. $PM_{2.5}$ absorbs and scatters sunlight, which alters surface temperature and affects precipitation in various ways [9]. The interaction between climate factors and $PM_{2.5}$ may impact NPP, making the relationship between $PM_{2.5}$ emission and NPP a crucial part of this study.

Previous research has shown that global warming may have an impact on the productivity of terrestrial ecosystems [10]. Since the beginning of the industrial revolution, the continuous increase in fossil fuel emissions has led to a sharp rise in the concentration of pollutants such as CO_2 and $PM_{2.5}$ in the atmosphere. A significant increase in CO_2 concentrations will result in global warming [11]. Current research indicates that the Earth may warm by 2–7 °C by the end of this century. Global warming will cause changes in plant photosynthesis, plant respiration, and organic matter decomposition, which may impact NPP [12]. However, the mechanism of NPP changes caused by global warming remains unclear. Given China's increasing pressure to reduce CO_2 emissions, investigating the mechanism of NPP fluctuations caused by climate change can help us better mitigate the adverse effects of climate warming in the future and formulate policies to address climate warming.

Numerous studies have found that rising levels of CO_2 caused by global warming will lead to a decrease in precipitation [13], and temperature and precipitation synergistically affect the accumulation of NPP [14]. $PM_{2.5}$ and CO_2 are the primary substances that cause global warming. $PM_{2.5}$ changes the net photosynthetic rate by affecting plant photosynthesis, thereby affecting ecosystem NPP [15–18]. NPP is an essential variable to characterize plant activities and a crucial factor in determining ecosystem carbon sinks and regulating ecological processes [19]. Estimating vegetation NPP on a large spatial and temporal scale using multisource remote sensing data and comprehensive spatial pattern and dynamic analysis with GIS technology has become an important method to quantify NPP [20,21]. Nonetheless, uncertainties remain in the relationship between NPP and climate change to global warming.

This study employed MAT, MAP, $PM_{2.5}$, CO_2, and NPP data from 1137 sites across China from 2001 to 2017 to investigate (1) the spatiotemporal trends of NPP, MAT, MAP, $PM_{2.5}$ concentration, and CO_2 emissions; (2) the relationship between NPP and MAT, MAP, $PM_{2.5}$ concentration, and CO_2 emissions; and (3) the impact of different levels (high/low) of MAT, MAP, $PM_{2.5}$ concentration, and CO_2 emissions on NPP.

2. Results

From 2001 to 2017, MAT and MAP exhibited a significant upward trend (Figure 1A,B). The concentration of $PM_{2.5}$ increased from 2001 to 2011, and then began to decline after 2011 (Figure 1C). CO_2 emissions consistently increased during this period (Figure 1D). NPP decreased from 2001 to 2011 initially, and then increased from 2012 to 2017, exhibiting an opposite trend to that of $PM_{2.5}$. Over time, NPP displayed a decreasing-then-increasing trend, with its lowest value being in 2011 (Figure 2). As time progressed, NPP and climatic factors showed clear regional differences (Figures 3 and 4). In general, productivity and climate factors exhibited significant habitat heterogeneity.

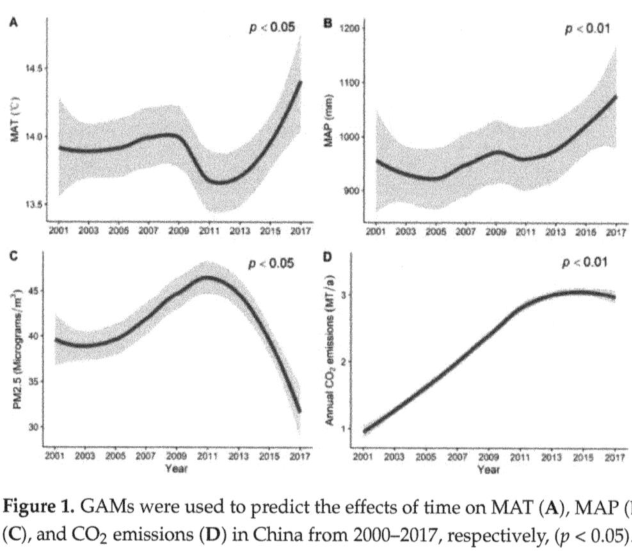

Figure 1. GAMs were used to predict the effects of time on MAT (**A**), MAP (**B**), $PM_{2.5}$ concentration (**C**), and CO_2 emissions (**D**) in China from 2000–2017, respectively, ($p < 0.05$).

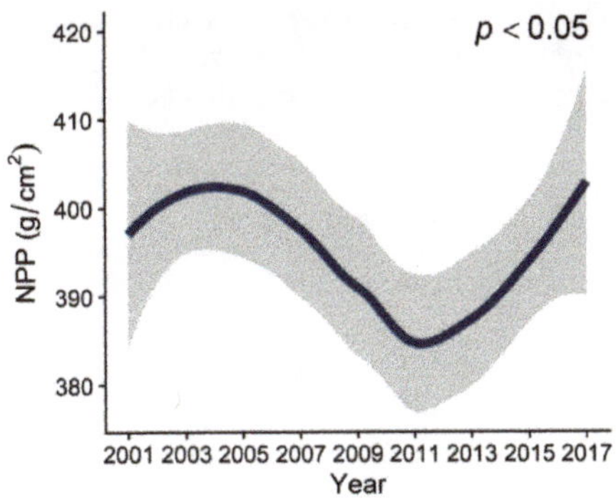

Figure 2. GAMs were used to predict the effects of time on NPP ($p < 0.05$). NPP showed a trend of decreasing first and then of rising over time. In 2011, the value of NPP reached the lowest.

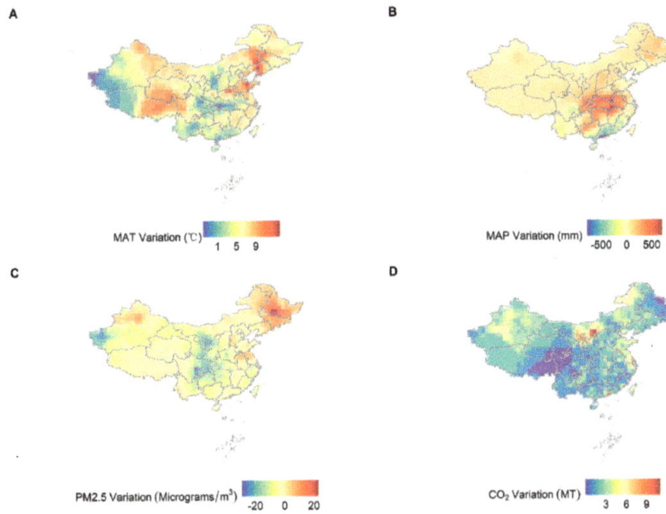

Figure 3. Variations in MAT, MAP, $PM_{2.5}$ concentrations and CO_2 emissions in China from 2001 to 2017. The spatial resolution is 1 km. (**A**) MAT; (**B**) MAP; (**C**) $PM_{2.5}$ concentration; (**D**) CO_2 emissions.

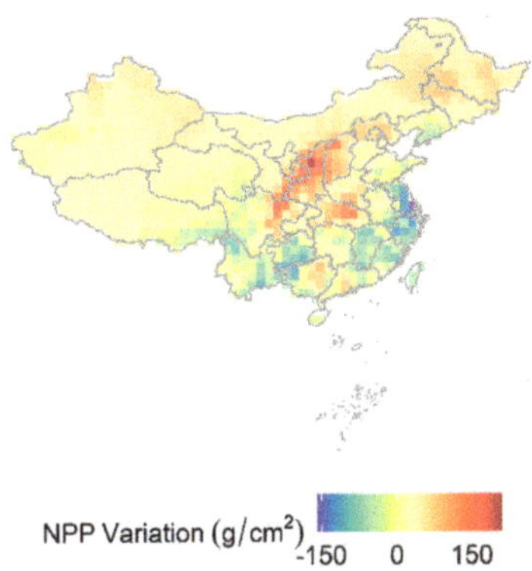

Figure 4. Variations in NPP in China from 2001 to 2017. The spatial resolution is 1 km. On the whole, there is a large number of regions in southern China where NPP is declining gradually.

MAT and MAP were significantly positively correlated with NPP ($R^2 = 0.23$, $p < 0.001$; $R^2 = 0.24$, $p < 0.001$) (Figure 5A,B). However, $PM_{2.5}$ concentrations and CO_2 emissions were significantly negatively correlated with NPP ($R^2 = 0.04$, $p < 0.001$; $R^2 = 0.06$, $p < 0.001$) (Figure 5C,D).

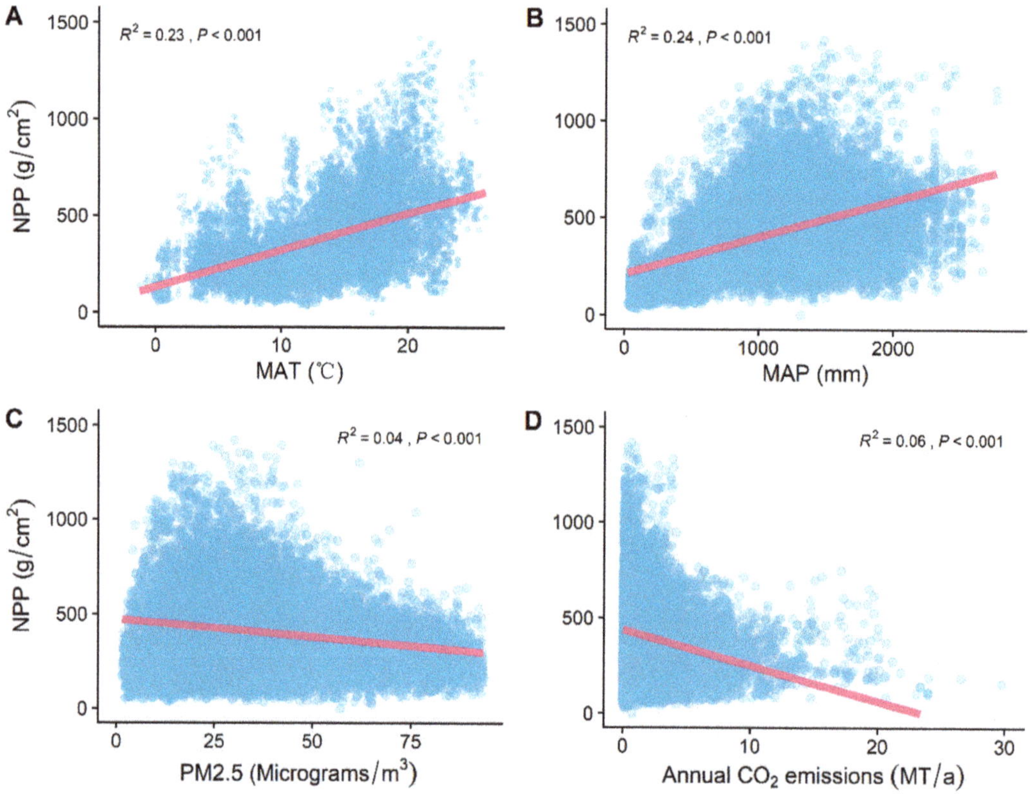

Figure 5. General linear correlation analysis of (**A**) MAT, (**B**) MAP, (**C**) $PM_{2.5}$ concentrations and (**D**) CO_2 emissions with NPP. R^2 represents how well the model fits the variables studied, and the *p* value represents the significance level.

Although MAT, MAP and NPP has shown a positive correlation over time, the positive correlation is gradually weakening. $PM_{2.5}$ concentrations, CO_2 emissions and NPP have been negatively correlated. However, the negative correlation is gradually increasing (Figure 6A,B). High levels of MAT and MAP have a positive effect on NPP. However, high levels of $PM_{2.5}$ concentration and CO_2 emissions have a negative effect on NPP (Figure 7).

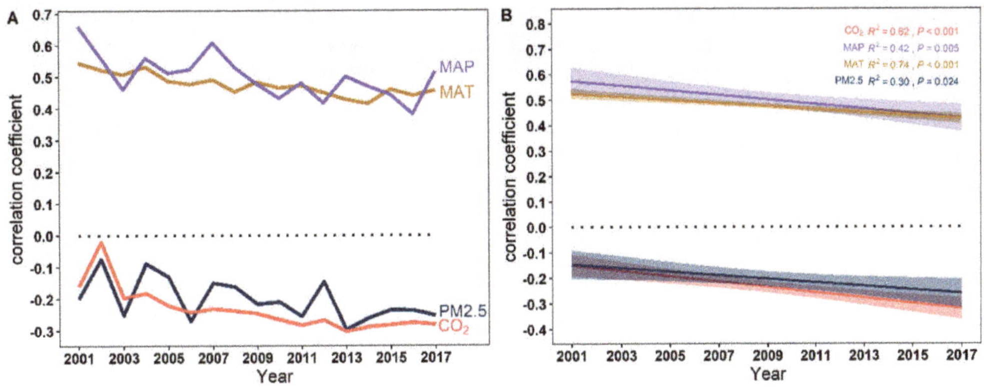

Figure 6. (**A**) The correlation between environmental factors (CO_2 emission, MAP, MAT, and $PM_{2.5}$ concentration) and NPP over time from 2001 to 2017. (**B**) The plots of the GAMs' smooth function for indicating the effects of time (year) on the correlation coefficient.

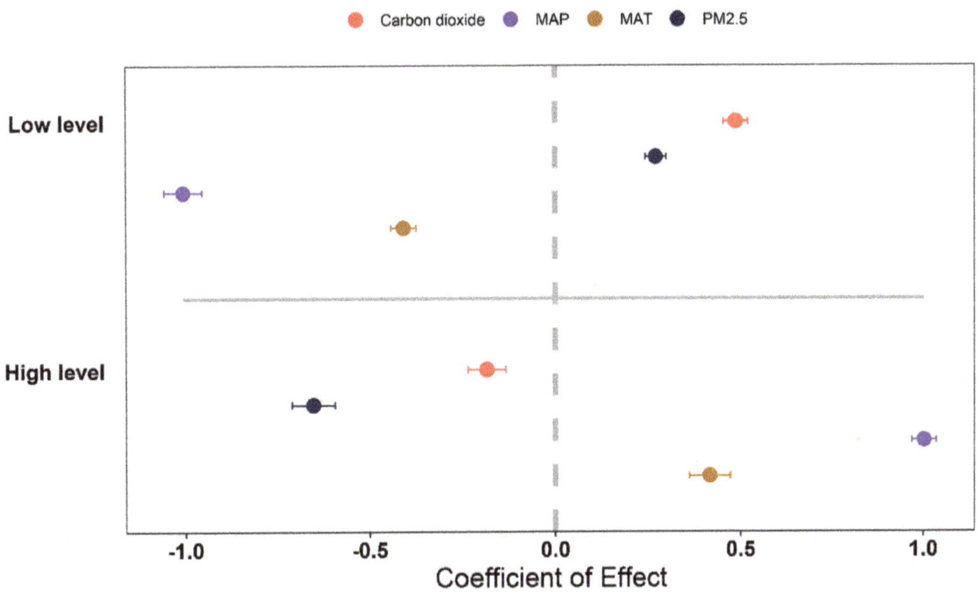

Figure 7. The effects of different environmental factors on high and low levels on NPP were analyzed, and the mean confidence interval of ±95% of the effect size is reported. The dotted line indicates that the effect size is zero. The environmental factors considered were CO_2 emissions, MAP, MAT, and $PM_{2.5}$ concentration, and were classified into three levels: high, medium, and low. The high and low levels were the experimental groups, while the medium level was the control group. For MAT, the low level was less than or equal to 10 °C, the medium level was greater than 10 °C but less than 20 °C, and the high level was greater than or equal to 20 °C. For MAP, the low level was less than or equal to 400 mm, the medium level was greater than 400 mm but less than 800 mm, and the high level was greater than or equal to 800 mm. For $PM_{2.5}$ concentration, the low level was less than or equal to 35 μg/m^3, the medium level was greater than 35 μg/m^3 but less than 75 μg/m^3, and the high level was greater than or equal to 75 μg/m^3. For CO_2 emissions, the low level was less than or equal to 2 MT/a, the medium level was higher than 2 MT/a but less than 5 MT/a, and the high level was greater than or equal to 5 MT/a.

3. Discussion

We investigated the spatiotemporal changes in forest NPP in China from 2001 to 2017. NPP was calculated based on the CASA model, although its prediction accuracy is not perfect due to the differences in spatial resolution and the actual and simulated values of variable factors [22]. Nevertheless, it is widely recognized and applied in numerous macroecological studies [23–25].

Changes in various environmental factors, such as temperature, precipitation, $PM_{2.5}$ concentration, and CO_2 emissions, caused by global warming can impact the interannual variation in NPP [6,26]. Greenhouse gas emissions have become an urgent ecological problem globally, making it crucial to explore the relationship between various climate factors and NPP to understand the impacts of climate warming on ecosystem functions [27].

Temperature is one of the main climatic factors that affect ecosystem productivity [28,29]. In our study, MAT exhibited an increasing trend from 2001 to 2017. NPP demonstrated a significant positive correlation with temperature, which is consistent with the results of Gu et al. [30]. Rising temperatures can increase the diversity and richness of functional genes of photosynthetic microorganisms in soil, as well as the activity of soil microorganisms [31]. Under rising temperatures, nutrient availability and the utilization efficiency required for forest growth can also improve, promoting soil nutrient mineralization and accelerating nutrient release, which can contribute to the improvement of vegetation NPP [32]. Temperature increase can also accelerate plant photosynthesis, stimulate plant growth, improve carbon absorption, and thus increase vegetation NPP [33–35].

Precipitation has also been found to be one of the climatic factors affecting ecosystem productivity [36]. NPP significantly increased with an increase in MAP [12]. The availability of nutrients in soil increases with increasing water content, and the effect of precipitation on soil nutrient availability can affect plant photosynthesis, thus improving vegetation NPP [37,38].

$PM_{2.5}$ has attracted widespread attention in recent times [38]. However, few studies have investigated the influence of $PM_{2.5}$ on vegetation NPP. We found a significant negative correlation between $PM_{2.5}$ concentration and NPP, which may have been due to plants adsorbing particulate matter through leaf holes or adsorbing particulate matter onto the leaf surface. As a result, plants reduce the absorption efficiency of external oxygen, water, and sunlight during photosynthesis, thus reducing the NPP of plants [39].

Since the industrial revolution, the significant increase in the use of fossil fuels has led to a continuous increase in CO_2 emissions, which is a primary factor contributing to global warming. The increase in CO_2 emissions significantly affects vegetation NPP [40]. We found a significant negative correlation between CO_2 emissions and NPP. With an increase in CO_2 emissions, the carbon sink will fail prematurely, and leaves will age quickly, leading to a decrease in plant NPP [41,42]. The interspecific survival–growth tradeoff theory suggests that, in the case of higher CO_2 concentrations, faster plant growth would be accompanied by a higher mortality risk, thus affecting plant NPP [43]. The above results are consistent with the findings of numerous studies in recent years [5,44,45]. However, due to the significant physiological spatiotemporal differences of plant individuals, the calculation of NPP by satellite also involves some uncertainty.

The results of this study indicate that high emissions of $PM_{2.5}$ and CO_2 have serious negative effects on NPP. Therefore, we should pay close attention to the negative effects of increasing $PM_{2.5}$ and CO_2 emissions on plant NPP. It is crucial to deeply understand the driving mechanism of the terrestrial carbon cycle, increase carbon fixation, reduce carbon emissions, and formulate air pollution prevention measures in a timely way to mitigate the negative impact of climate warming on the ecosystem.

4. Study Area and Methods

4.1. Data Collection

MAT and MAP data for China from 2000 to 2017 were extracted from the WorldClim global climate database with a spatial resolution of 1 km [45]. $PM_{2.5}$ data with a 1 km spatial resolution were extracted from a published database [46]. CO_2 data were extracted from a database derived from ODIAC (https://download.csdn.net/download/sooluo/73733136?utm_source=bbsseo (accessed on 20 October 2022) with a spatial resolution of 1 km.

NPP data from 2000 to 2017 were obtained at a 250 × 250 m resolution from NASA (https://search.earthdata.nasa.gov/search) (accessed on 10 October 2022). The Carnegie-Ames-Stanford Approach (CASA) model was used to estimate NPP as follows [22]:

$$NPP(x,t) = APAR(x,t) \times \varepsilon(x,t) \quad (1)$$

where $APAR(x, t)$ represents the photosynthetically active radiation (PAR, in units of MJ/m^2) absorbed at pixel x in month t, and $\varepsilon(x, t)$ represents the actual light energy utilization at pixel x in month t (g C/MJ).

4.2. Data Analysis

To examine the contribution of climatic factors (MAT, MAP, $PM_{2.5}$, and CO_2 emissions) to the spatial variation in annual NPP, a linear regression model was used, with R^2 being used to assess the model's goodness-of-fit. Linear regression was conducted using the R package 'lme4' [45].

To evaluate the effects of time (year) on the variables, we used Generalized Additive Models (GAMs), which utilize both parametric and non-parametric components to reduce the model risks inherent in linear models. All calculations were conducted within the R environment using the 'mgcv' package [47].

Furthermore, we examined the effects of different concentrations of climate factors on NPP by dividing all the factors into three parts according to the calculated value, with the middle part serving as the reference variable. The effect size of climatic variables was analyzed based on the following model:

$$\ln RR = \ln(X_e/X_c) \quad (2)$$

where X_e and X_c represent the mean values of the variables.

Author Contributions: Conceptualization, J.G.; experimental implementation, J.G.; validation, J.G.; formal analysis, J.G.; writing—original draft preparation, J.G.; writing—review and editing, J.G., J.W. and Y.L.; visualization, J.G.; project administration, J.G. and Y.L.; funding acquisition, J.G. and Y.L. All authors have read and agreed to the published version of the manuscript.

Funding: The study was supported by the Xinjiang Normal University Landmark Achievements Cultivation Project, China (grant number: no number), Natural Science Foundation of Xinjiang Uygur Autonomous Region (no. 2022D01A213) and Tianchi Talent Program.

Data Availability Statement: Not applicable.

Conflicts of Interest: The authors declare no conflict of interest. The funders had no role in the design of the study; in the collection, analyses, or interpretation of data; in the writing of the manuscript; or in the decision to publish the results.

References

1. Sallaba, F.; Lehsten, D.; Seaquist, J.; Sykes, M.T. A Rapid NPP Meta-Model for Current and Future Climate and CO_2 Scenarios in Europe. *Ecol. Model.* **2015**, *302*, 29–41. [CrossRef]
2. Wieder, W.; Cleveland, C.; Smith, W.; Todd-Brown, K. Future Productivity and Carbon Storage Limited by Terrestrial Nutrient Availability. *Nat. Geosci.* **2015**, *8*, 441–444. [CrossRef]
3. Gang, C.; Zhou, W.; Zhaoqi, W.; Chen, Y.; Li, J.; Chen, J.; Qi, J.; Odeh, I.; Groisman, P. Comparative Assessment of Grassland NPP Dynamics in Response to Climate Change in China, North America, Europe and Australia from 1981 to 2010. *J. Agron. Crop Sci.* **2014**, *201*, 57–68. [CrossRef]
4. Liu, Y.; Wang, Q.; Zhang, Z.; Tong, L.; Wang, Z.; Li, J. Grassland Dynamics in Responses to Climate Variation and Human Activities in China from 2000 to 2013. *Sci. Total Environ.* **2019**, *690*, 27–39. [CrossRef] [PubMed]
5. Chen, Y.; Chen, L.; Cheng, Y.; Ju, W.; Chen, H.Y.H.; Ruan, H. Afforestation Promotes the Enhancement of Forest LAI and NPP in China. *For. Ecol. Manag.* **2020**, *462*, 117990. [CrossRef]

6. Naeem, S.; Zhang, Y.; Tian, J.; Qamer, F.M.; Latif, A.; Paul, P.K. Quantifying the Impacts of Anthropogenic Activities and Climate Variations on Vegetation Productivity Changes in China from 1985 to 2015. *Remote Sens.* **2020**, *12*, 1113. [CrossRef]
7. Bi, X.; Li, B.; Zhang, L.; Nan, B.; Zhang, X.; Yang, Z. Response of Grassland Productivity to Climate Change and Anthropogenic Activities in Arid Regions of Central Asia. *PeerJ* **2020**, *8*, e9797. [CrossRef]
8. Fang, K.; Wang, T.; He, J.; Wang, T.; Xie, X.; Tang, Y.; Shen, Y.; Xu, A. The Distribution and Drivers of $PM_{2.5}$ in a Rapidly Urbanizing Region: The Belt and Road Initiative in Focus. *Sci. Total Environ.* **2020**, *716*, 137010. [CrossRef]
9. Liu, S.; Xing, J.; Zhao, B.; Wang, J.; Wang, S.; Zhang, X.; Ding, A. Understanding of Aerosol–Climate Interactions in China: Aerosol Impacts on Solar Radiation, Temperature, Cloud, and Precipitation and Its Changes Under Future Climate and Emission Scenarios. *Curr. Pollut. Rep.* **2019**, *5*, 36–51. [CrossRef]
10. Xu, X.; Niu, S.; Sherry, R.A.; Zhou, X.; Zhou, J.; Luo, Y. Interannual Variability in Responses of Belowground Net Primary Productivity (NPP) and NPP Partitioning to Long-Term Warming and Clipping in a Tallgrass Prairie. *Glob. Chang. Biol.* **2012**, *18*, 1648–1656. [CrossRef]
11. Ruddiman, W.F. The Anthropocene. *Annu. Rev. Earth Planet. Sci.* **2013**, *41*, 45–68. [CrossRef]
12. Azhdari, Z.; Rafeie Sardooi, E.; Bazrafshan, O.; Zamani, H.; Singh, V.P.; Mohseni Saravi, M.; Ramezani, M. Impact of Climate Change on Net Primary Production (NPP) in South Iran. *Environ. Monit. Assess.* **2020**, *192*, 409. [CrossRef]
13. Sun, Y.; Yang, Y.; Zhao, X.; Tang, Z.; Wang, S.; Fang, J. Global Patterns and Climatic Drivers of Above- and Belowground Net Primary Productivity in Grasslands. *Sci. China Life Sci.* **2021**, *64*, 739–751. [CrossRef] [PubMed]
14. Liu, H.; Mi, Z.; Lin, L.; Wang, Y.; Zhang, Z.; Zhang, F.; Wang, H.; Liu, L.; Zhu, B.; Cao, G.; et al. Shifting Plant Species Composition in Response to Climate Change Stabilizes Grassland Primary Production. *Proc. Natl. Acad. Sci. USA* **2018**, *115*, 4051–4056. [CrossRef]
15. Yang, D.; Zhang, H.; Li, J. Changes in Concentrations of Fine and Coarse Particles under the CO_2-Induced Global Warming. *Atmos. Res.* **2019**, *230*, 104637. [CrossRef]
16. Zhang, Z.; Wang, X.; Cheng, S.; Guan, P.; Zhang, H.; Shan, C.; Fu, Y. Investigation on the Difference of $PM_{2.5}$ Transport Flux between the North China Plain and the Sichuan Basin. *Atmos. Environ.* **2022**, *271*, 118922. [CrossRef]
17. Zhang, C.; Ren, W. Complex Climatic and CO_2 Controls on Net Primary Productivity of Temperate Dryland Ecosystems over Central Asia during 1980–2014. *J. Geophys. Res. Biogeosciences* **2017**, *122*, 2356–2374. [CrossRef]
18. Meng, L.; Wu, J.; Feng, Y.; Niu, B.; He, Y.; Zhang, X. Climate Variability Rather Than Livestock Grazing Dominates Changes in Alpine Grassland Productivity Across Tibet. *Front. Ecol. Evol.* **2021**, *9*, 631024.
19. Wei, X.; Yang, J.; Luo, P.; Lin, L.; Lin, K.; Guan, J. Assessment of the Variation and Influencing Factors of Vegetation NPP and Carbon Sink Capacity under Different Natural Conditions. *Ecol. Indic.* **2022**, *138*, 108834. [CrossRef]
20. Cao, S.; Sanchez-Azofeifa, G.A.; Duran, S.M.; Calvo-Rodriguez, S. Estimation of Aboveground Net Primary Productivity in Secondary Tropical Dry Forests Using the Carnegie-Ames-Stanford Approach (CASA) Model. *Environ. Res. Lett.* **2016**, *11*, 075004. [CrossRef]
21. Wang, Y.; Guo, C.; Chen, X.; Jia, L.; Guo, X.; Chen, R.; Zhang, M.; Chen, Z.; Wang, H. Carbon Peak and Carbon Neutrality in China: Goals, Implementation Path and Prospects. *China Geol.* **2021**, *4*, 720–746. [CrossRef]
22. Du, Z.; Liu, X.; Wu, Z.; Zhang, H.; Zhao, J. Responses of Forest Net Primary Productivity to Climatic Factors in China during 1982–2015. *Plants* **2022**, *11*, 2932. [CrossRef] [PubMed]
23. Hao, L.; Wang, S.; Cui, X.; Zhai, Y. Spatiotemporal Dynamics of Vegetation Net Primary Productivity and Its Response to Climate Change in Inner Mongolia from 2002 to 2019. *Sustainability* **2021**, *13*, 13310. [CrossRef]
24. Zhang, M.; Liu, X.; Nazieh, S.; Wang, X.; Nkrumah, T.; Hong, S. Spatiotemporal Distribution of Grassland NPP in Gansu Province, China from 1982 to 2011 and Its Impact Factors. *PLoS ONE* **2020**, *15*, e0242609. [CrossRef]
25. Li, H.; Wu, Y.; Liu, S.; Xiao, J. Regional Contributions to Interannual Variability of Net Primary Production and Climatic Attributions. *Agric. For. Meteorol.* **2021**, *303*, 108384. [CrossRef]
26. Peng, J.; Dan, L. Impacts of CO_2 Concentration and Climate Change on the Terrestrial Carbon Flux Using Six Global Climate–Carbon Coupled Models. *Ecol. Model.* **2015**, *304*, 69–83. [CrossRef]
27. Ju, W.M.; Chen, J.M.; Harvey, D.; Wang, S. Future Carbon Balance of Chinas Forests under Climate Change and Increasing CO_2. *J. Environ. Manag.* **2007**, *85*, 538–562. [CrossRef]
28. Jiang, D.; Sui, Y.; Lang, X. Timing and Associated Climate Change of a 2 °C Global Warming. *Int. J. Climatol.* **2016**, *36*, 4512–4522. [CrossRef]
29. Baig, S.; Medlyn, B.E.; Mercado, L.M.; Zaehle, S. Does the Growth Response of Woody Plants to Elevated CO_2 Increase with Temperature? A Model-Oriented Meta-Analysis. *Glob. Chang. Biol.* **2015**, *21*, 4303–4319. [CrossRef]
30. Gu, F.; Zhang, Y.; Huang, M.; Tao, B.; Guo, R.; Yan, C. Effects of Climate Warming on Net Primary Productivity in China during 1961–2010. *Ecol. Evol.* **2017**, *7*, 6736–6746. [CrossRef]
31. Ofiti, N.O.E.; Zosso, C.U.; Soong, J.L.; Solly, E.F.; Torn, M.S.; Wiesenberg, G.L.B.; Schmidt, M.W.I. Warming Promotes Loss of Subsoil Carbon through Accelerated Degradation of Plant-Derived Organic Matter. *Soil Biol. Biochem.* **2021**, *156*, 108185. [CrossRef]
32. Maslov, M.N.; Maslova, O.A. Nitrogen Limitation of Microbial Activity in Alpine Tundra Soils along an Environmental Gradient: Intra-Seasonal Variations and Effect of Rising Temperature. *Soil Biol. Biochem.* **2021**, *156*, 108234. [CrossRef]

33. Michaletz, S.; Cheng, D.; Kerkhoff, A.; Enquist, B. Convergence of Terrestrial Plant Production across Global Climate Gradients. *Nature* **2014**, *512*, 39–43. [CrossRef] [PubMed]
34. Acuña-Rodríguez, I.S.; Newsham, K.K.; Gundel, P.E.; Torres-Díaz, C.; Molina-Montenegro, M.A. Functional Roles of Microbial Symbionts in Plant Cold Tolerance. *Ecol. Lett.* **2020**, *23*, 1034–1048. [CrossRef] [PubMed]
35. Walker, T.W.N.; Kaiser, C.; Strasser, F.; Herbold, C.W.; Leblans, N.I.W.; Woebken, D.; Janssens, I.A.; Sigurdsson, B.D.; Richter, A. Microbial Temperature Sensitivity and Biomass Change Explain Soil Carbon Loss with Warming. *Nat. Clim. Chang.* **2018**, *8*, 885–889. [CrossRef] [PubMed]
36. Guo, X.; Zuo, X.; Yue, P.; Li, X.; Hu, Y.; Chen, M.; Yu, Q. Direct and Indirect Effects of Precipitation Change and Nutrients Addition on Desert Steppe Productivity in Inner Mongolia, Northern China. *Plant Soil* **2022**, *471*, 527–540. [CrossRef]
37. Wang, X.; Wang, J.; Zhang, L.; Lv, C.; Liu, L.; Zhao, H.; Gao, J. Climatic Factors Determine the Distribution Patterns of Leaf Nutrient Traits at Large Scales. *Plants* **2022**, *11*, 2171. [CrossRef]
38. Wang, Y.; Li, H.; Feng, J.; Wang, W.; Liu, Z.; Huang, L.; Yaluk, E.; Lu, G.; Manomaiphiboon, K.; Gong, Y.; et al. Spatial Characteristics of $PM_{2.5}$ Pollution among Cities and Policy Implication in the Northern Part of the North China Plain. *Atmosphere* **2021**, *12*, 77. [CrossRef]
39. Lin, Y.; Yuan, X.; Zhai, T.; Wang, J. Effects of Land-Use Patterns on $PM_{2.5}$ in Chinas Developed Coastal Region: Exploration and Solutions. *Sci. Total Environ.* **2020**, *703*, 135602. [CrossRef]
40. Piao, S.; Sitch, S.; Ciais, P.; Friedlingstein, P.; Peylin, P.; Wang, X.; Ahlström, A.; Anav, A.; Canadell, J.G.; Cong, N.; et al. Evaluation of Terrestrial Carbon Cycle Models for Their Response to Climate Variability and to CO_2 Trends. *Glob. Chang. Biol.* **2013**, *19*, 2117–2132. [CrossRef]
41. Maschler, J.; Bialic-Murphy, L.; Wan, J.; Andresen, L.C.; Zohner, C.M.; Reich, P.B.; Lüscher, A.; Schneider, M.K.; Müller, C.; Moser, G.; et al. Links across Ecological Scales: Plant Biomass Responses to Elevated CO_2. *Glob. Chang. Biol.* **2022**, *28*, 6115–6134. [CrossRef] [PubMed]
42. Keenan, T.F.; Luo, X.; Kauwe, M.G.; Medlyn, B.E.; Prentice, I.C.; Stocker, B.D.; Smith, N.G.; Terrer, C.; Wang, H.; Zhang, Y.; et al. RETRACTED ARTICLE: A constraint on historic growth in global photosynthesis due to increasing CO2. *Nature* **2021**, *600*, 253–258. [CrossRef] [PubMed]
43. Brienen, R.J.W.; Caldwell, L.; Duchesne, L.; Voelker, S.; Barichivich, J.; Baliva, M.; Ceccantini, G.; Di Filippo, A.; Helama, S.; Locosselli, G.M.; et al. Forest Carbon Sink Neutralized by Pervasive Growth-Lifespan Trade-Offs. *Nat. Commun.* **2020**, *11*, 4241. [CrossRef]
44. Needham, J.F.; Chambers, J.; Fisher, R.; Knox, R.; Koven, C.D. Forest Responses to Simulated Elevated CO_2 under Alternate Hypotheses of Size- and Age-Dependent Mortality. *Glob. Chang. Biol.* **2020**, *26*, 5734–5753. [CrossRef] [PubMed]
45. Gao, J.; Ji, Y.; Zhang, X. Net primary productivity exhibits a stronger climatic response in planted versus natural forests. *For. Ecol. Manag.* **2023**, *529*, 120722. [CrossRef]
46. Zhong, J.; Zhang, X.; Gui, K.; Liao, J.; Fei, Y.; Jiang, L.; Guo, L.; Liu, L.; Che, H.; Wang, Y.; et al. Reconstructing 6-Hourly $PM_{2.5}$ Datasets from 1960 to 2020 in China. *Earth Syst. Sci. Data* **2022**, *14*, 3197–3211. [CrossRef]
47. Huang, W.; Ding, Y.; Wang, S.; Song, C.; Wang, F. Growth and Development Responses of the Rhizome-Root System in Pleioblastus pygmaeus to Light Intensity. *Plants* **2022**, *11*, 2204. [CrossRef]

plants

MDPI

Article

Temporal Changes in Litterfall and Nutrient Cycling from 2005–2015 in an Evergreen Broad-Leaved Forest in the Ailao Mountains, China

Shiyu Dai [1], Ting Wei [1], Juan Tang [1], Zhixiong Xu [2] and Hede Gong [1,*]

1 School of Geography and Ecotourism, Southwest Forestry University, Kunming 650224, China
2 Ailaoshan Station for Subtropical Forest Ecosystem Studies, Chinese Academy of Sciences, Jingdong, Puer City 676209, China
* Correspondence: gonghede3@163.com

Abstract: The study of litter can provide an important reference for understanding patterns of forest nutrient cycling and sustainable management. Here, we measured litterfall (leaves, branches, etc.) from a wet, evergreen, broad-leaved forest in Ailao Mountains of southwestern China on a monthly basis for 11 years (2005–2015). We measured the total biomass of litter fall as well as its components, and estimated the amount of C, N, P, K, S, Ca, and Mg in the amount of litterfall. We found that: The total litter of evergreen, broadleaved forest in Ailao Mountains from 2005 to 2015 was 7.70–9.46 t/ha, and the output of litterfall differed between years. This provides a safeguard for the soil fertility and biodiversity of the area. The total amount of litterfall and its components showed obvious seasonal variation, with most showing a bimodal pattern (peak from March to May and October to November). The majority of litterfall came from leaves, and the total amount as well as its components were correlated with meteorological factors (wind speed, temperate and precipitation) as well as extreme weather events. We found that among years, the nutrient concentration was sorted as C > Ca > N > K > Mg > S > P. The nutrient concentration in the fallen litter and the amount of nutrients returned showed a decreasing trend, but the decreasing rate was slowed through time. Nutrient cycling was influenced by meteorological factors, such as temperature, precipitation, and wind speed, but the nutrient utilization efficiency is high, the circulation capacity is strong, and the turnover time is short. Our results showed that although there was nutrient loss in this evergreen, broad-leaved forest, the presence of forest litterfall can effectively curb potential ecological problems in the area.

Keywords: litterfall production; elemental composition; nutrient cycle; subtropical forest; Ailao Mountain

Citation: Dai, S.; Wei, T.; Tang, J.; Xu, Z.; Gong, H. Temporal Changes in Litterfall and Nutrient Cycling from 2005–2015 in an Evergreen Broad-Leaved Forest in the Ailao Mountains, China. *Plants* **2023**, *12*, 1277. https://doi.org/10.3390/plants12061277

Academic Editor: Jianhui Huang

Received: 19 January 2023
Revised: 6 March 2023
Accepted: 8 March 2023
Published: 10 March 2023

1. Introduction

In forests, organic matter from plants that is returned to the soil surface (e.g., fallen leaves, branches, floral and fruit parts) is generally referred to as litterfall [1,2]. The amount and quality of forest litterfall plays an important role in the development of soil and the cycling of nutrients. For example, growing plants absorb nutrients needed for their own growth from the soil, e.g., carbon, nitrogen, phosphorus, potassium, and other elements, and then return those elements back to the soil in the form of litterfall decomposition [3]. Among the different C fluxes of the forest ecosystem, canopy litterfall is the main aboveground organic C flux that reaches the soil, affecting C cycling as well as maintaining soil fertility globally [4]. Hence, litterfall acts as an important link between the aboveground production of trees and the soil organic C stock. At the same time, the litterfall production changes with climate, forest type, stand age, and season [5–7]. As a result, the quantity and quality of litterfall, as well as the environmental factors that influence them, regulate how these material cycles within ecosystems. However, there is variation in the turnover rates of different elements in a given ecosystem, as well as variation in turnover across regions in different climatic zones. Thus, it is useful to study the nature of forest litterfall dynamics through time in order to gain a deeper understanding of patterns of variation in litterfall among seasons and across years, and how this contributes to variation in forest nutrient cycling.

Plants periodically shed parts of their biomass as litterfall, which transfers C and nutrients from plants back into the soils and is a key biogeochemical process within forests [4,8]. However, litterfall in forests is variable throughout the year, with less litterfall during the growing season than the non-growing season, The main seasonal pattern of presentation is that: unimodal, bimodal, or irregular pattern [9]. In deciduous forests, litterfall happens during the non-growing season when low temperatures stimulate plant leaf synthesis of abscisic acid, resulting in a high levels of leaf fall, this has been confirmed in many studies [10–12]. In addition to variation across the season and across different forest types, litterfall is also variable among years as a result of variation in climatic conditions (e.g., wind and snow) and forest age. As global climate change continues, studies on the within and among year variation in litterfall, and its role in nutrient cycling, will provide important baseline knowledge for understanding how these will change in the future [13–16].

Subtropical forests have high primary productivity and are also hotspots for biodiversity research., which play an important role in carbon storage in global terrestrial ecosystems [17]. The montane, moist, evergreen, broad-leaved forest in the Ailao Mountain Nature Reserve in Yunnan is currently the largest and most well-preserved subtropical evergreen broad-leaved forest in China. It is one of the valuable zonal vegetation. This is particularly urgent for understanding patterns of litterfall and nutrient cycling in primary forests which are rapidly disappearing. Hence, we can also better understand and utilize natural resources, thus improving the stability and sustainability of the ecosystem.

Here, we measured the total amount of litterfall, as well as its nutrient concentration, from monthly samples collected over an 11-year period (2005–2015) in the primeval forest of Ailao Mountain National Nature Reserve in Yunnan Province, southwestern China. We compared this variation at monthly, seasonal, and annual periods and examined how they were correlated with precipitation, temperature, wind speed, and extreme weather. Results from our study show considerable variation within and among years in the quantity and quality of litterfall in this forest, providing baseline data for studying forest nutrient cycles.

2. Results

2.1. *Dynamic Characteristics of Litterfall and Its Component Output*

2.1.1. Interannual Dynamics of Litterfall and Its Component Output

Between 2005 to 2015, we found that the amount of litterfall per year ranged from 7.70 to 9.46 t/ha a (Figure 1, Table 1), with an annual average of 8.11 ± 0.73 t/ha a. Across the observation period, we found that deciduous leaf litter was the greatest component of the overall leaf litter (representing 42–62% of the litterfall). Other components of litterfall, including fruit/flower drop, bark, moss/lichen, and other debris, were observed to an intermediate (21–26% of the litterfall) extent, and litterfall from branches the least (17–32% of the litterfall) (Table 1).

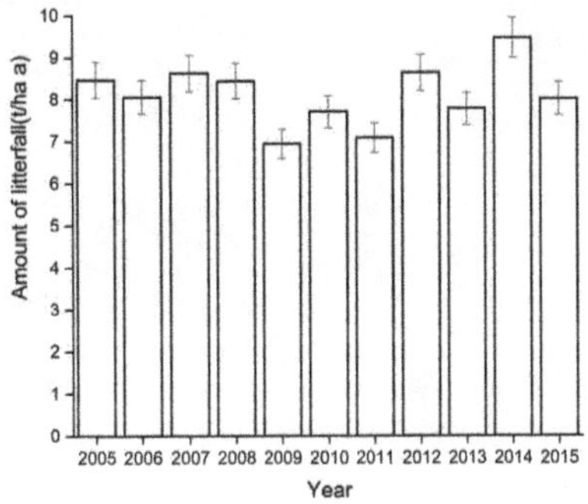

Figure 1. Interannual dynamics of litterfall from 2005 to 2015.

Table 1. Annual total litterfall productions and components.

Year	Branches (t/ha a)		Leaf (t/ha a)		Others (t/ha a)		Total (t/ha a)
2005	2.32 ± 0.80 Aa	26.9%	4.36 ± 0.52 Bb	52.0%	1.76 ± 0.12 Aa	21.1%	8.36 ± 0.88
2006	1.60 ± 0.20 Aa	19.5%	4.40 ± 0.84 Bb	54.6%	2.04 ± 0.24 Ba	25.9%	8.08 ± 1.16
2007	2.00 ± 0.60 Aa	22.9%	4.52 ± 0.64 Bb	52.8%	2.12 ± 0.16 Ba	24.3%	8.60 ± 1.00
2008	1.88 ± 0.36 Aa	23.2%	4.52 ± 0.8 Bb	52.5%	2.08 ± 0.28 Ba	24.3%	8.48 ± 1.12
2009	1.32 ± 0.20 Aa	19.7%	4.00 ± 0.48 Bb	59.1%	1.44 ± 0.12 Aa	21.2%	6.88 ± 0.60
2010	1.36 ± 0.24 Aa	17.6%	4.80 ± 0.84 Cb	61.5%	1.52 ± 0.20 Aa	20.9%	7.76 ± 1.20
2011	1.28 ± 0.16 Aa	18.6%	4.24 ± 0.56 Bb	59.5%	1.56 ± 0.08 Aa	21.8%	7.12 ± 0.52
2012	1.60 ± 0.20 Aa	18.9%	4.80 ± 0.84 Cc	55.6%	2.20 ± 0.28 Ca	25.5%	8.64 ± 1.04
2013	1.32 ± 0.20 Aa	17.3%	4.48 ± 0.60 Bc	57.5%	1.96 ± 0.28 Bb	25.2%	7.80 ± 0.80
2014	2.12 ± 0.52 Aa	22.5%	5.20 ± 1.12 Cb	54.1%	2.24 ± 0.28 Ca	23.3%	9.40 ± 1.44
2015	2.52 ± 1.40 Ba	32.0%	3.28 ± 0.44 Ab	41.7%	2.04 ± 0.32 Ca	26.3%	8.08 ± 1.92
Total average	1.76 ± 0.2	21.7%	4.42 ± 0.18	52.0%	1.91 ± 0.06	23.6%	8.11 ± 0.27

Note: Different uppercase letters indicate significant differences in litterfall output between different years of the same component ($p < 0.05$), and different lowercase letters indicate significant differences in litterfall output of different components in the same year ($p < 0.05$).

2.1.2. Monthly Variation in the Output of Litterfall and Its Components within a Year

There was clear seasonal variation in the total litterfall in this forest (Figure 2a), but the variation was not always consistent among years. For example, most years had multiple peaks of litterfall, with one peak around April (ranging from March–May) and another later in the October–November. There was variation in these peaks, however. One year, 2015, was unique with the highest observed litterfall of the whole time series occurring in January. When we analyzed litterfall in its components, we found that leaf litter (Figure 2b), as the most abundant component, largely mirrored that of the total litterfall biomass. The other two components of litterfall, branches (Figure 2c), and 'other' (Figure 2d) were more variable throughout the year. The main source of litterfall production is leaves and branches, with the highest production of branches in January and February, and the highest production of leaves in other months (Table 2).

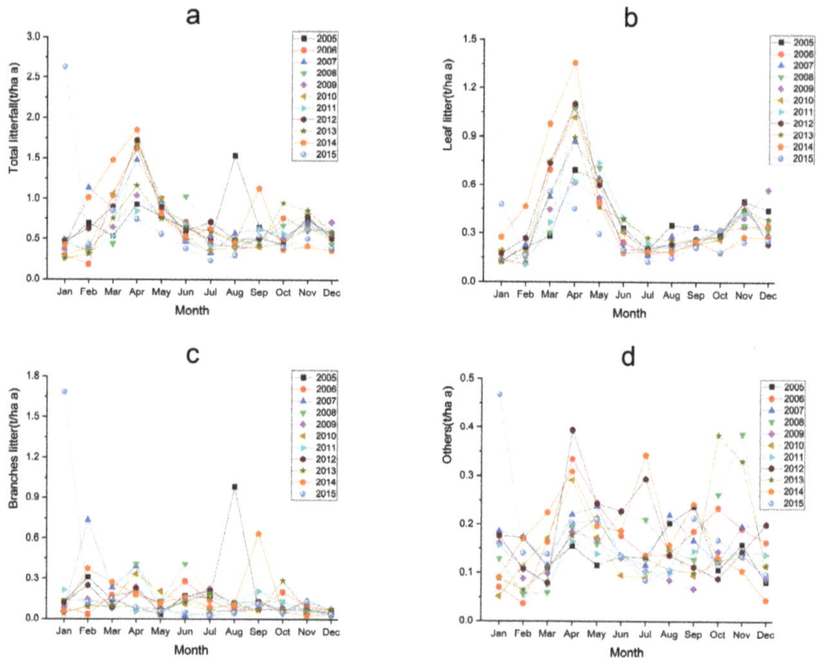

Figure 2. Monthly variation of litterfall and its component output in evergreen broadleaf forest in Ailao Mountain. (**a**) Total litterfall, (**b**) Leaf litter, (**c**) Branches litter, (**d**) Others.

Table 2. Intra-year variation of litterfall productions and components.

Month	Branches (t/ha a)		Leaf (t/ha a)		Others (t/ha a)		Total (t/ha a)
January	0.33 ± 0.54 Aa	41.04%	0.26 ± 0.15 Aa	32.10%	0.22 ± 0.18 Aa	26.86%	0.82 ± 0.76
February	0.3 ± 0.63 Aa	41.17%	0.28 ± 0.18 Aa	37.74%	0.15 ± 0.11 Aa	21.09%	0.73 ± 0.8
March	0.2 ± 0.23 Aa	17.22%	0.77 ± 0.26 Bb	67.93%	0.17 ± 0.1 Aa	14.85%	1.14 ± 0.4
April	0.29 ± 0.53 Aa	15.62%	1.22 ± 0.35 Cb	66.49%	0.33 ± 0.15 Ca	17.89%	1.85 ± 0.75
May	0.12 ± 0.21 Aa	10.82%	0.78 ± 0.24 Bb	66.57%	0.27 ± 0.13 Ba	22.61%	1.16 ± 0.37
Jun	0.21 ± 0.57 Aa	26.68%	0.4 ± 0.18 Aa	47.65%	0.22 ± 0.19 Aa	25.67%	0.84 ± 0.81
July	0.16 ± 0.32 Aa	26.58%	0.28 ± 0.09 Aa	40.76%	0.22 ± 0.34 Aa	32.66%	0.66 ± 0.6
August	0.22 ± 1.13 Aa	31.69%	0.31 ± 0.17 Aa	42.13%	0.19 ± 0.25 Aa	26.18%	0.74 ± 1.45
September	0.2 ± 0.33 Aa	27.19%	0.36 ± 0.12 Aa	45.86%	0.21 ± 0.23 Aa	26.95%	0.79 ± 0.45
October	0.14 ± 0.29 Aa	19.67%	0.38 ± 0.12 Ab	49.55%	0.24 ± 0.25 Aa	30.78%	0.77 ± 0.43
November	0.1 ± 0.16 Aa	11.98%	0.55 ± 0.17 Aa	59.74%	0.26 ± 0.36 Ba	28.27%	0.92 ± 0.46
December	0.07 ± 0.07 Aa	10.62%	0.49 ± 0.2 Ab	66.43%	0.17 ± 0.19 Aa	22.96%	0.73 ± 0.3
Total average	0.19 ± 0.5	23.36%	0.51 ± 0.28	51.91%	0.22 ± 0.18	24.73%	0.93 ± 0.74

Note: Different uppercase letters indicate significant differences in litterfall output between different months of the same component ($p < 0.05$), and different lowercase letters indicate significant differences in litterfall output of different components in the same month ($p < 0.05$).

2.1.3. Correlation Analysis of Litterfall and Its Components with Climatic Factors

Overall, we found that the total litterfall was negatively correlated with the average wind speed, but positively correlated with temperature and precipitation (Table 3). These trends were mirrored by the 'other' category of litterfall, while only leaf litterfall was positively correlated with average monthly temperature (Table 3). For monthly data, we found that the total amount of litterfall and meteorological factors was positively correlated with monthly precipitation in the first 1–2 months, but there were no other positive correlations between litterfall production and monthly climatic variation (Table 4).

Table 3. Correlation coefficients between litterfall productions of different components and various meteorological factors.

	Average Monthly Wind Speed	Monthly Precipitation	Monthly Precipitation Maximum	Average Monthly Temperature
Total	−0.127 *	0.166 **	0.158 **	0.139 *
Branches litter	−0.049	0.065	0.064	0.042
Leaf litter	−0.092	0.108	0.099	0.165 **
Others	−0.185 **	0.197 **	0.177 **	0.132 *

Note: * Significantly correlated at level 0.05 (two-tailed). ** Significant correlation at level 0.01 (double-tailed).

Table 4. Correlation coefficients between meteorological factors from different months and monthly total litterfall productions.

Month	Average Monthly Wind Speed	Monthly Precipitation	Monthly Precipitation Maximum	Average Monthly Temperature
January	−0.565	0.996 **	0.963 **	−0.268
February	0.028	0.773 **	0.717 *	−0.424
March	0.125	−0.332	0.085	0.497
April	0.419	−0.183	−0.291	0.406
May	0.103	0.190	0.012	−0.430
June	−0.468	−0.038	−0.012	−0.529
July	−0.028	−0.396	−0.240	0.303
August	−0.337	0.375	−0.121	0.240
September	−0.143	0.473	0.436	0.231
October	−0.308	0.384	0.317	−0.173
November	−0.056	0.291	0.255	−0.005
December	0.023	−0.245	−0.151	−0.178

Note: * Significantly correlated at level 0.05 (two-tailed). ** Significant correlation at level 0.01 (double-tailed).

2.2. *Dynamic Characteristics of Litterfall Nutrient Concentration*

2.2.1. Interannual Variation Characteristics of Litterfall Nutrient Concentration

We found high variation in the nutrient concentration of litterfall among years and in different components of the litterfall (Table 5). In general, the concentration of C, Ca, and Mg decreased from 2005 to 2010 and 2015, while the concentration of N, P, K, and S increased across this same period. While the nutrient concentration of litterfall was C > Ca > N > K > Mg > S > P in 2005 and 2010, the abundance of N increased relative to the other elements, such that C > N > Ca > K > Mg > S > P in 2015. Across years, we found higher nutrient concentration in leaves compared to branches.

Table 5. Characteristics of annual average nutrient concentration of litterfall in different years.

Nutrient Element	2005		2010		2015	
	Branches (g/kg)	Leaf (g/kg)	Branches (g/kg)	Leaf (g/kg)	Branches (g/kg)	Leaf (g/kg)
C	526.44 ± 2.94 Cd	546.25 ± 2.45 Dc	489.31 ± 3.28 Ad	507.25 ± 2.78 Bc	486.63 ± 2.52 Ad	500.7 ± 2.7 Bd
N	9.13 ± 0.22 Ab	13.23 ± 0.27 Cb	10.22 ± 0.56 Bb	13.81 ± 0.64 Cb	8.52 ± 0.49 Ab	14.7 ± 0.5 Dc
P	0.48 ± 0.01 Aa	0.72 ± 0.02 Ba	0.49 ± 0.02 Aa	0.74 ± 0.04 Ba	0.5 ± 0.03 Aa	0.84 ± 0.04 Ca

Table 5. *Cont.*

Nutrient Element	2005		2010		2015	
	Branches (g/kg)	Leaf (g/kg)	Branches (g/kg)	Leaf (g/kg)	Branches (g/kg)	Leaf (g/kg)
K	1.99 ± 0.11 Aa	4.95 ± 0.17 Ca	1.8 ± 0.12 Aa	4.55 ± 0.23 Ca	2.66 ± 0.37 Ba	5.32 ± 0.17 Da
S	0.87 ± 0.02 Ba	1.24 ± 0.02 Ca	0.92 ± 0.03 Ba	1.25 ± 0.02 Ca	0.74 ± 0.04 Aa	1.25 ± 0.04 Ca
Ca	15.19 ± 0.43 Dc	13.08 ± 0.2 Bb	14.38 ± 0.81 Cc	12.19 ± 0.28 Bb	11.78 ± 0.6 Bc	10.67 ± 0.53 Ab
Mg	1.55 ± 0.05 Aa	2.55 ± 0.03 Ba	1.54 ± 0.05 Aa	2.47 ± 0.09 Ba	1.45 ± 0.09 Aa	2.47 ± 0.06 Ba
Total average	555.65 ± 3.78	582.02 ± 3.16	518.66 ± 4.87	542.26 ± 4.08	512.28 ± 4.14	535.95 ± 4.04

Note: Different uppercase letters in the same row indicate significant differences between different years ($p < 0.05$), and different lowercase letters in the same column indicate significant differences between different nutrient elements in the same year ($p < 0.05$).

2.2.2. Characteristics of Intra-Year Variation of Litterfall Nutrient Concentration

We illustrate the within year variation of each element in the litterfall in Figure 3, showing that each element has a signature variation. C concentration is higher in the first half of the year and declines to its lowest level in late autumn and early winter (Figure 3a). N, P, and S showed similar variation across the year, first declining through the first months of the year, peaking in the middle of the growing season, and declining again towards fall and winter (Figure 3b–d). The last three elements showed less distinct patterns and fluctuated around mean values throughout the year (Figure 3e–g).

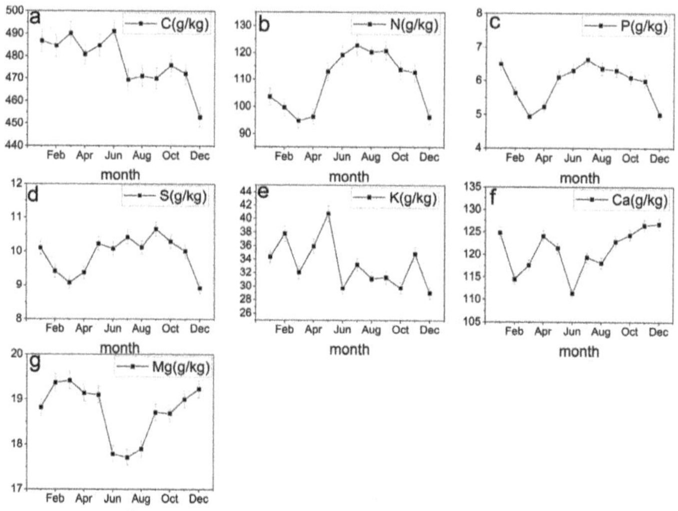

Figure 3. Within-year change of C, N, P, K, S, Ca and Mg concentrations in litterfall. (**a**) Within-year change of C element concentration. (**b**) Within-year change of N element concentration. (**c**) Within-year change of P element concentration. (**d**) Within-year change of S element concentration. (**e**) Within-year change of K element concentration. (**f**) Within-year change of Ca element concentration. (**g**) Within-year change of Mg element concentration.

2.3. *Dynamic Characteristics of Litterfall Nutrient Element Return*

2.3.1. Interannual Return Characteristics of Litterfall Nutrient Elements

We found that the return of leaf litter (except Ca) was higher than that of branches litter, and the return of C was much higher than that of other nutrient elements. In all, the returns were roughly sorted as C > N > Ca > K > Mg > S > P (Table 6). Overall, we found a downward trend of the annual mean return of nutrients from 2005 to 2015. In branches, the return of C, N, and S decreased through time, while the return of P and K increased; the return of Ca and Mg first decreased and then increased.

Table 6. Characteristics of annual average nutrient return of litterfall in different years.

Nutrient Element	2005		2010		2015	
	Branches (g/kg)	**Leaf (g/kg)**	**Branches (g/kg)**	**Leaf (g/kg)**	**Branches (g/kg)**	**Leaf (g/kg)**
C	542.27 ± 25.34 Bc	553.96 ± 24.60 Bc	491.33 ± 9.67 Ad	509.16 ± 16.44 Ad	481.84 ± 14.29 Ac	500.20 ± 10.59 Ae
N	10.30 ± 2.07 Ab	13.72 ± 1.66 Bb	9.98 ± 2.42 Ac	17.60 ± 5.16 Dc	8.50 ± 2.13 Ab	15.81 ± 2.30 Cd
P	0.51 ± 0.15 Aa	0.66 ± 0.12 Ba	0.66 ± 0.40 Ba	1.02 ± 0.50 Ca	0.73 ± 0.33 Ba	0.98 ± 0.24 Ca
K	2.17 ± 1.14 Aa	3.73 ± 0.86 Ba	3.25 ± 2.90 Bb	4.95 ± 2.96 Cb	3.79 ± 2.49 Ba	5.13 ± 2.23 Cb
S	0.93 ± 0.21 Ba	1.18 ± 0.09 Ca	0.76 ± 0.17 Aa	1.27 ± 0.17 Ca	0.74 ± 0.19 Aa	1.25 ± 0.15 Ca
Ca	11.16 ± 2.32 Cb	9.29 ± 1.24 Bb	8.81 ± 3.84 Bc	7.53 ± 2.81 Ab	10.43 ± 4.39 Cb	7.98 ±1.99 Ac
Mg	1.37 ± 0.36 Ba	1.96 ± 0.31 Ca	1.07 ± 0.37 Aa	1.80 ± 0.40 Ca	1.29 ± 0.44 Ba	1.88 ± 0.37 Ca
Total average	568.71 ± 26.20	584.5 ± 24.48	515.86 ± 17.83	543.33 ± 18.50	507.32 ± 18.05	533.23 ± 11.47

Note: Different uppercase letters in the same row indicate significant differences between different years ($p < 0.05$), and different lowercase letters in the same column indicate significant differences between different nutrient elements in the same year ($p < 0.05$).

2.3.2. Characteristics of Intra-Year Return of Litterfall Nutrient Elements

In Figure 4, we show that the return of each element varies greatly throughout the year. The return of C is the largest, with a peak May. The return of N and S showed a multimodal distribution at the beginning of the year, mid-year, and at the end of the year, but there was a sudden decrease in July. K and P showed a peak in August. The return of Ca and Mg first decreased until around July and then increased.

In Table 7, we show that some, but not all, element returns are affected by meteorological factors. The returns of C, N, and S were not correlated with meteorological factors (temperature, precipitation, wind speed). The return of P and K was significantly negatively correlated with wind speed. The return of K was also positively correlated with precipitation, while the return of Ca was negatively correlated with precipitation. Finally, the return of Mg was negatively correlated with temperature and precipitation.

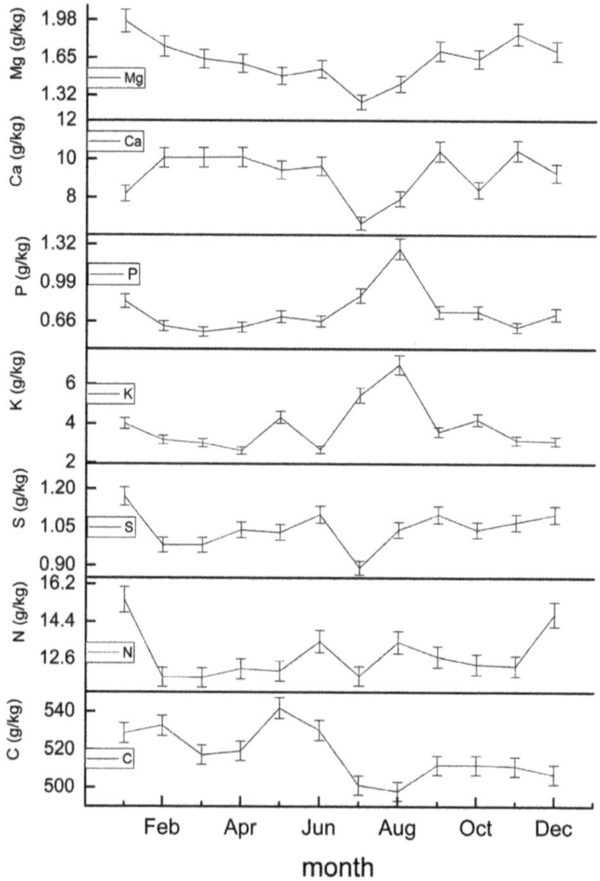

Figure 4. Within-year nutrient return of each element.

Table 7. Correlation coefficient between meteorological factors and nutrient return.

Nutrient Element	Temperature		Precipitation		Wind Speed	
	p	r^2	*p*	r^2	*p*	r^2
C	0.604	0.028	0.098	0.249	0.069	0.294
N	0.185	0.169	0.558	0.035	0.737	0.012
P	0.259	0.125	0.052	0.327	0.023 *	0.421
K	0.165	0.183	0.028 *	0.397	0.014 *	0.466
S	0.267	0.122	0.262	0.124	0.795	0.007
Ca	0.435	0.062	0.045 *	0.345	0.222	0.145
Mg	0.001 **	0.655	0.008 **	0.527	0.247	0.131

Note: * indicates significant correlation ($p < 0.05$), ** indicates very significant correlation ($p < 0.01$).

2.4. Biological Cycle of Nutrient Elements in Evergreen Broad-Leaved Forests of Mount Ailao

Nutrient cycling refers to the absorption of nutrients from the soil by plant, some of which is used for plant growth, while the rest is returned to the soil through litterfall, secretions and rainwater. This is given by absorption, retention, and return of the three links, where the cycle balance formula is: absorption = retention + return [18,19]. In our study, we were only able calculate the return of litterfall, thus underestimating the total cycle. Nevertheless, we can use a utilization coefficient, circulation coefficient, and turnover time to estimate elements of the cycle [20]. The nutrient utilization coefficients were distributed from 0.23 to 0.29, with an average value of 0.25, which showed Ca > Mg > S, C > N, K > P. The circulation coefficients were distributed from 0.42 to 0.84, with an average value was 0.53, showing P > K > N > C > S > Mg > Ca. The turnover time was distributed from 8.40a to 14.14a, with an average value of 10.50a, which was manifest as Ca > Mg > S > C > N > K > P (Table 8).

Table 8. Biological cycle of different nutrients.

Nutrient Element	R_e	R_g	T_t (a)
C	0.25	0.50	9.93
N	0.24	0.52	9.32
P	0.23	0.84	8.40
K	0.24	0.53	9.09
S	0.25	0.49	10.30
Ca	0.29	0.42	14.14
Mg	0.28	0.44	12.80
average	0.25	0.53	10.50

3. Materials and Methods

3.1. Overview of the Study Area

Our study area was located in the Xujiaba area (24°32′ N, 102°01′ E) within the Ailaoshan National Nature Reserve, Jingdong Yi Autonomous County, Puer City, Yunnan Province. The study area occurred at an altitude of 2400–2600 m, and the soil was fertile, acidic, yellow-brown soil [21]. The climate at the Ailao Mountain Forest Ecosystem Research Station is southwestern monsoon, with distinct dry and wet seasons (annual average temperature is 11.3 °C and the annual average precipitation is 1931 mm.), at the transition from central subtropical to south Asian tropics. Our study site is within the largest area of the primitive Zhongshan wet, evergreen, broad-leaved forest preserved in China, with a closed canopy and layered shrub layer with abundant epiphytes on the trees. The dominant tree species at the site include *Machilus bombycina, Populus rotundifolia, Schima noronhae, Castanopsis rufescens*, and *Castanopsis delavayi* [22].

3.2. Litterfall Sampling and Collection

We established a litterfall collection grid within a 1-ha long term observation plot of the forest. Specifically, we divided the plot into 100 10 m × 10 m subplots. From those 100 subplots, we randomly selected 25 and placed a 1 m^2 litterfall collection basket in each. We constructed litterfall collection baskets out of steel frame boxes 1 m × 1 m × 0.25 m covered with 0.5 mm nylon mesh. We inserted the four corners of each basket into the soil such that the bottom of the basket was about 0.5 m from the ground.

We collected litterfall from each basket at the end of each month from 2005 to 2015. We sorted litterfall into categories, including branches, leaves, fallen flowers and fruits, bark, moss and lichen, and other debris, drying each in an oven at 65 °C to a constant weight, and then recorded the dry weight of each component.

3.3. Meteorological Data Observation

Meteorological data, including precipitation, temperature, and wind speed, were collected from the Ailao Mountain Meteorological Station. Data were averaged monthly and according to season (Figure 5). The wet season is from June to October, while the dry season is from November to May.

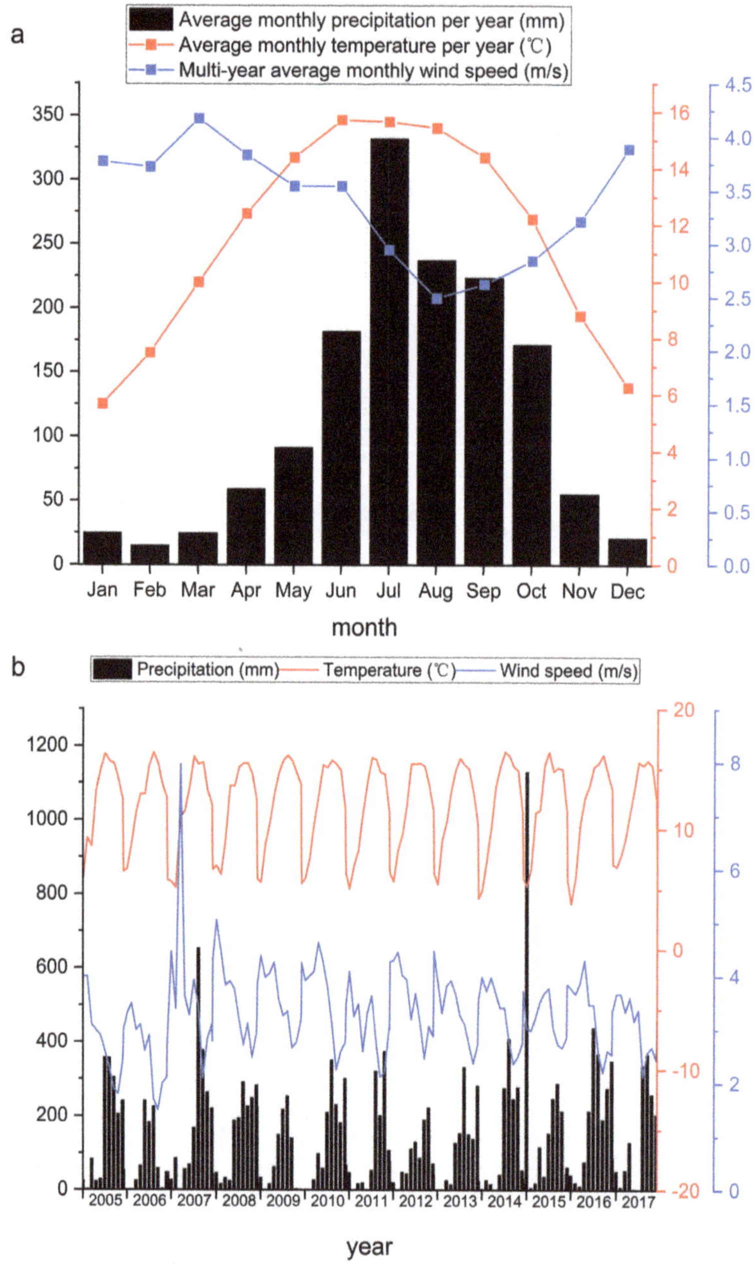

Figure 5. (**a**) Ailao Mountain Meteorological Station has average monthly precipitation, average monthly temperature for many years, average monthly wind speed for many years, (**b**) temperature, precipitation and wind speed from 2005 to 2017.

3.4. *Measuring Nutrients in the Litterfall*

We estimated important elements in the litterfall after drying by first grounding the littler into a fine powder, which was subsequently sieved through a 250-μm mesh. We determined carbon (C) and nitrogen (N) using a carbon analyzer (EA3000 EuroVector, Milan, Italy) [23,24]. To prepare samples for phosphorus (P) and potassium (K), we first digested samples in H_2O_2-H_2SO_4. We determined phosphorus (P) concentration using molybdenum antimony colorimetry, potassium (K) was measured using plasma atomic emission spectrophotometer [2], and the concentration of sulfur (S), calcium (Ca) and magnesium (Mg) was measured using a flame photometer and spectrophotometer [25].

3.5. Statistical Analysis of Data

We averaged the total litterfall output across the 25 collection baskets and took monthly and annual (sum of the 12 months) data for analyses. After using the Shapiro–Wilk test to test the normality of the data, one-way ANOVA and LSD were used to compare the difference in the amount of litterfall in different parts of different years and its components, the concentration of nutrient elements, and the amount of return. The ANOVA is mainly used to test whether there is a significant difference in the mean of the components of litterfall between the interannual and intra-annual periods. The use of multiple comparisons can support a better understanding of the differences between them. In addition, we also use SPSS26.0 to linear fit the environmental variables (temperature, precipitation, wind speed) and the output of components of litterfall and nutrient concentration, and then explore the correlation between them.

We calculated the annual return of litterfall nutrients as follows:

$$\mathrm{La} = \sum_{i=1}^{12}\sum_{j=1}^{25} LijCij/100 \tag{1}$$

where L_a is the annual amount of nutrients returned by litterfall; L_{ij} is the litterfall amount of the *j*th component in the *i*th month (kg/m^2); C_{ij} is the nutrient concentration (g/kg) of the *j*th component of litterfall in the *i*th month [26].

The biocirculating coefficient mainly includes nutrient utilization coefficient, cycle coefficient, and turnover time. The nutrient utilization coefficient is the ratio of the elements absorbed by the plant per unit time and unit area to the existing elements of the plant, and the calculation method of nutrient utilization is mainly based on the Chapin index.

$$E = \mathrm{A_p}/\mathrm{M} \tag{2}$$

It can be seen from Equation (2) that E is Chapin index, M is plant biomass, and $\mathrm{A_p}$ is nutrient storage (t/ha). In essence, Chapin index is the average content of plant body nutrients, which reflects the amount of nutrients consumed by the plant construction unit per unit biomass. However, Chapin index has a bias in overestimating the nutrient utilization efficiency of trees, and formula (2), revised from the perspective of nutrient cycle to obtain formula (3), can also better reflect the nutrient utilization status of trees.

$$R_\mathrm{e} = \mathrm{F_a}/\mathrm{A_p} \tag{3}$$

where R_e is the utilization coefficient, $\mathrm{F_a}$ is the nutrient uptake (t/ha a), and $\mathrm{A_p}$ is the nutrient storage (t/ha).

The nutrient cycle coefficient is a kind of index proposed based on the concept of the biological cycle, also known as the biological return coefficient. The method has certain limitations in reflecting the overall situation of forest nutrient cycling, since the calculation of the nutrient cycling coefficient does not involve the decomposition of forest litterfall, and the decomposition of litterfall is an important link in the nutrient cycle. The calculation is performed as follows:

$$R_g = \mathrm{F_d}/\mathrm{F_a} \tag{4}$$

where R_g is the cycle coefficient, $\mathrm{F_a}$ is the nutrient uptake (t/ha a), and $\mathrm{F_d}$ is the nutrient return (t/ha a).

Turnover time is the time it takes for a nutrient element to go through one cycle.

$$T_t = F_d / A_p \quad (5)$$

where T_t is the turnover time, A_p is the nutrient storage (t/ha), and F_d is the nutrient return (t/ha a) [27].

4. Discussion

Litterfall volume is a component of the forest ecosystem biomass, which reflects the primary productivity level as well as the functions of the forest ecosystem [25,28]. The research shows that average annual litterfall of evergreen broad-leaved forest is 6.96 t/ha [29], the average annual litterfall of Yuanjiang savanna ecosystem is 2.5–3 t/ha [30], and the average litterfall of Chinese grassland is 0.59 t/ha [31]. It can be seen that fallen forest materials play a very important role in the global ecosystem. Our results showed that the average annual litterfall from 2005 to 2015 was 8.11 t/ha. This is similar to that observed in evergreen broadleaved forests in other subtropical regions (e.g., Dinghushan South Asian tropical evergreen broadleaved forest (7–11 t/ha) [32], and Xiaokeng subtropical evergreen broadleaf forest (7.99–8.45 t/ha) [33]. Furthermore, Guan Xin summarized the research results of central, subtropical, evergreen broadleaved forests and found that the annual litterfall recovery ranged from 3.90 to 7.72 t/ha [34]. Thus, the amount of litterfall observed at our site was intermediate between the South Asian tropical monsoon, evergreen, broad-leaved forest and the central, subtropical, evergreen, broad-leaved forest; its litterfall production is much higher than other ecosystems.

When compared to evergreen, broadleaf forests in other subtropical regions, we found that the ratio of leaf litter to total litter was less than that of Wuyishan rice oak forest (77.03 $\pm$ 1.93%) [35], Guangxi Longgang National Nature Reserve (85%) [36], Baishan Zu evergreen broadleaf forest (51.34%) [37] and Zhejiang Ningbo Tiantong Mountain evergreen broadleaved forest (50.7%) [38]. The high variability we observed within and among years for total litter and its components was consistent with the results of Zou Bingzhang [39]. Likewise, we found that most litterfall came from leaves with less from other sources, which was consistent with the results of Wan Chunhong [40]. The size and dynamic changes of forest litterfall output are influenced by many factors and are the result of a combination of factors [41,42]. We found that the overall amount of litterfall on the ground is high, which represents positive feedback for the forest ecosystem. It provides abundant food sources for forest organisms, especially fruits and flowers, which play an important role in the survival and reproduction of rodents. This also indirectly ensures the survival and reproduction of birds that feed on rodents. On the other hand, the litterfall decomposes and releases nutrients, keeping the fertility of the research site at a high level, as well as changing the physical properties of the forest soil. Litterfall also has a strong water retention capacity, which can reduce water evaporation and maintain sufficient water storage on the forest surface, which is important for water conservation and maintaining soil environment stability [43,44]. The presence of litterfall plays a foundational role in the entire forest ecosystem. We also found that litterfall dynamics correlated with a number of features of the environment, including wind speed, precipitation, and temperature. In the one exceptionally unusual month year in our survey (January 2015), branch content was high as a result of an unusual amount of snow in this period (1129.2 mm), which created considerable tree and branch fall in this period. This indicates that, after the interference of extreme ice and snow weather, the leaves were violently shaken by strong external forces, resulting in non physiological shedding.

The nutrient concentration of litterfall is related to the characteristics of the plants and the soil nutrient content [45]. We found that the nutrient concentration of litterfall was roughly the same in different years and different litterfall components. Specifically, we found that C concentration was the highest, Ca and N concentration were second, and K, Mg, S, and P concentrations were relatively low, consistent with the results of Chen Jinlei and Xue Fei [46,47]. Our finding that different nutrients varied through time is similar to the research results of Xue Fei and Zhao Chang [45,47]. Our results show that many elements are correlated with temperature and precipitation, which may be influenced by the relationship between these elements and plant growth [48]. For example, K is highly mobile, and P is easily affected by multiple factors, such as vegetative growth rhythms, rainfall leaching, microbial degradation, etc.

The return of litterfall nutrients to the soil is influenced by a combination of factors. We found that the return of nutrient elements in litterfall showed a decreasing trend through the year. Our annual nutrient return size of litterfall is roughly C > N > Ca > K > Mg > S > P, which differs from the results of Liu Yi, but is similar to those of Liu Lei and Gao Shilei [49–51]. Furthermore, the annual return amount of litterfall C in our study area (6.21 t/ha) was within the range of litterfall C return observed (0.05~7.50 t/ha) across the world's forested ecosystems [52]. Likewise, our results regarding the return of C, N, and P were higher than those of Mijiao natural forest in Sanming City, Fujian Province and Wuyi Mountain evergreen, broadleaved forest [53,54]. This suggests that evergreen, broadleaf forest literfall in at our study site in the Ailao Mountain plays a very important role in the carbon cycle of the soil. Additionally, the litterfall results in modifications in forest ecosystems, particularly in subtropical, evergreen, broad-leaved forests because the amount of litterfall can regulate the micro-climate in the soil, affecting the decomposition rate with the changes in microbial community, and soil microorganisms in the soil affect soil respiration [55].

The utilization coefficient, circulation coefficient, and turnover time are all important parameters in the nutrient cycling process, and the nutrient cycling parameters also vary due to the difference between nutrient uptake and return by different forest types [56]. The utilization coefficient is the ratio of absorption to storage, reflecting the storage rate of ecosystem elements; the larger the coefficient, the greater the storage capacity of plants and the lower the utilization efficiency. The nutrient utilization coefficient in our study area was 0.25, which was lower than that found in the karst peak cluster depression in Huanjiang, Guangxi (0.35) and the four-year-old Mazhan Acacia plantation (0.51) in the state-owned peak forest farm in Nanning City, Guangxi [20,57]. This indicates that the forest in our study system had high nutrient utilization efficiency and low storage capacity. The circulation coefficient is the ratio of plant return to absorption, and reflects the size of the residual amount of the element during the cycle; the larger the coefficient, the faster the rate of element circulation and the greater the fluidity. The cycling coefficient in our study (0.53) was higher than that of Gongga Mountain Natural Forest (0.474) [58], but lower than that of Dinghu Mountain Horsetail Pine Forest (0.68) and *Pinus tabulosis* forest (0.71–0.85) in the loess hilly area [19,59]. This indicates that the forest in the study area had high nutrient cycling capacity. Turnaround time is the ratio of a plant's total nutrient storage to return, indicating the time it takes for a nutrient element to go through the cycle. The longer the turnaround time, the longer the nutrients stay in plants. In this study, we found that the turnover time ranged from 8.40 a to 14.14 a (average value of 10.50 a), which was manifested as Ca > Mg > S > C > N > K > P, suggesting that Ca and Mg were inactive elements and P was the active element. Meantime, this also indicates that the plants in the research area absorb nutrients quickly, grow quickly, and have high yield and large biomass.

5. Conclusions

In this paper, we present the results of an 11-year study investigating the dynamic changes of litterfall output, nutrient concentration, and return in a wet, evergreen, broadleaf forest in the Ailao Mountains of China to draw the following conclusions:

From 2005 to 2015, the total litter of evergreen, broadleaved forests in the Ailao Mountains was 7.70–9.46 t/ha a, and the interannual fluctuation of litterfall was large, with an average of 8.11 $\pm$ 0.73 t/ha, which was higher than that of Central Asian, thermal, evergreen, broadleaved forests. The presence of a large amount of litterfall provides nutrients to the study area, promoting the development and stability of the study area's ecosystem and ensuring the fertility of the soil and biodiversity. The output of litterfall was significantly different between different years. There are significant interannual and seasonal variations in the amount of litterfall, mostly bimodal (peak from March to May, October to November), with higher levels of leaf litter than other components in each year, some of which was correlated with meteorological factors ($p < 0.05$). The article only discussed the meteorological factor as the cause of the litterfall, indicating that the production of litterfall is the result of multiple factors. We generally found the nutrient concentration was sorted as C > Ca > N > K > Mg > S > P, except for the slight difference in Ca and N, this also conforms to the general pattern of changes in forest nutrient concentrations. The nutrient concentration and returned amount of litterfall are related to the growth of trees, as well as variation in meteorological factors, such as temperature, precipitation, and wind speed. Our results showed that although there was nutrient loss in the evergreen, broad-leaved forest area of the Ailao Mountains, forest litterfall could still maintain soil fertility in the area, maintaining the normal operation of the entire forest ecosystem.

Author Contributions: S.D. conceived and designed the study, oversaw data collection, Z.X. provide experimental materials and assist in the inspection and analysis of related chemical elements, and conducted all litterfall analyses; S.D., T.W., J.T. and H.G. analyzed the data and co-wrote the paper. All authors have read and agreed to the published version of the manuscript.

Funding: This research received no external funding.

Data Availability Statement: The data are not publicly available due to the dataset is proprietary and the author currently does not have the rights to make the data public.

Acknowledgments: The data of this paper are collected and obtained by the staff of the Ailao Mountain Ecological Station of Xishuangbanna Botanical Garden of the Chinese Academy of Sciences for many years, especially the determination of nutrients and the acquisition of field materials. Here, I would like to especially thank Li Dawen for the sampling work.

Conflicts of Interest: The authors declare no conflict of interest.

References

1. Nakagawa, M.; Ushio, M.; Kume, T.; Nakashizuka, T. Seasonal and long-term patterns in litterfall in a Bornean tropical rainforest. *Ecol. Res.* **2019**, *34*, 31–39. [CrossRef]
2. Wang, Z.B.; Ji, M.; Li, Y.X.; Zhang, H.D.; Li, Y.L.; Gong, S.F.; Liu, J.; Bai, G.W. Effects of nitrogen addition and litter management on soil chemistry of larch plantations in North China. *West. China For. Sci.* **2021**, *50*, 26–32+40.
3. Spohn, M.; Berg, B. Import and release of nutrients during the first five years of plant litter decomposition. *Soil Biol. Biochem.* **2023**, *176*, 108878. [CrossRef]
4. Neumann, M.L.; Ukonmaanaho, J.; Johnson, S.; Benham, L.; Vesterdal, R.; Novotný, A.; Verstraeten, L.; Lundin, A.; Thimonier, P.; Michopoulos, H. Hasenauer. Quantifying Carbon and Nutrient Input From Litterfall in European Forests Using Field Observations and Modeling. *Glob. Biogeochem. Cycles* **2018**, *32*, 784–798. [CrossRef]
5. Kitayama, K.; Ushio, M.; Aiba, S.I. Temperature is a dominant driver of distinct annual seasonality of leaf litter production of equatorial tropical rain forests. *J. Ecol.* **2020**, *109*, 727–736. [CrossRef]
6. Jia, B.R.; Sun, H.R.; Yu, W.Y.; Zhou, G.S. Quantifying the interannual litterfall variations in China's forest ecosystems. *J. Plant Ecol.* **2020**, *13*, 266–272. [CrossRef]
7. Zhu, X.A.; Zou, X.; Lu, E.F.; Deng, Y.; Luo, Y.; Chen, H.; Liu, W.J. Litterfall biomass and nutrient cycling in karst and nearby non-karst forests in tropical China: A 10-year comparison. *Sci. Total Environ.* **2020**, *758*, 143619. [CrossRef]
8. Sayer, E.J.; Tanner, E.V.J. Experimental investigation of the importance of litterfall in lowland semi-evergreen tropical forest nutrient cycling. *J. Ecol.* **2010**, *98*, 1052–1062. [CrossRef]
9. Xing, J.M.; Wang, K.Q.; Song, Y.L.; Zhang, Y.J.; Zhang, Z.M.; Pan, T.S. Characteristics of litter return and nutrient dynamic change in four typical forests in the subalpine of central Yunnan province. *J. Cent. South Univ. For. Technol.* **2021**, *41*, 134–144.
10. Torres, J.R.; Sánchez-Mejía, Z.M.; Arreola-Lizárraga, J.A.; Galindo-Félix, J.I.; Mascareño-Grijalva, J.J.; Rodríguez-Pérez, G. Environmental factors controlling structure, litter productivity, and phenology of mangroves in arid region of the Gulf of California. *Acta Oecologica* **2022**, *117*, 103861. [CrossRef]
11. Singh, P.; Ghosh, A.K.; Kumar, S.; Kumar, M.; Sinha, P.K. Influence of input litter quality and quantity on carbon storage in post-mining forest soil after 14 years of reclamation. *Ecol. Eng.* **2022**, *178*, 106575. [CrossRef]
12. Zhang, Y.F.; Xu, D.M.; Jiang, K.L.; Luo, Z.R. Seasonal variation and driving factors of evergreen broadleaf forest in Baishan ancestral forest. *Zhejiang For. Sci. Technol.* **2021**, *41*, 42–49.
13. Ge, X.G.; Wang, C.G.; Wang, L.L.; Zhou, B.Z.; Cao, Y.H.; Xiao, W.F.; Li, M.H. Drought changes litter quantity and quality, and soil microbial activities to affect soil nutrients in moso bamboo forest. *Sci. Total Environ.* **2022**, *838*, 156351. [CrossRef]

14. Fan, C.N.; Guo, Z.L.; Zhen, J.P.; Li, B.; Yang, B.G.; Yue, L.; Yu, H.B. Quantity and dynamics of litter in natural secondary forest in Mopanshan. *Acta Ecol. Sin.* **2014**, *34*, 633–641.
15. de Morais, T.M.O.; Berenguer, E.; Barlow, J.; França, F.; Lennox, G.D.; Malhi, Y.; Chesini, R.L.; Maria, M.d.S.M.; Ferreira, J. Leaf-litter production in human-modified Amazonian forests following the El Niño-mediated drought and fires of 2015–2016. *For. Ecol. Manag.* **2021**, *496*, 119441. [CrossRef]
16. Railoun, M.; Simaika, J.; Jacobs, S. Leaf litter production and litter nutrient dynamics of invasive Acacia mearnsii and native tree species in riparian forests of the Fynbos biome, South Africa. *For. Ecol. Manag.* **2021**, *498*, 119515. [CrossRef]
17. Li, J.B. Dynamics of Litter and Nutrient in 4 Types of Subtropical. Ph.D. Thesis, Central South University of Forestry and Technology, Changsha, China, December 2011.
18. Qin, W.M.; He, B.; Qin, S.Y.; Huang, S.F.; Mo, Y.F.; Qin, Y.H. Study on nutrient element biological cycling in Acacia pachypod plantation. *J. Soil Water Conserv.* **2007**, *4*, 103–107.
19. Zhang, X.; Zheng, S.; Shangguan, Z. Comparison of nutrient distribution and biological cycle between Pinus tabulosis plantation and natural forest in loess hilly area. *Acta Ecol. Sin.* **2006**, *26*, 373–382. [CrossRef]
20. Yu, Y.F.; He, T.G.; Peng, W.X.; Song, T.Q.; Zeng, F.P.; Du, H.; Han, C.; Li, S.S. Nutrient cycling characteristics of different types of forests in karst peak depression. *Acta Ecol. Sin.* **2015**, *35*, 7531–7542.
21. Su, Y.; Wang, P. Vertical variation characteristics of temperature and precipitation in Taijiding Mountain at the northern end of Ailao Mountain. *J. Baoshan Univ.* **2021**, *40*, 59–65.
22. Gong, H.D.; Yang, G.P.; Lu, Z.Y.; Liu, Y.H. Tree species diversity and spatial distribution pattern of evergreen broad-leaved forest in Ailao Mountain. *Biodivers. Sci.* **2011**, *19*, 143–150.
23. Laskowski, R.; Niklinska, M.; Maryanski, M. The dynamics of chemical elements in forest litter. *Ecology* **1995**, *76*, 1393–1406. [CrossRef]
24. Kappelle, M.; Leal, M.E. Changes in leaf Morphology and foliar nutrient status along a successional gradient in Costa Rican upper montane Quercus forest. *Biotropica* **1996**, *28*, 331–344. [CrossRef]
25. Liski, J.; Nissinen, A.; Erhard, M.; Taskinen, O. Climatic effects on litter decomposition from arctic tundra to tropical rainforest. *Glob. Chang. Biol.* **2003**, *9*, 575–584. [CrossRef]
26. Zhang, Y.J.; Song, Y.L.; Wang, K.Q. Ecostoichiometry characteristics of each organ in the arbor layer of alpine forest in Central Asia. *Chin. J. Ecol.* **2019**, *38*, 1669–1678.
27. Liu, Z.W.; Zhao, X.G. Research on Nutrient Cycle Characteristic Parameters of Forest Ecosystem. *J. Northwest For. Univ.* **2001**, *4*, 21–24.
28. Sundarapandian, S.M.; Swamy, P.S. Litter production and leaf-litter decomposition of selected tree species in tropical forests at Kodayar in the western ghats, India. *For. Ecol. Manag.* **1999**, *123*, 231–244. [CrossRef]
29. Liao, J.; Wang, X.G. An overview of the research on forest litter. *South China For. Sci.* **2000**, *1*, 31–34.
30. Jin, Y.Q.; Li, J.; Liu, C.G.; Liu, Y.T.; Zhang, Y.P.; Song, Q.H.; Sha, L.Q.; Chen, A.G.; Yang, D.X.; Li, P.G. Response of net primary productivity to precipitation exclusion in a savanna ecosystem. *For. Ecol. Manag.* **2018**, *429*, 69–76. [CrossRef]
31. Wen, D.; He, N.P. Spatial Distribution Pattern and Its Controlling Factors of Existing Stocks of Forest and Grassland Litter in China. *Acta Ecol. Sin.* **2016**, *36*, 2876–2884.
32. Tu, M.Z.; Yao, W.H.; Weng, H.; Li, Z.A. Characteristics of Litterfall in a Subtropical Evergreen Broad-Leaved Forest in Dinghushan Mountain. *Acta Pedol. Sin.* **1993**, *1*, 34–42.
33. Lan, C.C. Litter Amount and Nutrient Dynamics of Xiaokeng Subtropical Evergreen Broad-Leaved Forest. Master's Thesis, Anhui Agricultural University, Hefei, China, 2008.
34. Guan, X.; Huang, K.; Yan, S.K.; Wang, S.L. Monthly dynamic dataset of litter recovery and existing stock in central subtropical evergreen broadleaf forests from 2005–2015. *China Sci. Data* **2021**, *6*, 205–212.
35. Huang, S.D. Degradation dynamics and its relationship with meteorological factors in Wuyi Mountain Mique forest community. *J. Southwest For. Univ. (Nat. Sci.)* **2022**, *42*, 76–83.
36. Guo, Y.L.; Li, D.X.; Wang, B.; He, Y.L.; Xiang, W.S.; Jiang, Y.L.; Li, X.K. Composition and temporal and spatial dynamics of litter components in northern tropical karst seasonal rainforest. *Biodivers. Sci.* **2017**, *25*, 265–274. [CrossRef]
37. Hu, L.Z.; Chen, D.L.; Zhu, H.L.; Zhang, Y.H.; Ding, B.Y. Dying rhythm and composition of litter in evergreen broadleaf forest of Baishan ancestral ancestor. *J. Zhejiang Univ. (Agric. Life Sci.)* **2011**, *37*, 533–539.
38. Zhao, Y.Z.; Wang, X.P.; Zhang, Y.G. Relationship between litter quantity and meteorological factors in Tiantong evergreen broadleaf forest. *J. Beijing Agric. Univ.* **2017**, *32*, 73–77.
39. Zou, B.Z. Relationship between litter amount and soil nutrient in major subtropical forest types. *Fujian For. Sci. Technol.* **2019**, *46*, 8–12, 45.
40. Wan, C.H.; Tao, C.; Yang, X.B.; Huang, J.; Feng, D.D.; Yang, Q.; Zhou, W.S. Litter yield and its influencing factors in different forest types in Hainan Island. *Chin. J. Trop. Biol.* **2014**, *5*, 153–161.
41. Fang, J.Y.; Li, Y.D.; Zhu, B.; Liu, G.H.; Zhou, G.Y. Community structure, species diversity and status of the Jianfengling mountain rainforest in Hainan Island. *Biodivers. Sci.* **2004**, *1*, 29–43.
42. Shi, J.Z.; Xu, H.; Lin, M.X.; Li, Y.D. Yield and dynamics of litter in tropical montane rainforest in Jianfengling, Hainan. *Plant Sci. J.* **2019**, *37*, 593–601.

43. Ilek, A.; Kucza, J.; Szostek, M. The effect of stand species composition on water storage capacity of the organic layers of forest soils. *Eur. J. For. Res.* **2015**, *134*, 187–197. [CrossRef]
44. Borchard, N.; Adolphs, T.; Beulshausen, F.; Ladd, B.; Gießelmann, U.C.; Hegenberg, D.; Möseler, B.M.; Amelung, W. Carbon accrual rates, vegetation and nutrient dynamics in a regularly burned coppice woodland in Germany. *GCB Bioenergy* **2016**, *9*, 1140–1150. [CrossRef]
45. Zhao, C.; Long, J.; Li, J.; Liao, H.K.; Liu, L.F.; Zhang, M.J.; Hua, J. Existential and nutrient characteristics of litter in different slopes and decomposition layers of the Maolan Karst primary forest. *Chin. J. Ecol.* **2018**, *37*, 295–303.
46. Chen, J.L.; Zhang, S.J.; Li, L. Current stock and nutrient characteristics of forest litter layer in different subtropical vegetation restoration stages. *Acta Ecol. Sin.* **2020**, *40*, 4073–4086.
47. Xue, F.; Long, C.L.; Liao, Q.L.; Xiong, L. Current inventory and nutrient characteristics of litter in different terrains of karst forest. *J. Northwest For. Univ.* **2021**, *36*, 28–35.
48. Li, A.; Zhang, Y.J.; Sun, J.X.; Zhang, J.W.; Gong, Z.H.; Pan, X.J.; Guo, W.J.; Liu, X.Q. Effects of plant growth regulators and temperature on Phalaenopsis double stem rate, flowering period and flower traits. *Chin. J. Trop. Crops* **2021**, *42*, 732–738.
49. Liu, Y.; Cao, Y.H.; Zhang, Y.F.; Xie, W.D. Nutrient release dynamics of litter from three stand types in tropical coastal sandy areas of South Asia. *J. Northwest AF Univ. (Nat. Sci. Ed.)* **2022**, *50*, 55–68.
50. Liu, L.; Shen, G.Z.; Chen, F.Q.; Luo, L.; Xie, Z.q.; Yu, J. Current stocks and nutrient cycling dynamics of four typical forest litter on the elevation gradient of Shennongjia. *Acta Ecol. Sin.* **2012**, *32*, 2142–2149. [CrossRef]
51. Gao, S.L.; He, Z.M.; Ding, G.C.; Lin, Y.; Lin, S.Z.; Liu, Z.M. Dynamic study on litter and nutrient return in Acacia plantation forest in coastal sandy sand area. *J. Southwest For. Univ.* **2015**, *35*, 17–23.
52. Li, H.T.; Yu, G.R.; Li, J.Y.; Liang, T.; Chen, Y.R. Dynamic decomposition of litter and phosphorus and potassium release rates in Jinggangshan forest. *Chin. J. Appl. Ecol.* **2007**, *2*, 233–240.
53. Zhang, N.; Yang, Z.J.; Xu, C.; Liu, X.F.; Xiong, D.C.; Lin, C.F. Effects of Central Subtropical Forest Conversion on Nutrient Return and Nutrient Use Efficiency of Litter. *Chin. J. Appl. Ecol.* **2022**, *33*, 321–328.
54. Huang, Y.R.; Gao, W.; Huang, S.D.; Lin, J.; Tan, F.L.; You, H.M.; Yang, L. Ecostoichiometric characteristics of carbon, nitrogen and phosphorus in three evergreen broad-leaved forests in Fujian. *Acta Ecol. Sin.* **2021**, *41*, 1991–2000.
55. Yan, W.; Peng, Y.; Zhang, C.; Chen, X. The manipulation of aboveground litter input affects soil CO_2 efflux in a subtropical liquidambar forest in China. *iForest Biogeosci. For.* **2019**, *12*, 181–186. [CrossRef]
56. Nie, D.P. Biological cycle of nutrients in forest ecosystems. *For. Res.* **1991**, *4*, 435–440.
57. He, B.; Qin, W.M.; Yu, H.G.; Liu, Y.H.; Qin, L.; Qin, Y.H. Biological cycling of nutrients in Acacia plantation at different ages. *Acta Ecol. Sin.* **2007**, *12*, 5158–5167.
58. Luo, J.; Chen, G.W.; Li, W.; He, Z.W. Biological cycle characteristics of nutrient elements in natural forest in Gongga Mountain. *J. Beijing For. Univ.* **2005**, *2*, 13–17.
59. Mo, J.M.; Brown, S.D.; Kong, G.H.; Zhang, Y.C.; Lenart, M. Distribution and biological cycle characteristics of nutrients in Pinus massoniana forest in Dinghu Mountain. *Acta Ecol. Sin.* **1999**, *5*, 47–52.

plants

MDPI

Article

Linking Leaf Functional Traits with Soil and Climate Factors in Forest Ecosystems in China

Xingyu Zhou [1,†], Jiaxun Xin [2,†], Xiaofei Huang [1,3], Haowen Li [2], Fei Li [1,3,*] and Wenchen Song [2,*]

1 College of Nuclear Technology and Automation Engineering, Chengdu University of Technology, Chengdu 610059, China
2 College of Life and Environmental Sciences, Minzu University of China, No. 27 Zhongguancun South Street, Haidian, Beijing 100081, China
3 Applied Nuclear Techniques in Geosciences Key Laboratory of Sichuan, Chengdu University of Technology, Chengdu 610059, China
* Correspondence: lifeimvp@sina.com (F.L.); songw@muc.edu.cn (W.S.)
† These authors contributed equally to this work.

Abstract: Plant leaf functional traits can reflect the adaptive strategies of plants to environmental changes. Exploring the patterns and causes of geographic variation in leaf functional traits is pivotal for improving ecological theory at the macroscopic scale. In order to explore the geographical variation and the dominant factors of leaf functional traits in the forest ecosystems of China, we measured 15 environmental factors on 16 leaf functional traits in 33 forest reserves in China. The results showed leaf area (LA), carbon-to-nitrogen ratio (C/N), carbon-to-phosphorus ratio (C/P), nitrogen-to-phosphorus ratio (N/P), phosphorus mass per area (Pa) and nitrogen isotope abundance (δ^{15}N)) were correlated with latitude significantly. LA, Pa and δ^{15}N were also correlated with longitude significantly. The leaf functional traits in southern China were predominantly affected by climatic factors, whereas those in northern China were mainly influenced by soil factors. Mean annual temperature (MAT), mean annual precipitation (MAP) and mean annual humidity (MAH) were shown to be the important climate factors, whereas available calcium (ACa), available potassium (AK), and available magnesium (AMg) were shown to be the important climate factors that affect the leaf functional traits of the forests in China. Our study fills the gap in the study of drivers and large-scale geographical variability of leaf functional traits, and our results elucidate the operational mechanisms of forest–soil–climate systems. We provide reliable support for modeling global forest dynamics.

Keywords: leaf functional traits; climate; soil; geographical variation; δ^{13}C; δ^{15}N; forest ecosystem

Citation: Zhou, X.; Xin, J.; Huang, X.; Li, H.; Li, F.; Song, W. Linking Leaf Functional Traits with Soil and Climate Factors in Forest Ecosystems in China. *Plants* **2022**, *11*, 3545. https://doi.org/10.3390/plants11243545

Academic Editors: Jie Gao, Weiwei Huang, Johan Gielis and Peijian Shi

Received: 22 November 2022
Accepted: 12 December 2022
Published: 15 December 2022

1. Introduction

Leaf functional traits can not only reflect plant growth, metabolism and reproduction [1] but also represent plant adaptation strategies to different ecological environments [2,3]. In recent decades, patterns of geographic variation in leaf functional traits at large spatial scales have been paid much attention [4–6]. Understanding the geographical variation in leaf functional traits and its relationship with environmental factors can improve the predictions of vegetation changes [7–10], the large-scale mapping of plant function types [11], analysis of the community structure [12] and dynamic modeling of global vegetation [13,14]. At present, the mainstream view is that the variation in leaf functional traits is mainly affected by climate factors at large scale [15–21]. Temperature and precipitation were shown to be the most important climate factors, and they have been well known as the dominant factors that affect leaf functional traits, such as leaf area (LA), special leaf area (SLA), nitrogen mass (Nm), leaf phosphorus mass (Pm) and nitrogen-to-phosphorus ratio (N/P) [16,22–24]. Humidity is also one of the climate factors affecting leaf functional traits (e.g., Nm, N/P) [25,26]. Furthermore, evapotranspiration has been found to explain 30% of the variation in leaf area [27], and it is one of the limiting factors for some leaf functional traits (e.g., leaf dry matter content (LDMC), Pm, SLA, Nm) [28].

In recent years, soil factors have been shown to be one of the main factors affecting the functional traits at large scale [4,17–19,29,30], because the plant–soil interactions are typically the major determinants of changes in processes and functions in forest ecosystems [31]. Leaf functional traits, such as leaf mass per area (LMA), Nm and Pm, are strongly correlated with soil factors, such as available phosphorus (AP), available potassium (AK) available nitrogen (AN) and pH [16,32–34]. The carbon-to-nitrogen ratio of leaves (C/N) is associated with the soil salinity [35]. Besides, total nitrogen content (TN) is the main factor affecting LA and SLA, while the total organic carbon (TOC) in soil mainly affects the Pm and Nm of leaves [36]. Soil microorganisms play an important ecological role in soil formation and the protection and regulation of nutrient cycling [37], further changing the physiological adaptation of plants and making soil factors affect leaf functional traits in forest ecosystems [38].

The large latitudinal span of forest ecosystems in China covers all four major forest ecosystem types from north to south, namely, boreal coniferous forests, temperate deciduous broadleaf forests, subtropical evergreen broadleaf forests and tropical forests, making it a valuable sample place for studying geographic variation in leaf functional traits at large scale. Therefore, the variation patterns of leaf functional traits along the gradients of environmental factors (e.g., temperature, precipitation, soil elements and topography) have been widely studied in several climatic regions of China, such as the Tibetan Plateau region [39], the Sanjiangyuan region of northeast China [40], the Karst region of southwest China [35,41], the arid and semi-arid region of northwest China [15,23,42], the deciduous broadleaf forest region of southeastern China [15,43] and the Loess Plateau region of central China [36]. However, at the national scale, only a few reports have explored geographical variation, and most of these studies focused on several species [16,36,44]. Herein, an important question was asked: What are the main factors affecting leaf functional traits in China? The answer to this question will contribute to the global understanding of leaf functional trait variation.

To address this question, we explored the following two hypotheses:

(a) Forest leaf functional traits are mainly correlated with geographic change in climate factors [45];

(b) The geographic change of forest leaf functional traits is mainly influenced by shifts in soil factors [46];

In this study, we sampled soils and leaves from 33 forest reserves in China to test which of the two hypotheses are more realistic. The aim of this study is to link leaf functional traits with soil and climate factors in forest ecosystems in China.

2. Results

2.1. Geographical Variation

The leaf area (LA) and phosphorus mass per area (Pa) decreased with increasing latitude (Figure 1a,b); however, the respective curves did not show a good fit ($R^2 = 0.37$, $p < 0.01$; $R^2 = 0.37$, $p < 0.01$, respectively). The carbon-to-phosphorus ratio (C/P), nitrogen-to-phosphorus ratio (N/P) and nitrogen isotope abundance ($\delta^{15}N$) increased with increasing latitude (Figure 1d–f). $\delta^{15}N$ showed the strongest correlation with latitude ($R^2 = 0.53$, $p < 0.01$), whereas those of C/P and N/P were weak, albeit significant ($R^2 = 0.24$, $p < 0.01$; $R^2 = 0.36$, $p < 0.01$, respectively). In the range of 20–40° N, carbon-to-nitrogen ratio (C/N) decreased with increasing latitude, and in the range of 40–55° N, C/N increased slowly with increasing latitude ($R^2 = 0.24$, $p < 0.05$) (Figure 1c).

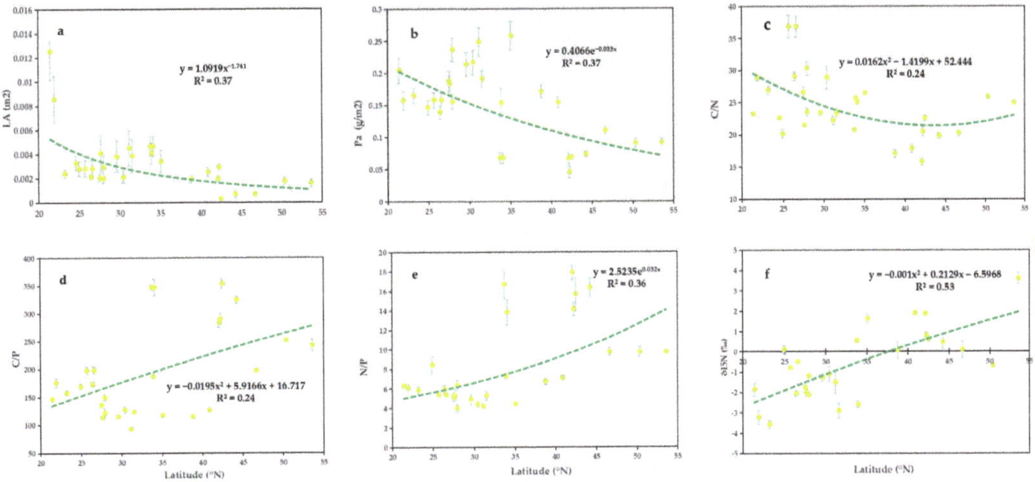

Figure 1. Relationships between latitude and **(a)** LA, **(b)** Pa, **(c)** C/N, **(d)** C/P, **(e)** N/P and **(f)** δ^{15}N, respectively.

The LA of leaves from the south was, on average, higher than that of leaves in the north (Figure 2a), which was consistent with the conclusions drawn in Figure 2a. The LA of leaves from the south had a significant but weak negative correlation with longitude (R^2 = 0.33, p < 0.01). Leaves from the north showed a significant and stronger correlation with longitude (R^2 = 0.62, p < 0.01). Between 105° E and 120° E, LA was negatively correlated with longitude, and between 120° E and 130° E, LA was positively correlated with longitude. The phosphorus mass per area (Pa) of southern leaves was generally higher than that of leaves in the north, which is consistent with the conclusion obtained from Figure 1b (Figure 2b). With increasing longitude, the Pa of the southern leaves first decreased and then increased, with the lowest value at 110° E (R^2 = 0.46, p < 0.05). The Pa of the northern leaves increased with longitude and then decreased, with the highest value at 112 °E (R^2 = 0.30, p < 0.05). The degree of variation of δ^{15}N with longitude was not significant in the south and the north, but as a whole, δ^{15}N showed a significant positive correlation with longitude (R^2 = 0.32, p < 0.01) (Figure 2c).

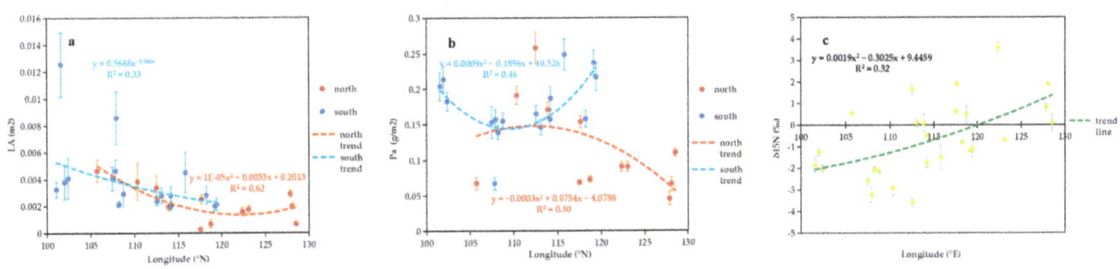

Figure 2. Relationships between longitude and **(a)** LA, **(b)** Pa and **(c)** δ^{15}N, respectively.

2.2. Environmental Factors Affecting Functional Traits of Leaves

The RDA showed that the first axis explained 41.74% of the relationship of leaf functional traits and environmental factors, and the second axis explained 20.84%. The mean annual precipitation (MAP), mean annual temperature (MAT), mean annual humidity (MAH), available magnesium (AMg), available potassium (AK) and pH were more strongly correlated with leaf functional traits, followed by correlations with available aluminum (AAl), available phosphorus (AP) and electrical conductivity (EC). AMg, AK, pH, EC and

AP, all of which were positively correlated with the first axis, and AMg, AK and pH which were closely associated with the first axis. MAP, MAH, MAT and AAl were negatively correlated with the first axis, with MAT being the most closely related to the first axis. Total potassium (TK) was closely and positively correlated with the second axis, and total organic carbon (TOC) and AP were closely and negatively correlated with the second axis (Figure 3).

We found that the factors influencing the functional traits of southern leaves differed significantly from those of northern leaves (Figure 3). Southern leaf functional traits were mainly influenced by climate factors (MAP, MAH, MAT) and to a lesser extent by soil factors (TK, AAl, TOC). Northern leaf functional traits were mainly influenced by soil factors (AP, TP [total phosphorus], TN [total nitrogen], EC, pH, AK, ACa, AMg) and to a lesser extent by climate factors (MAE). The area of the ellipse where the functional traits of southern leaves were located was significantly smaller than that of the ellipse where the functional traits of the northern leaves were located, thus it can be assumed that the intra-group variability among the functional traits of the southern leaves is smaller than that of the northern leaves.

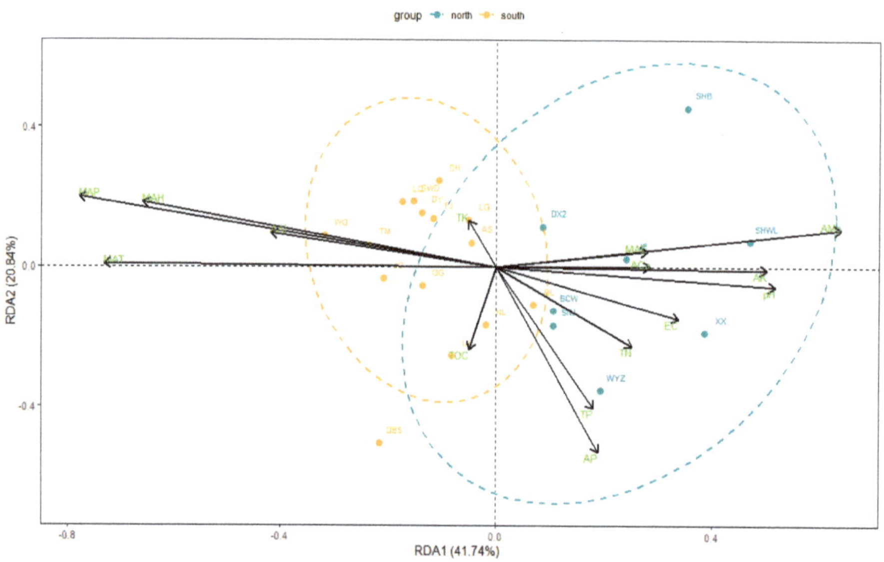

Figure 3. RDA between leaf functional traits and environmental factors.

We performed pairwise Pearson's correlation analyses between 16 leaf functional trait variables and 14 environmental variables and produced heat maps (Figure 4). A total of 23 relationship pairs reached significance at the $p < 0.01$ level, and 25 relationship pairs reached significance at the $p < 0.05$ level. Horizontally, all 14 leaf functional traits, apart from potassium mass per area (Ka) and carbon mass (Cm), were significantly influenced by some of the many environmental variables, to varying degrees. Longitudinally, the remaining nine environmental variables, except for EC, TOC, TN, TP and AP, were significantly correlated with certain leaf trait indicators. Among all environmental variables, AMg, MAP, MAT and MAH had the most significant effects on leaf functional traits, and they were significantly correlated with 7, 9, 10 and 8 leaf functional trait variables, respectively. According to the division criteria of a previous study [47], among all significant correlations, only ACa showed strong positive correlations with leaf mass per area (LMA) and nitrogen mass per area (Na), and most of the remaining correlations were moderately correlated in degree, with R^2 values generally ranging from 0.4 to 0.7, and a small number of correlations were weak.

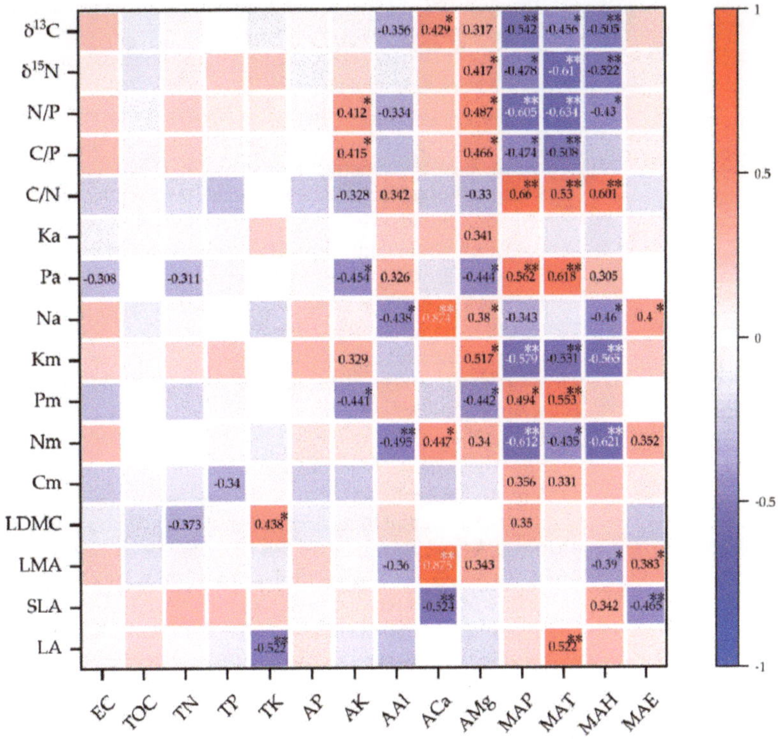

Figure 4. Heatmap of Pearson's correlations between environmental variables and leaf functional traits. Significant difference was indicated as follows: * < 0.05; ** < 0.01.

3. Discussion

3.1. Geographical Variation in Leaf Functional Traits

The relationship between leaf physiological traits and latitude was not linear in any of our analyses (Figure 1), and this finding was consistent with those of a previous study [20]. The possible reason is that the presence of local habitat heterogeneity reduces the climate variation along the latitudinal gradient [4]. Leaf area (LA) decreased with latitude (Figure 1a), which was in line with the findings of a previous study [48]. Latitude did not have a significant relationship with specific leaf area (SLA) or leaf dry matter content (LDMC), in contrast to previous observations [20,48]. A significant positive correlation between the nitrogen-to-phosphorus ratio (N/P) and latitude was observed (Figure 1e), which was consistent with the observations of one previous study [38] but not with global patterns [45]. The law of N/P change with latitude suggests that, in China, nitrogen is restricted by low latitudes and mainly by phosphorus in high latitudes [49]. The reason for this pattern may be that the dependence of ectomycorrhizal fungi (EMF) on trees is higher at high latitudes than at low latitudes [38]. EMF can inhibit the mycorrhizal root colonization of neighboring arbuscular mycorrhizal herbs by promoting litter accumulation and limiting nutrient access [50,51], and they help trees absorb more N while enhancing their competitiveness, thus leading to higher N/P with increasing latitude [52–55]. From a general perspective, $\delta^{13}C$ showed no significant correlation with latitude, although fitting curves differed between north and south [56]. Leaf $\delta^{15}N$ increased with increasing latitude (Figure 1f), probably because EMF colonization is also positively correlated with latitude [56]. EMF can supply relatively ^{15}N-enriched N to their hosts in the rhizosphere [57–59]; however, this was contrary to the global trend [46], and it also differed from the trend of increasing and then decreasing with latitude in the southern hemisphere [60].

Overall, most leaf functional traits showed no significant longitudinal trends and varied at random, as observed previously [61]. This may be due to the low number of our sampling sites and the lack of a significant trend in hydrothermal conditions between sample sites in the longitude direction [62]. The LA of leaves in southern China decreased significantly with increasing longitude (Figure 2a), which was also found in a study on southern *Taxus mairei* [61]. However, this trend cannot be explained exclusively by precipitation patterns [63]. In contrast, the LA of northern leaves first increased and then decreased with longitude ascending height (Figure 2a). A possible reason for this trend is that, over the longitude of northern sampling sites, the latitude first decreases and then increases, roughly corresponding to decreasing and then increasing warmth. From an overall perspective, the correlation between $\delta^{13}C$ and longitude is not significant, which is consistent with the conclusions of previous studies [56,64]. $\delta^{15}N$ showed a significant positive correlation with longitude (Figure 2c). A different study also found that $\delta^{15}N$ was positively associated with longitude, but the correlation was not significant [56].

3.2. *Factors Influencing Leaf Functional Traits*

3.2.1. Climate Factors

Mean annual temperature (MAT) showed a significant positive correlation with leaf area (LA) and phosphorus mass (Pm) (Figure 4), as observed previously [45,65]. Long-term monitoring results showed that leaf size increases with increasing temperature, which reflects plant adaptation [66]. The increase of MAT is conducive to the litter decomposition and nutrient circulation, increasing the mineralization rate of nitrogen in the soil, which favors the increase of available nitrogen (AN) in the soil and nitrogen mass (Nm) in the leaves [67]. However, a significant negative correlation between MAT and Nm was found in the present study (Figure 4), which probably occurred due to the presence of other soil elements limiting the growth of Nm [67]. Leaf carbon isotope abundance ($\delta^{13}C$) is markedly affected by MAT [38,64,68]. Temperature-related variables were found to exert stronger effects on $\delta^{13}C$ than precipitation-related factors [12,69]. However, the direction and degree of influence of MAT differed between studies [64,70–72]. In the present study, MAT had a significant negative correlation with $\delta^{13}C$. MAT was previously shown to have a significant positive correlation with carbon mass (Cm) [73]; however, no such correlation was observed in the current study (Figure 5). No correlation between SLA and MAT was found, by contrast to other studies [20,74].

Mean annual precipitation (MAP) had a significant negative correlation with $\delta^{13}C$ and LMA, as observed previously [64,75,76]. MAP is the strongest predictor of leaf $\delta^{13}C$ among global climate variables, and it explains approximately half of the global variation in leaf $\delta^{13}C$ [77]. When the soil water content and air humidity decrease due to insufficient precipitation, plants may reduce stomatal conductance or stomatal density, leading to improved water-use efficiency and positive leaf $\delta^{13}C$ in plants [24,78–80]. Nm also had a significant negative correlation with MAP (Figure 4). This may be because, under water shortage, plants increase the allocation of N to the leaves, increase osmotic pressure in the cells, reduce the consumption of water by operating at lower stomatal conductance and improve water retention [81,82]. Leaf $\delta^{15}N$ was negatively correlated with MAP (Figure 4), likely because high moisture levels reduce rhizomicrobial activity and the ability of mycorrhiza to obtain nutrition from decomposing soil organic matter [46,83,84]. However, no correlation of MAP and SLA (or LA) was observed, in contrast to previous studies [65,74,85,86]. This may be because MAP values at all sampling sites were high (>299 mm) and thus did not elicit plant stress.

Increasing water-table depth negatively and directly affects SLA [87], and mean annual evapotranspiration (MAE) is negatively correlated with water-table depth at a country-wide scale [88]. Thus, MAE should be positively correlated with SLA. However, in the current study, MAE and SLA were negatively correlated (Figure 4), which is contrary to

the prediction. This discrepancy occurs may be due to the different scales of regional studies [89]. MAH was significantly negatively correlated with $\delta^{13}C$ (Figure 4), and Liu et al. obtained similar results using *Quercus variabilis* [75].

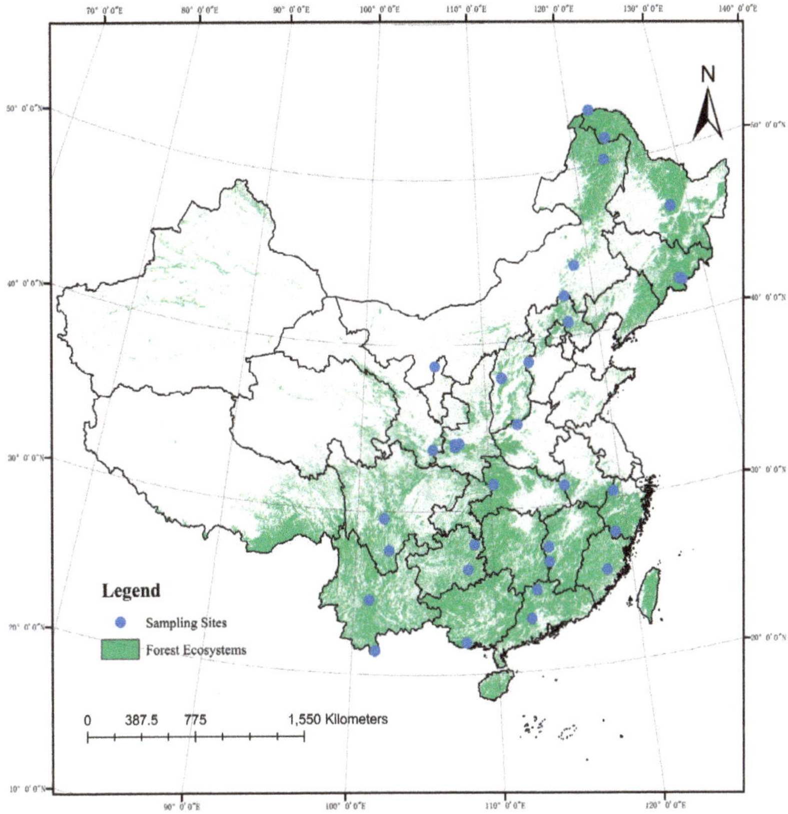

Figure 5. Geographic distribution of the sampling sites.

3.2.2. Soil Mineral Elements

AK had a significant positive effect on N/P (Figure 4), possibly because EMF colonization increases with increasing latitude [55,56], promoting tree roots to deposit their exudates into the soil to facilitate mineral transformation [90,91], thereby increasing AK content [92–94]. Higher AK concentrations are beneficial to soil EMF diversity [24,95,96], and EMF are considered to have a facilitating effect on N/P [55]. Nm was significantly positively influenced by ACa (Figure 4). Studies conducted in karst areas similarly found that alkaline soils with Ca^{2+} accumulation promote plant Nm [45]. AMg was positively correlated with $\delta^{15}N$ and ACa with $\delta^{13}C$ and LMA, which was observed previously.

4. Conclusions

This study found that both climate factors and soil factors significantly affected the leaf functional traits of forests in China; these influences have obvious geographic variability. The leaf functional traits in southern China were predominantly affected by climate factors, whereas those in northern China were mainly influenced by soil factors. Mean annual precipitation (MAP), mean annual temperature (MAT) and mean annual humidity (MAH) were the major climate factors affected leaf functional traits, available magnesium (AMg), available potassium (AK) and available calcium (ACa) were the major soil factors. In this study, it is believed that climatic and geological variation processes dominate the geographical pattern of forest functional traits in China, and their impacts must be comprehensively considered in future studies.

5. Materials and Methods

5.1. Soil and Leaf Sampling

Leaf and soil samples were collected from 33 mountain forest reserves in China (Figure 1). These forest reserves are located in a latitude range of 21.40°–53.56° N and a longitude range of 101.03°–128.52° E. These areas were characterized by rich vegetation communities (tropical forest, subtropical forest, temperate deciduous broadleaf forest, temperate mixed coniferous–broadleaf forest and boreal forest), mean annual precipitation of 299–2210, mean annual temperature of −5.5–22.7 °C, mean annual humidity of 46.4%–80% and mean annual evapotranspiration of 604–1276 mm. In each forest reserve, 9–15 sampling plots (10 m × 10 m) were randomly selected along the same aspect of the mountain, and 5 topsoil samples (5 cm depth) were randomly collected in each plot and immediately stored in precooled polyethylene bags, at each plot (see Wang et al. [97] for details). Within each plots, 5–10 major species were selected and 20–50 leaf samples on individuals from 2–5 adult healthy trees of each species were collected (see Song and Zhou [56] for details). All meteorological data in the present study were downloaded from the National Meteorological Science Data Center of China (http://data.cma.cn; accessed on 14 April 2019).

5.2. Soil and Leaf Analysis

Basic information on environmental variables and leaf functional traits is shown in Table 1. Soil pH and EC were measured using specific electrodes (Sartorius PB-10, Göttingen, Germany) and a soil suspension (soil and deionized water at 1:2.5). TOC was measured using a Shimadzu TOC Analyzer (TOC-Vcsh; Kyoto, Japan). The SmartChem Discrete Auto Analyzer (SmartChem 200; WESTCO Scientific Instruments Inc., Connecticut, USA) was used to measure TN and TP. Available Ca, Mg, Al, P and K were extracted using Mehlich-III solution and were measured through inductively coupled plasma optical emission spectrometry (ICP-OES, Optima 2100 DV; Perkin-Elmer, Waltham, MA, USA). TK was also measured by ICP-OES after sample digestion. The elemental content of C, N, P and K in the leaves was measured using TOC-Vcsh, SmartChem 200 and Optima 2100 DV instruments, respectively.

Table 1. Leaf functional traits and environmental variables.

Category	Abbreviation	Meaning	Unit
Leaf functional traits	LA	leaf area	m^2
	SLA	specific leaf area	$m^2 \cdot kg$
	LMA	leaf mass per area	$kg \cdot m^{-2}$
	LDMC	leaf dry matter content	$mg \cdot g^{-1}$
	Cm	carbon mass	$g \cdot kg^{-1}$
	Nm	nitrogen mass	$g \cdot kg^{-1}$
	Pm	phosphorus mass	$g \cdot kg^{-1}$
	Km	potassium mass	$g \cdot kg^{-1}$
	Na	nitrogen mass per area	$g \cdot m^{-2}$
	Pa	phosphorus mass per area	$g \cdot m^{-2}$
	Ka	potassium mass per area	$g \cdot m^{-2}$
	C/N	carbon-to-nitrogen ratio	-
	C/P	carbon-to-phosphorus ratio	-
	N/P	nitrogen-to-phosphorus ratio	-
	$\delta^{15}N$	nitrogen isotope abundance	‰
	$\delta^{13}C$	carbon isotope abundance	‰

Table 1. *Cont.*

Category	Abbreviation	Meaning	Unit
Soil characteristics	pH	potential of hydrogen	-
	EC	electrical conductivity	$\mu S \cdot cm^{-2}$
	TOC	total organic carbon	ppm
	TN	total nitrogen	ppm
	TP	total phosphorus	ppm
	TK	total potassium	ppm
	AP	available phosphorus	ppm
	AK	available potassium	ppm
	AAl	available aluminum	ppm
	ACa	available calcium	ppm
	AMg	available magnesium	ppm
Climatic variables	MAP	mean annual precipitation	mm
	MAT	mean annual temperature	°C
	MAH	mean annual humidity	%
	MAE	mean annual evapotranspiration	mm

After retrieving the leaf samples from the sample points, we performed the following experimental steps: (a) Samples were sorted and washed in distilled water. (b) Leaf thickness, root length and plant height were measured using a Vernier caliper. (c) The leaf area was calculated through Photoshop pixel analysis. (d) The samples were placed in a constant-temperature drying oven for 8 h at 75 °C. (f) An analytical balance was used to record plant mass.

To measure chemical properties, plants were ground with mortar and were sieved through 200 mesh. The values of C%, N%, $\delta^{13}C$ and $\delta^{15}N$ were measured using a Finnigan M.A.T 253 Isotope Ratio Mass Spectrometer and Flash 2000 EA-HT Elemental Analyzer (Thermo Fisher Scientific, Waltham, MA, USA). The measurement precisions for $\delta^{13}C$ and $\delta^{15}N$ were $< \pm 0.1‰$ and $< \pm 0.2‰$, respectively (see Song and Zhou [56] for further details).

The determination of $\delta^{13}C$ and $\delta^{15}N$ was based on the international standard Peedee Belemnite (PDB) formation and was calculated according to the following Equations (1) and (2):

$$\delta^{13}C = \left[\frac{(^{13}C/^{12}C)_{Sample}}{(^{13}C/^{12}C)_{PDB}} - 1 \right] \cdot 1000‰ \quad (1)$$

$$\delta^{15}N = \left[\frac{(^{15}N/^{15}N)_{Sample}}{(^{15}N/^{15}N)_{PDB}} - 1 \right] \cdot 1000‰ \quad (2)$$

where $\delta^{13}C$ is the thousand percent deviation of sample $^{13}C/^{12}C$ from the standard sample; $(^{13}C/^{12}C)$Sample is the $^{13}C/^{12}C$ of the leaf sample, and $(^{13}C/^{12}C)$PDB is $^{13}C/^{12}C$ in Peedee Belemnite (South Carolina).

5.3. Statistical Analyses

We first performed a DCA test on the leaf functional trait data, and the DCA1 value of axis lengths was 0.459, considerably smaller than 3.0, implying that the data distribution was consistent with the linear model. We therefore used an RDA to test correlations of leaf functional trait indices and environmental factors [98–100]. The 'vegan' package in RStudio (Integrated Development for R, RStudio Inc., Boston, MA, USA) was used to calculate RDA ordinations. The Pearson's correlation coefficient between leaf functional traits and environmental variables was calculated using IBM SPSS Statistics (IBM Inc., New York, NY, USA), and a respective heatmap was produced using Origin 2022 (OriginLab Inc., Northampton, MA, USA). The distribution map of forest reserves was produced using ArcGIS 10.4 (ESRI Inc., Redlands, CA, USA). Statistical analyses were performed using

Excel 2019 (Microsoft Inc., Redmond, WA, USA). Values are presented as means ± standard error of the mean.

Author Contributions: Conceptualization, W.S. and F.L.; methodology, X.Z.; software, J.X.; formal analysis, J.X., H.L. and X.H.; data curation, W.S.; writing—original draft preparation, J.X. and H.L.; writing—review and editing, W.S.; project administration, X.Z.; funding acquisition, F.L. All authors have read and agreed to the published version of the manuscript.

Funding: This research was funded by National Natural Science Foundation of China, grant number 42127807.

Data Availability Statement: The data produced in this study are available in Refs. [38,56,97].

Conflicts of Interest: The authors declare no conflict of interest.

References

1. Luo, L.; Shen, G.Z.; Xie, Z.Q.; Yu, J. Leaf functional traits of four typical forests along the altitudinal gradients in Mt. Shennongjia. *Acta Ecol. Sin.* **2011**, *31*, 6420–6428.
2. Cochrane, A.; Hoyle, G.L.; Yates, C.J.; Neeman, T.; Nicotra, A.B. Variation in plant functional traits across and within four species of western Australian *Banksia* (Proteaceae) along a natural climate gradient. *Austral Ecol.* **2016**, *41*, 886–896. [CrossRef]
3. Xiao, H.G.; Wang, C.Y.; Liu, J.; Wang, L.; Du, D.L. Insights into the differences in leaf functional traits of heterophyllous *Syringa oblata* under different light intensities. *J. For. Res.* **2015**, *26*, 613–621. [CrossRef]
4. Gong, H.D.; Gao, J. Soil and climatic drivers of plant SLA (specific leaf area). *Glob. Ecol. Conserv.* **2019**, *20*, e00696. [CrossRef]
5. Reichstein, M.; Michael, B.; Mahecha, M.D.; Kattge, J.; Baldocchi, D.D. Linking plant and ecosystem functional biogeography. *Proc. Natl. Acad. Sci. USA* **2014**, *111*, 13697–13702. [CrossRef]
6. Violle, C.; Reich, P.B.; Pacala, S.W. The emergence and promise of functional biogeography. *Proc. Natl. Acad. Sci. USA* **2014**, *111*, 13690–13696. [CrossRef] [PubMed]
7. Chave, J.; Coomes, D.; Jansen, S.; Lewis, S.L.; Swenson, N.G.; Zanne, A.E. Towards a worldwide wood economics spectrum. *Ecol. Lett.* **2009**, *12*, 351–366. [CrossRef] [PubMed]
8. Moles, A.T.; Ackerly, D.D.; Tweddle, J.C.; Dickie, J.B.; Smith, R.; Leishman, M.R.; Mayfield, M.M.; Pitman, A.; Wood, J.T.; Westoby, M. Global patterns in seed size. *Global Ecol. Biogeogr.* **2007**, *16*, 109–116. [CrossRef]
9. Moles, A.T.; Warton, D.I.; Warman, L.; Swenson, N.G.; Laffan, S.W.; Zanne, A.E.; Pitman, A.; Hemmings, F.A.; Leishman, M.R. Global patterns in plant height. *J. Ecol.* **2009**, *97*, 923–932. [CrossRef]
10. Reich, P.B.; Oleksyn, J. Global patterns of plant leaf N and P in relation to temperature and latitude. *Proc. Natl. Acad. Sci. USA* **2004**, *101*, 11001–11006. [CrossRef]
11. Hansen, M.C.; DeFries, R.S.; Townshend, J.R.G.; Carroll, M.; Dimiceli, C.; Sohlberg, R.A. Global percent tree cover at a spatial resolution of 500 meters: First results of the MODIS vegetation continuous fields algorithm. *Earth Interact.* **2003**, *7*, 1–15. [CrossRef]
12. Moles, A.T.; Perkins, S.E.; Shawn, W.; Laffan, S.W.; Flores-Moreno, H.; Awasthy, M.; Tindall, M.L.; Sack, L.; Pitman, A.; Kattge, J.; et al. Which is a better predictor of plant traits: Temperature or precipitation? *J. Veg. Sci.* **2014**, *25*, 1167–1180. [CrossRef]
13. Bondeau, A.; Cramer, W.; Smith, P.C.; Schaphoff, S.; Lucht, W.; Zaehle, S.; Gerten, D. Modelling the role of agriculture for the 20th century global terrestrial carbon balance. *Glob. Ecol. Biogeogr.* **2007**, *13*, 679–706. [CrossRef]
14. Verheijen, L.M.; Brovkin, V.; Aerts, R.; Bönisch, G.; Cornelissen, J.H.C.; Kattge, J.; Reich, P.B.; Wright, I.J.; van Bodegom, P.M. Impacts of trait variation through observed trait climate relationships on performance of an Earth system model: A conceptual analysis. *Biogeosciences* **2013**, *10*, 5497–5515. [CrossRef]
15. Yao, T.T.; Meng, T.T.; Ni, J.; Yan, S.; Feng, X.H.; Wang, G.H. Leaf functional trait variation and its relationship with plant phylogenic background and the climate in Xinjiang Junggar Basin, NW China. *Chin. J. Biodivers. Sci.* **2010**, *18*, 188. [CrossRef]
16. Wei, M.; Wang, S.; Wu, B.D.; Jiang, K.; Zhou, J.W.; Wang, C.Y. Variability of leaf functional traits of invasive tree *Rhus typhina* L. in North China. *Chin. J. Cent. South Univ.* **2020**, *27*, 155–163. [CrossRef]
17. Freschet, G.T.; Cornelissen, J.H.; Van Logtestijn, R.S.; Aerts, R. Evidence of the plant economics spectrum in a subarctic flora. *J. Ecol.* **2010**, *98*, 362–373. [CrossRef]
18. Sundqvist, M.K.; Giesler, R.; Wardle, D.A. Within- and across species responses of plant traits and litter decomposition to elevation across contrasting vegetation types in subarctic tundra. *PLoS ONE* **2011**, *6*, e27056. [CrossRef]
19. Mason, N.W.; Richardson, S.J.; Peltzer, D.A.; de Bello, F.; Wardle, D.A.; Allen, R.B. Changes in coexistence mechanisms along a long-term soil chronosequence revealed by functional trait diversity. *J. Ecol.* **2012**, *100*, 678–689. [CrossRef]
20. Gong, H.D.; Cui, Q.J.; Gao, J. Latitudinal, soil and climate effects on key leaf traits in northeastern China. *Glob. Ecol. Conserv.* **2020**, *22*, e00904. [CrossRef]
21. Zhang, S.H.; Zhang, Y.; Xiong, K.N.; Yu, Y.H.; Min, X.Y. Changes of leaf functional traits in karst rocky desertification ecological environment and the driving factors. *Glob. Ecol. Conserv.* **2020**, *24*, e01381. [CrossRef]
22. Wright, I.J.; Reich, P.B.; Westoby, M.; Ackerly, D.D.; Baruch, Z.; Bongers, F.; Ackerly, D.D.; Baruch, Z.; Bongers, F.; Cavender-Bares, J.; et al. The worldwide leaf economics spectrum. *Nature* **2004**, *428*, 821–827. [CrossRef]

23. Li, S.; Gou, W.; Wang, H.; White, J.F.; Wu, G.; Su, P. Trade-Off Relationships of Leaf Functional Traits of *Lycium ruthenicum* in Response to Soil Properties in the Lower Reaches of Heihe River, Northwest China. *Diversity* **2021**, *13*, 453. [CrossRef]
24. Cernusak, L.A.; Ubierna, N.; Winter, K.; Holtum, J.A.M.; Marshall, J.D.; Farquhar, G.D. Environmental and physiological determinants of carbon isotope discrimination in terrestrial plants. *New Phytol.* **2013**, *200*, 950–965. [CrossRef] [PubMed]
25. Sellin, A.; Tullus, A.; Niglas, A.; Õunapuu, E.; Karusion, A.; Lõhmus, K. Humidity-driven changes in growth rate, photosynthetic capacity, hydraulic properties and other functional traits in silver birch (*Betula pendula*). *Ecol. Res.* **2013**, *28*, 523–535. [CrossRef]
26. Richards, J.H.; Henn, J.J.; Sorenson, Q.M.; Adams, M.A.; Smith, D.D.; McCulloh, K.A.; Givnish, T.J. Mistletoes and their eucalypt hosts differ in the response of leaf functional traits to climatic moisture supply. *Oecologia* **2021**, *195*, 759–771. [CrossRef] [PubMed]
27. Guerin, G.R.; Gallagher, R.V.; Wright, I.J.; Andrew, S.C.; Falster, D.S.; Wenk, E.; Munroe, S.E.M.; Lowe, A.J.; Sparrow, B. Environmental associations of abundance-weighted functional traits in Australian plant communities. *Basic Appl. Ecol.* **2022**, *58*, 98–109. [CrossRef]
28. Dong, Y.; Liu, Y. Response of Korean pine's functional traits to geography and climate. *PLoS ONE* **2017**, *12*, e0184051. [CrossRef]
29. Pan, F.; Liang, Y.; Wang, K.; Zhang, W. Responses of fine root functional traits to soil nutrient limitations in a karst ecosystem of Southwest China. *Forests* **2018**, *9*, 743. [CrossRef]
30. Qi, D.; Wieneke, X.; Zhou, X.; Jiang, X.; Xue, P. Succession of plant community composition and leaf functional traits in responding to karst rocky desertification in the Wushan County in Chongqing, China. *Community Ecol.* **2017**, *18*, 157–168. [CrossRef]
31. Monika, R.; Kusum, A.; Ayyandar, A.; Juha, M.A.; Ujjwal, K.; Barbara, S.; Levente, H.; Erika, M.; Rajiv, P. Relative contribution of plant traits and soil properties to the functioning of a temperate forest ecosystem in the Indian Himalayas. *Catena* **2020**, *194*, 10467. [CrossRef]
32. Wang, X.P.; Ye, M.R.; Zhang, X.P.; Xu, R.S.; Xu, D.Q. Changes in leaf functional traits of *Houttuynia cordata* in response to soil environmental factors in Anqing city of Anhui Province in China. *Environ. Pollut. Bioavail.* **2019**, *31*, 240–251. [CrossRef]
33. Ordoñez, J.C.; Van Bodegom, P.M.; Witte, J.P.M.; Wright, I.J.; Reich, P.B.; Aerts, R. A global study of relationships between leaf traits, climate and soil measures of nutrient fertility. *Glob. Ecol. Biogeogr.* **2009**, *18*, 137–149. [CrossRef]
34. Chen, Y.; Han, W.; Tang, L.; Tang, Z.; Fang, J. Leaf nitrogen and phosphorus concentrations of woody plants differ in responses to climate, soil and plant growth form. *Ecography* **2013**, *36*, 178–184. [CrossRef]
35. Liu, R.; Liang, S.; Long, W.; Jiang, Y. Variations in leaf functional traits across ecological scales in riparian plant communities of the Lijiang river, Guilin, Southwest China. *Trop. Conserv. Sci.* **2018**, *11*, 1940082918804680. [CrossRef]
36. Duan, X.; Jia, Z.; Li, J.; Wu, S. The influencing factors of leaf functional traits variation of *Pinus densiflora* Sieb. et Zucc. *Glob. Ecol. Conserv.* **2022**, *38*, e02177. [CrossRef]
37. Zeilinger, S.; Gupta, V.K.; Dahms, T.E.S.; Silva, R.N.; Singh, H.B.; Upadhyay, R.S.; Gomes, E.V.; Tsui, C.K.M.; Nayak, S.C. Friends or foes? Emerging insights from fungal interactions with plants. *FEMS Microbiol. Rev.* **2015**, *40*, 182–207. [CrossRef]
38. Zheng, L.M.; Song, W.C. Phosphorus Limitation of Trees Influences Forest Soil Fungal Diversity in China. *Forests* **2022**, *13*, 223. [CrossRef]
39. Hu, M.Y.; Zhang, L.; Luo, T.X.; Shen, W. Variations in leaf functional traits of *Stipa purpurea* along a rainfall gradient in Xizang, China. *Chin. J. Plant Ecol.* **2012**, *36*, 136. [CrossRef]
40. Song, L.L.; Tian, Q.; Li, G.; Li, Z.X.; Liu, X.Y.; Gui, J.; Li, Y.C.; Cui, Q.; Zhao, Y. Variation in characteristics of leaf functional traits of alpine vegetation in the Three-River Headwaters Region, China. *Ecol. Indic.* **2022**, *145*, 109557. [CrossRef]
41. Zhang, C.; Zeng, F.; Zeng, Z.; Du, H.; Su, L.; Zhang, L.; Zhang, H. Impact of Selected Environmental Factors on Variation in Leaf and Branch Traits on Endangered Karst Woody Plants of Southwest China. *Forests* **2022**, *13*, 1080. [CrossRef]
42. Liu, G.; Freschet, G.T.; Pan, X.; Cornelissen, J.H.; Li, Y.; Dong, M. Coordinated variation in leaf and root traits across multiple spatial scales in Chinese semi-arid and arid ecosystems. *New Phytol.* **2010**, *188*, 543–553. [CrossRef] [PubMed]
43. Wang, J.; Zhu, J.; AI, X.R.; Yao, L.; Huang, X.; Wu, M.L.; Hong, J.F. Effects of topography on leaf functional traits across plant life forms in Xingdou Mountain, Hubei, China. *Chin. J. Plant Ecol.* **2019**, *43*, 447. [CrossRef]
44. Duan, Y.Y.; Song, L.J.; Niu, S.Q.; Huang, T.; Yang, G.H.; Hao, W.F. Variation in leaf functional traits of different-aged *Robinia pseudoacacia* communities and relationships with soil nutrients. *Chin. J. Appl. Ecol.* **2017**, *28*, 28–36. [CrossRef]
45. Zhang, K.R.; Cheng, X.L.; Dang, H.S.; Zhang, Q.F. Biomass:N:K:Ca:Mg:P ratios in forest stands world-wide: Biogeographical variations and environmental controls. *Glob. Ecol. Biogeogr.* **2020**, *29*, 2176–2189. [CrossRef]
46. Amundson, R.; Austin, A.T.; Schuur, E.A.; Yoo, K.; Matzek, V.; Kendall, C.; Uebersax, A.; Brenner, D.L.; Baisden, W.T. Global patterns of the isotopic composition of soil and plant nitrogen. *Glob. Biogeochem. Cy.* **2003**, *17*, 1031–1038. [CrossRef]
47. Rumsey, D.J. *Statistics for Dummies*, 2nd ed.; Wiley Publishing: Hoboken, NJ, USA, 2011; p. 284.
48. Luo, Y.K.; Hu, H.F.; Zhao, M.Y.; Li, H.; Liu, S.S.; Fang, J.Y. Latitudinal pattern and the driving factors of leaf functional traits in 185 shrub species across eastern China. *J. Plant Ecol.* **2019**, *12*, 67–77. [CrossRef]
49. Wright, I.J.; Reich, P.B.; Cornelissen, J.H.C.; Falster, D.S.; Garnier, E.; Hikosaka, K.; Lamont, B.B.; Lee, W.; Oleksyn, J.; Osada, N.; et al. Assessing the generality of global leaf trait relationships. *New Phytol.* **2005**, *166*, 485e496. [CrossRef]
50. Becklin, K.M.; Pallo, M.L.; Galen, C. Willows indirectly reduce arbuscular mycorrhizal fungal colonization in understorey communities. *J. Ecol.* **2012**, *100*, 343–351. [CrossRef]
51. Ferlian, O.; Goldmann, K.; Eisenhauer, N.; Tarkka, M.T.; Buscot, F.; Heintz-Buschart, A. Distinct effects of host and neighbour tree identity on arbuscular and ectomycorrhizal fungi along a tree diversity gradient. *ISME Commun.* **2021**, *1*, 40. [CrossRef]

52. Toju, H.; Kishida, O.; Katayama, N.; Takagi, K. Networks Depicting the Fine-Scale Co-Occurrences of Fungi in Soil Horizons. *PLoS ONE* **2016**, *11*, e0165987. [CrossRef] [PubMed]
53. Luo, S.; Schmid, B.; De Deyn, G.B.; Yu, S.X. Soil microbes promote complementarity effects among co-existing trees through soil nitrogen partitioning. *Funct. Ecol.* **2018**, *32*, 1879–1889. [CrossRef]
54. Boberg, J.B.; Finlay, R.D.; Stenlid, J.; Ekblad, A.; Lindahl, B.D. Nitrogen and carbon reallocation in fungal mycelia during decomposition of boreal forest litter. *PLoS ONE* **2014**, *9*, e92897. [CrossRef]
55. Du, E.; van Doorn, M.; de Vries, W. Spatially divergent trends of nitrogen versus phosphorus limitation across European forests. *Sci. Total Environ.* **2021**, *771*, 145391. [CrossRef] [PubMed]
56. Song, W.C.; Zhou, Y.J. Linking leaf $\delta^{15}N$ and $\delta^{13}C$ with soil fungal biodiversity, ectomycorrhizal and plant pathogenic abundance in forest ecosystems of China. *Catena* **2021**, *200*, 105176. [CrossRef]
57. He, X.H.; Xu, M.G.; Qiu, G.Y.; Zhou, J.B. Use of ^{15}N stable isotope to quantify nitrogen transfer between mycorrhizal plants. *J. Plant Ecol.* **2009**, *2*, 107–118. [CrossRef]
58. Angst, G.; Mueller, K.E.; Eissenstat, D.M.; Trumbore, S.; Freeman, K.H.; Hobbie, S.E.; Chorover, J.; Oleksyn, J.; Reich, P.B.; Mueller, C.W. Soil organic carbon stability in forests: Distinct effects of tree species identity and traits. *Glob. Chang. Biol.* **2019**, *25*, 1529–1546. [CrossRef]
59. Song, W.C.; Tong, X.J.; Liu, Y.H.; Li, W.K. Microbial community, newly sequestered soil organic carbon, and $\delta^{15}N$ variations driven by tree roots. *Front. Microbiol.* **2020**, *11*, 314. [CrossRef]
60. Schulze, E.D.; Nicolle, D.; Boerner, A.; Lauerer, M.; Aas, G. Stable carbon and nitrogen isotope ratios of *Eucalyptus* and *Acacia* species along a seasonal rainfall gradient in Western Australia. *Trees* **2014**, *28*, 1125–1135. [CrossRef]
61. Luo, Q.Q.; Zhou, Z.C.; Deng, Z.F.; Yu, L.H.; Sun, J.J.; Xu, G.B. Variation law of phenotypic traits and nitrogen and phosphorus stoichiometric characteristics of leaf of natural populations of *Taxus wallichiana* var. *mairei*. *Chin. J. Plant Res. Environ.* **2021**, *10*, 27–35. [CrossRef]
62. Zhang, K.; Hou, J.H.; He, N.P. Leaf functional trait distribution and controlling factors of *Pinus tabuliformis*. *Acta Ecol. Sin.* **2017**, *37*, 736–749. [CrossRef]
63. Liu, M.X. Studies on physiological and leaf morphological traits for photosynthesis on different slopes in a subalpine meadow. *Acta Ecol. Sin.* **2017**, *37*, 8526–8536. [CrossRef]
64. Li, S.J.; Zhang, Y.F.; Chen, T. Relationships between foliar stable carbon isotope composition and environmental factors and leaf element contents of *Pinus tabulaeformis* in northwestern China. *Chin. J. Plant Ecol.* **2011**, *35*, 596–604. [CrossRef]
65. Zhu, Y.H.; Kang, H.Z.; Xie, Q.; Wang, Z.; Yin, S.; Liu, C.J. Pattern of leaf vein density and climate relationship of *Quercus variabilis* populations remains unchanged with environmental changes. *Trees* **2012**, *26*, 597–607. [CrossRef]
66. Hudson, J.M.; Henry, G.H.; Cornwell, W.K. Taller and larger: Shifts in Arctic tundra leaf traits after 16 years of experimental warming. *Glob. Chang. Biol.* **2011**, *17*, 1013–1021. [CrossRef]
67. Noguchi, K.; Terashima, I. Responses of spinach leaf mitochondria to low N availability. *Plant Cell Environ.* **2006**, *29*, 710–719. [CrossRef]
68. Rao, Z.G.; Guo, W.K.; Cao, J.T.; Shi, F.X.; Jiang, H.; Li, C.Z. Relationship between the stable carbon isotopic composition of modern plants and surface soils and climate: A global review. *Earth-Sci. Rev.* **2017**, *165*, 110–119. [CrossRef]
69. Rumman, R.; Atkin, O.K.; Bloomfield, K.J.; Eamus, D. Variation in bulk-leaf ^{13}C discrimination, leaf traits and water-use efficiency-trait relationships along a continental-scale climate gradient in Australia. *Glob. Chang. Biol.* **2017**, *24*, 1186–1200. [CrossRef]
70. Su, B.; Han, X.G.; Li, L.H.; Huang, J.H.; Bai, Y.F.; Qu, C.M. Responses of $\delta^{13}C$ value and water use efficiency of plant species to environmental gradients along the grassland zone of Northeast China Transect. *Acta Phytoecol. Sin.* **2000**, *24*, 648–655.
71. Marshall, J.D.; Zhang, J.W. Carbon isotope discrimination and water use efficiency in native plants of the North Central Rockies. *Ecology* **2018**, *75*, 1887–1895. [CrossRef]
72. Anderson, W.T.; Bernasconi, S.M.; Mckenzie, J.A.; Saurer, M. Oxygen and carbon isotopic record of climatic variability in tree ring cellulose (*Picea abies*): An example from central Switzerland (1913–1995). *J. Geophys. Res.* **1998**, *103*, 31625–31636. [CrossRef]
73. Xing, K.X.; Niinemets, Ü.; Rengel, Z.; Onoda, Y.; Xia, J.Z.; Chen, H.Y.; Zhao, M.F.; Han, W.X.; Li, H.B. Global patterns of leaf construction traits and their covariation along climate and soil environmental gradients. *New Phytol.* **2021**, *232*, 1648–1660. [CrossRef] [PubMed]
74. Poorter, H.; Niinemets, U.; Poorter, L.; Wright, I.J.; Villar, R. Causes and consequences of variation in leaf mass per area (LMA): A meta-analysis. *New Phytol.* **2010**, *182*, 565e588. [CrossRef] [PubMed]
75. Liu, J.F.; Zhang, Y.T.; Ni, Y.Y.; Huang, Y.N.; Jiang, Z.P. Latitudinal trends in foliar $\delta^{13}C$ and $\delta^{15}N$ of *Quercus variabilis* and their influencing factors. *Chin. J. Appl. Ecol.* **2018**, *29*, 1373–1380. [CrossRef]
76. Dong, L.L.; Liu, S.R.; Shi, Z.M.; Feng, Q.H. Relationships between leaf traits of *Castanopsis* species and the environmental factors in the North-South transect of eastern China. *Jpn. J. For. Res.* **2009**, *22*, 463–469.
77. Diefendorf, A.F.; Mueller, K.E.; Wing, S.L.; Koch, P.L.; Freeman, K.H.; John, M.H. Global patterns in leaf ^{13}C discrimination and implications for studies of past and future climate. *Proc. Natl. Acad. Sci. USA* **2010**, *107*, 5738–5743. [CrossRef]
78. Tieszen, L.L.; Boutton, T.W. *Stable Carbon Isotopes in Terrestrial Ecosystem Research*; Springer: New York, NY, USA, 1989; pp. 167–195.
79. Du, B.M.; Kang, H.Z.; Zhu, Y.H.; Zhou, X.; Yin, S.; Burgess, P.J.; Liu, C.J. Variation of Oriental Oak (*Quercus variabilis*) Leaf $\delta^{13}C$ across Temperate and Subtropical China: Spatial Patterns and Sensitivity to Precipitation. *Forests* **2015**, *6*, 2296–2306. [CrossRef]

80. Li, M.X.; Peng, C.H.; Wang, M.; Yang, Y.Z.; Zhang, K.R.; Li, P.; Yang, Y.; Ni, J.; Zhu, Q.A. Spatial patterns of leaf $\delta^{13}C$ and its relationship with plant functional groups and environmental factors in China. *J. Geophys. Res-Biogeosci.* **2017**, *122*, 1564–1575. [CrossRef]
81. No'am, G.S.; Sinclair, T.R. Global environment change and simulated forage quality of wheat II. Water and nitrogen stress. *Field Crops Res.* **1995**, *40*, 29–37.
82. Wright, I.J.; Reich, P.B.; Westoby, M. Strategy shifts in leaf physiology, structure and nutrient content between species of high- and low-rainfall and high- and low-nutrient habitats. *Funct. Ecol.* **2001**, *15*, 423–434. [CrossRef]
83. Song, W.C.; Tong, X.J.; Zhang, J.S.; Meng, P.; Li, J. Autotrophic and heterotrophic components of soil respiration caused by rhizosphere priming effects in a plantation. *Plant Soil Environ.* **2017**, *63*, 295–299. [CrossRef]
84. Song, W.C.; Tong, X.J.; Zhang, J.S.; Meng, P.; Li, J. How a root-microbial system regulates the response of soil respiration to temperature and moisture in a plantation. *Pol. J. Environ. Stud.* **2018**, *27*, 2749–2756. [CrossRef] [PubMed]
85. Liu, C.C.; Liu, Y.G.; Guo, K. Ecophysiological adaptations to drought stress of seedlings of four plant species with different growth forms in karst habitats. *Chin. J. Plant Ecol.* **2011**, *35*, 1070–1082. [CrossRef]
86. Zhang, Z.F.; You, Y.M.; Huang, Y.Q.; Li, X.K.; Zhang, J.C.; Zhang, D.N.; He, C.X. Effects of drought stress on *Cyclobalanopsis glauca* seedlings under simulating karst environment condition. *Acta Ecol. Sin.* **2012**, *32*, 6318–6325. [CrossRef]
87. Lourenoç, J., Jr.; Newman, E.A.; Ventura, A.J.; Milanez, C.R.; Thomaz, L.D.; Wandekoken, D.T.; Enquist, B.J. Soil-associated drivers of plant traits and functional composition in Atlantic Forest coastal tree communities. *Ecosphere* **2021**, *12*, e03629. [CrossRef]
88. Li, L. Research on the Relationship between Evapotranspiration and Groundwater Based on Remote Sensing Technology. Master's Thesis, China University of Geosciences, Beijing, China, 2013.
89. Huo, L.L.; Chen, D.D.; Li, Q.; Zhang, G.L.; He, F.Q.; Shu, M.; Zhao, L. Study on the Relationship between Functional Traits of Grassland Plants and Evapotranspiration in the Sanjiangyuan. *Acta Agrestia Sin.* **2022**, *30*, 2182–2190. [CrossRef]
90. Müller, M.; Schickhoff, U.; Scholten, T.; Drollinger, S.; Böhner, J.; Chaudhary, R.P. How do soil properties affect alpine treelines? General principles in a global perspective and novel findings from Rolwaling Himal, Nepal. *Prog. Phys. Geogr. Earth Environ.* **2015**, *40*, 135–160. [CrossRef]
91. Piao, H.C.; Li, S.L.; Yan, Z.F.; Li, C. Understanding nutrient allocation based on leaf nitrogen isotopes and elemental ratios in the karst region of Southwest China. *Agric. Ecosyst. Environ.* **2020**, *294*, 106864. [CrossRef]
92. Fujii, K.; Shibata, M.; Kitajima, K.; Ichie, T.; Kitayama, K.; Turner, B.L. Plant–soil interactions maintain biodiversity and functions of tropical forest ecosystems. *Ecol. Res.* **2018**, *33*, 149–160. [CrossRef]
93. Sasse, J.; Martinoia, E.; Northen, T. Feed Your Friends: Do Plant Exudates Shape the Root Microbiome? *Trends Plant Sci.* **2018**, *23*, 25–41. [CrossRef]
94. Chalot, M.; Brun, A. Physiology of organic nitrogen acquisition by ectomycorrhizal fungi and ectomycorrhizas. *FEMS Microbiol. Rev.* **1998**, *22*, 21–44. [CrossRef] [PubMed]
95. Wu, H.L.; Xiang, W.H.; Ouyang, S.; Forrester, D.I.; Zhou, B.; Chen, L.X.; Ge, T.D.; Lei, P.F.; Chen, L.; Zeng, Y.L.; et al. Linkage between tree species richness and soil microbial diversity improves phosphorus bioavailability. *Funct. Ecol.* **2019**, *33*, 1549–1560. [CrossRef]
96. Laliberté, E.; Lambers, H.; Burgess, T.I.; Wright, S.J. Phosphorus limitation, soil-borne pathogens and the coexistence of plant species in hyperdiverse forests and shrublands. *New Phytol.* **2015**, *206*, 507–521. [CrossRef] [PubMed]
97. Wang, P.D.; Chen, Y.J.; Sun, Y.T.; Tan, S.; Zhang, S.Y.; Wang, Z.H.; Zhou, J.Z.; Zhang, G.; Shu, W.S.; Luo, C.L.; et al. Distinct biogeography of different fungal guilds and their associations with plant species richness in forest ecosystems. *Front. Ecol. Evol.* **2019**, *7*, 216. [CrossRef]
98. Paliy, O.; Shankar, V. Application of multivariate statistical techniques in microbial ecology. *Mol. Ecol.* **2016**, *25*, 1032–1057. [CrossRef] [PubMed]
99. Lepš, J.; Šmilauer, P. *Multivariate Analysis of Ecological Data Using CANOCO*; Cambridge Press: Cambridge, UK, 2003.
100. Legendre, P.; Gallagher, E.D. Ecologically meaningful transformations for ordination of species data. *Oecologia* **2001**, *129*, 271–280. [CrossRef]

plants

Article

Comparison of Two Simplified Versions of the Gielis Equation for Describing the Shape of Bamboo Leaves

Weihao Yao [1], Ülo Niinemets [2,3], Wenjing Yao [1,*], Johan Gielis [4], Julian Schrader [5], Kexin Yu [1] and Peijian Shi [1,*]

1 Bamboo Research Institute, College of Biology and the Environment, Nanjing Forestry University, Nanjing 210037, China
2 Institute of Agricultural and Environmental Sciences, Estonian University of Life Sciences, 51006 Tartu, Estonia
3 Estonian Academy of Sciences, 10130 Tallinn, Estonia
4 Department of Biosciences Engineering, University of Antwerp, B-2020 Antwerp, Belgium
5 School of Natural Sciences, Macquarie University, Sydney, NSW 2109, Australia
* Correspondence: yaowenjing@njfu.edu.cn (W.Y.); pjshi@njfu.edu.cn (P.S.)

Abstract: Bamboo is an important component in subtropical and tropical forest communities. The plant has characteristic long lanceolate leaves with parallel venation. Prior studies have shown that the leaf shapes of this plant group can be well described by a simplified version (referred to as SGE-1) of the Gielis equation, a polar coordinate equation extended from the superellipse equation. SGE-1 with only two model parameters is less complex than the original Gielis equation with six parameters. Previous studies have seldom tested whether other simplified versions of the Gielis equation are superior to SGE-1 in fitting empirical leaf shape data. In the present study, we compared a three-parameter Gielis equation (referred to as SGE-2) with the two-parameter SGE-1 using the leaf boundary coordinate data of six bamboo species within the same genus that have representative long lanceolate leaves, with >300 leaves for each species. We sampled 2000 data points at approximately equidistant locations on the boundary of each leaf, and estimated the parameters for the two models. The root–mean–square error (RMSE) between the observed and predicted radii from the polar point to data points on the boundary of each leaf was used as a measure of the model goodness of fit, and the mean percent error between the RMSEs from fitting SGE-1 and SGE-2 was used to examine whether the introduction of an additional parameter in SGE-1 remarkably improves the model's fitting. We found that the RMSE value of SGE-2 was always smaller than that of SGE-1. The mean percent errors among the two models ranged from 7.5% to 20% across the six species. These results indicate that SGE-2 is superior to SGE-1 and should be used in fitting leaf shapes. We argue that the results of the current study can be potentially extended to other lanceolate leaf shapes.

Keywords: leaf shape; percent error; *Pleioblastus*; polar angle; polar radius

Citation: Yao, W.; Niinemets, Ü.; Yao, W.; Gielis, J.; Schrader, J.; Yu, K.; Shi, P. Comparison of Two Simplified Versions of the Gielis Equation for Describing the Shape of Bamboo Leaves. *Plants* **2022**, *11*, 3058. https://doi.org/10.3390/plants11223058

Academic Editor: Yasutomo Hoshika

Received: 2 October 2022
Accepted: 10 November 2022
Published: 11 November 2022

1. Introduction

The subfamily Bambusoideae, which includes >1300 species covering 75 genera of Poaceae, are important components in many ecosystems, and are particularly abundant in subtropical and tropical areas [1]. As typical to Poaceae, leaves of all bamboo species have parallel venation, and most species have long lanceolate leaves. Lin et al. [2] reported that the leaf lamina width/length ratio (referred to as leaf width/length ratio for convenience hereinafter) ranged from 0.05 to 0.35 for 101 bamboo taxa, and the interspecific variation in leaf shape is mainly due to differences in the leaf width/length ratio. When the leaf width/length ratio is large, the leaf shape of some bamboo species (e.g., *Shibataea chinensis*) appears to be ovate. In fact, in bamboos, leaf width/length ratio provides an objective criterion to distinguish among lanceolate or linear leaves and ovate leaves [3]. Given the importance of leaf shape in the resource harvesting and evolution of plants, several indices (e.g., leaf width/length ratio, leaf dissection index, leaf roundness index, leaf ellipticalness index, and the fractal dimension of leaf boundary) were proposed to quantify the leaf shape geometrical characteristics, especially the tapering and curvature of a leaf's boundary [2,4–7]. However, the number of studies that have developed explicit models that can quantitatively describe leaf boundaries of the Poaceae species is very limited [8–11].

It would be highly beneficial to have a "universal" parametric model that can describe all natural geometries, such as the diverse leaf shapes across different plant groups; such an ambition stems from the successful use of general models in other natural science fields, especially in physics, where general laws have been defined since the Renaissance [12]. However, the variations in natural geometries, especially asymmetry, handedness, and spirality, have far exceeded what we can imagine based on the extant physical and mathematical knowledge. It is difficult to find a universal parametric model to describe all morphological variations in leaves across different plant groups. Fortunately, it is still hoped to find one that can apply to some groups. Gielis [13] proposed a polar coordinate equation, referred to as the Gielis equation hereinafter, which can simulate many geometries found in nature, although its capacity to describe actual biological objects has been seldom tested. The Gielis equation is a generalization of the superellipse equation [14], while the latter is a generalization of the ellipse equation. In recent years, several studies have demonstrated the validity of the Gielis equation for describing and fitting many natural geometries (see Ref. [15] and the references therein). The first practical application of the Gielis equation was a description of leaf shapes of four bamboo species from the genus *Indocalamus* [9], followed by Ref. [11], in which the leaf shapes of an additional 42 bamboo species were demonstrated to follow this equation.

The original Gielis equation has six empirical parameters, and its mathematical expression in the polar coordinate system is as follows:

$$r(\varphi) = \left(\left| \frac{1}{A} \cos\left(\frac{m}{4}\varphi\right) \right|^{n_2} + \left| \frac{1}{B} \sin\left(\frac{m}{4}\varphi\right) \right|^{n_3} \right)^{-\frac{1}{n_1}} \quad (1)$$

where r and φ are the polar radius and polar angle, respectively; A, B, n_1, n_2, and n_3 are parameters to be fitted; m is a positive integer that determines the number of angles of the Gielis curve within $[0, 2\pi]$. This equation can be re-expressed as [16,17]:

$$r(\varphi) = a\left(\left| \cos\left(\frac{m}{4}\varphi\right) \right|^{n_2} + \left| \frac{1}{k} \sin\left(\frac{m}{4}\varphi\right) \right|^{n_3} \right)^{-\frac{1}{n_1}} \quad (2)$$

where $a = A^{n_2/n_1}$ and $k = B/A^{n_2/n_3}$. To decrease the model's complexity and more effectively fit the empirical boundary data of bamboo leaves, Shi et al. [9] used a simplified version of Equation (2) by setting $m = 1$, $k = 1$ and $n_2 = n_3 = 1$, which is referred to as SGE-1 hereinafter:

$$r(\varphi) = a\left(\left| \cos\left(\frac{\varphi}{4}\right) \right| + \left| \sin\left(\frac{\varphi}{4}\right) \right| \right)^{-\frac{1}{n_1}} \quad (3)$$

The SGE-1 was confirmed to provide very good fits to empirical leaf boundary coordinate data for the studied 46 bamboo species (the coefficients of determination were all larger than 0.985) [9,11]. The model parameter n_1 characterizes the elongation change (accompanied with the change in tapering and curvature) of leaf shape, and it was significantly different among species, but it varied in a narrow range, from 0.02 to 0.10 [9,11]. However, the question is whether additional modifications of the Gielis equation can result in a model that describes the leaf shapes of bamboo with better goodness of fit, while keeping the number of fitted parameters low. Previously, the following simplified version of the original Gielis equation with an additional parameter n_2, which can render the equation to generate more diverse symmetrical geometries [18], was used and shown to perform similarly to SGE-1 in fitting the shapes of avian eggs [19]:

$$r(\varphi) = a\left(\left|\cos\left(\frac{\varphi}{4}\right)\right|^{n_2} + \left|\sin\left(\frac{\varphi}{4}\right)\right|^{n_2}\right)^{-\frac{1}{n_1}} \tag{4}$$

We refer to Equation (4) as SGE-2. When m is set to 5 instead of 2, this model version can describe the shapes of some sea stars, and the geometries of the outer rims of corolla tubes of *Vinca major* [17,20].

In the present work, we sampled 1996 leaves from six bamboo species from the genus *Pleioblastus*, and compared the predictions using SGE-1 and SGE-2 to test whether SGE-2 can improve the model prediction of bamboo leaf shapes.

2. Materials and Methods

2.1. Plant Materials and Leaf Collection

We sampled 1996 leaves of six *Pleioblastus* species (Figure 1 for the leaf samples) growing at the Nanjing Forestry University campus (118°48′35″ E, 32°4′67″ N) in late August 2021 when the leaves were fully mature in this season. For each species, we randomly sampled more than 300 leaves from different plant canopy positions without distinguishing among different canopy microenvironments and among leaf ages. For each species, leaves were sampled from 10 to 60 culms (Table 1 for sampling information). Although the accurate age of each culm cannot be determined, all species had been planted on this site more than 20 years ago. We argue that due to the large sample size, influences of sampling vertical positions, azimuth, leaf age, and culm age do not alter our results qualitatively. The leaves were wrapped in wet paper, and put into plastic self-sealing bags (45 cm × 34 cm) to reduce water loss. The bags with leaves were stored at 5 °C in a fridge for less than 24 h before scanning.

Figure 1. Outlines of leaf samples of six *Pleioblastus* species collected from the Nanjing Forestry University campus.

Table 1. Sampling information of the six bamboo species.

Species Code	Scientific Name	Number of Culms	Number of Leaves	Sampling Date
1	*Pleioblastus argenteostriatus*	60	335	27 August 2021
2	*Pleioblastus chino* var. *hisauchii*	15	336	21 August 2021
3	*Pleioblastus fortunei*	60	337	24 August 2021
4	*Pleioblastus kongosanensis* f. *aureostriatus*	60	336	22 August 2021
5	*Pleioblastus maculatus*	10	323	25 August 2021
6	*Pleioblastus viridistriatus*	60	329	23 August 2021

2.2. Data Acquisition

We scanned the fresh leaves with a photo scanner (V550, Epson, Batam, Indonesia) at 600 dpi resolution. The scanned color images were converted to black and white BMP files with Photoshop CS6, ver. 13.0 (Adobe, San Jose, CA, USA). Matlab (version $\geq$ 2009a; MathWorks, Natick, MA, USA) procedures developed by Refs. [10,21] were used to extract the planar coordinates of the boundary of each leaf. The boundary of each leaf was characterized by 2000 approximately equidistantly spaced coordinates using the "adjdata" function of the "biogeom" package (version 1.2.1) [15] in R (version 4.2.0) [22].

2.3. Data Fitting and Model Evaluation

We used two simplified versions of the Gielis equation, SGE-1 (Equation (3)) and SGE-2 (Equation (4)), to fit the boundary coordinates for each leaf using the "fitGE" function in the "biogeom" package (version 1.2.1) [15]. This function estimates the model parameters by minimizing the residual sum of squares (RSS) between the observed and predicted radii (r_i vs. $\hat{r}_i$) from the polar point to the leaf boundary:

$$\text{RSS} = \sum_{i=1}^{N} (r_i - \hat{r}_i)^2 \tag{5}$$

where N is the number of data points on the leaf's boundary (N = 2000 in our study). The root–mean–square error (RMSE) was calculated to characterize the goodness of fit of the nonlinear regression:

$$\text{RMSE} = \sqrt{\text{RSS}/N} \tag{6}$$

We used the paired t-test at the 0.05 significance level to compare the goodness of fits of the two models, SGE-1 and SGE-2. We further calculated the mean percent error (MPE) between the two groups of RMSEs:

$$\text{MPE} = \frac{1}{Q}\sum_{j=1}^{Q} \frac{\text{RMSE}_{1,j} - \text{RMSE}_{2,j}}{\text{RMSE}_{1,j}} \times 100\% \tag{7}$$

where j represents the j-th leaf, and Q represents the number of leaves for each species. MPE was used to assess whether the introduction of an additional parameter in SGE-1 to form SGE-2 enhances model predictability enough to compensate for the increase in model complexity. As a rule of thumb, a >5% MPE indicates that it is worth adding an additional parameter [23].

For the estimated values of n_1 and n_2 in SGE-2, we used one-way ANOVA followed the Tukey's HSD test [24] to examine whether the model parameters differed among any two species. Before comparing the parameter values among the species, the parameter values were log- or exp-transformed, depending on the shape of the parameter frequency distributions. For a right-skewed distribution (parameter n_1), a logarithmic transformation was used; for a left-skewed distribution (parameter n_2), an exponential transformation was used [25]. Estimated values of parameters and goodness of fit for models SGE-1 and SGE-2 for all the 1996 leaves are shown in Tables S1 and S2 in the online Supplementary Materials.

The statistical software R (version 4.2.0) [22] was used to carry out the statistical analyses and to draw figures.

3. Results

Both models provided good fits to the boundary of leaves in all studies species (Tables S1 and S2 in the online Supplementary Materials; see Figures 2 and 3 for the six leaf examples as intuitively shown in Figure 1). The RMSE varied among species, with the lowest RMSE observed for *Pleioblastus argenteostriatus* fitted with model SGE-2 and the highest RMSE observed for *P. viridistriatus* fitted with model SG-1 (Figure 4). Visually, the boundaries predicted by model SGE-2 more closely matched the actual leaf boundaries than those predicted by the model SGE-1 (Figure 3 versus Figure 2). This was confirmed by a comparison of the mean RMSEs among species. For all species, the RMSE for the model SGE-1 was greater than that for the model SGE-2 (all p values < 0.001; Figure 4). The mean percentage errors (MPEs) between the RMSEs for the two models (Equation (7)) were greater than 5% for all studied bamboo species (20.2%, 12.8%, 7.5%, 11.3%, 15.3%, and 8.5%, following species order in Table 1). That is, the introduction of n_2 into SGE-2 largely improved the goodness of fit. The parameters, $\log(n_1)$ or $\exp(n_2)$, varied among the six species, reflecting differences in leaf elongation and margin curvature (Figure 5). All means of the estimated values of n_2 of the six species were greater than 1, and most numerical values of n_2 (1559 out of 1996) were greater than 1.0 (Figure 5B), further suggesting that an additional parameter needs to be incorporated.

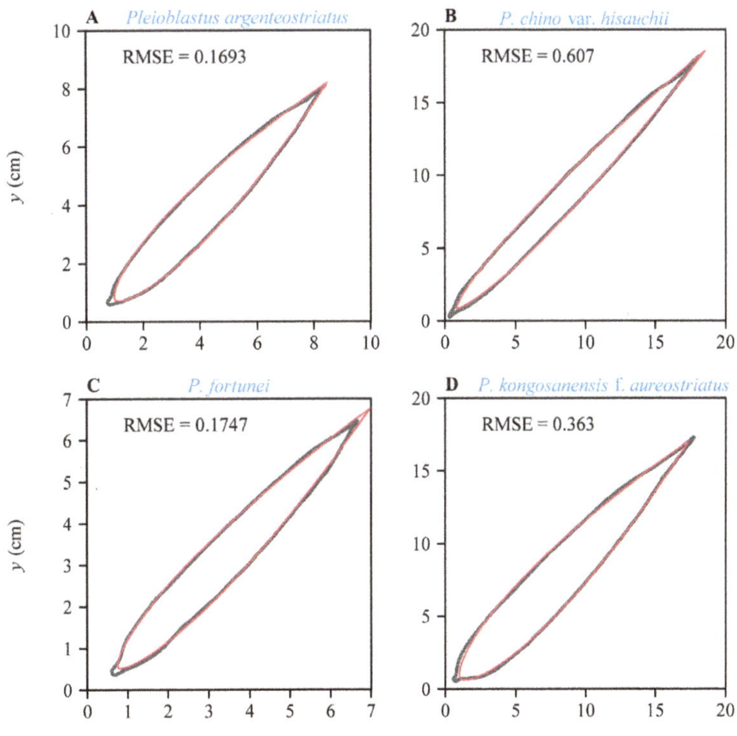

Figure 2. *Cont.*

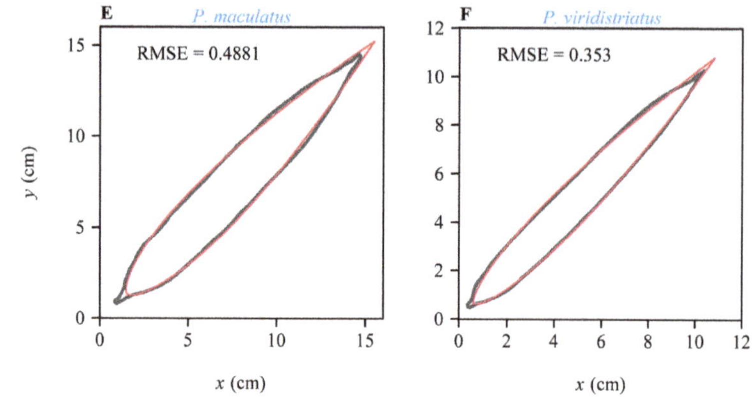

Figure 2. Illustration of the results of fitting the boundary coordinates of representative leaf samples for the six studied bamboo species (the same leaves as shown in Figure 1) using SGE-1. The gray curves are the actual scanned leaf boundaries; the red curves are the leaf boundaries predicted by the model SGE-1, i.e., Equation (3).

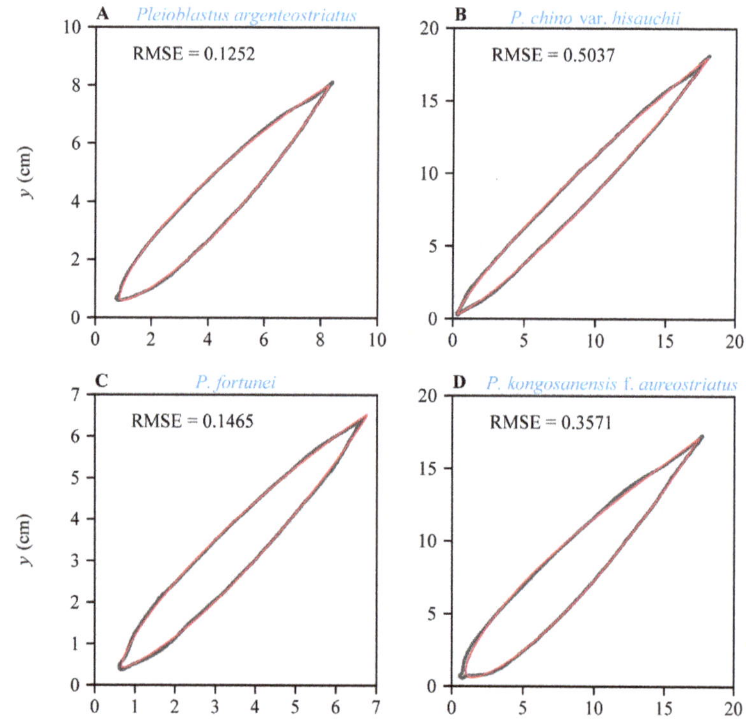

Figure 3. *Cont.*

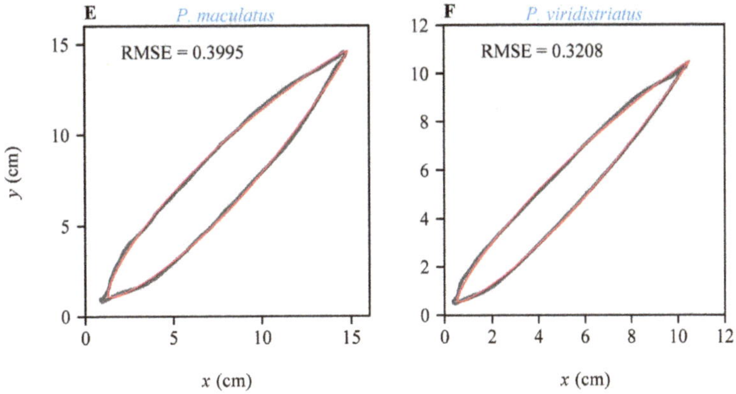

Figure 3. Illustration of the results of fitting the boundary coordinates of representative leaf samples for the six studied bamboo species (the same leaves as shown in Figure 1) using SGE-2. The gray curves are the actual scanned leaf boundaries; the red curves are the leaf boundaries predicted by the model SGE-2, i.e., Equation (4).

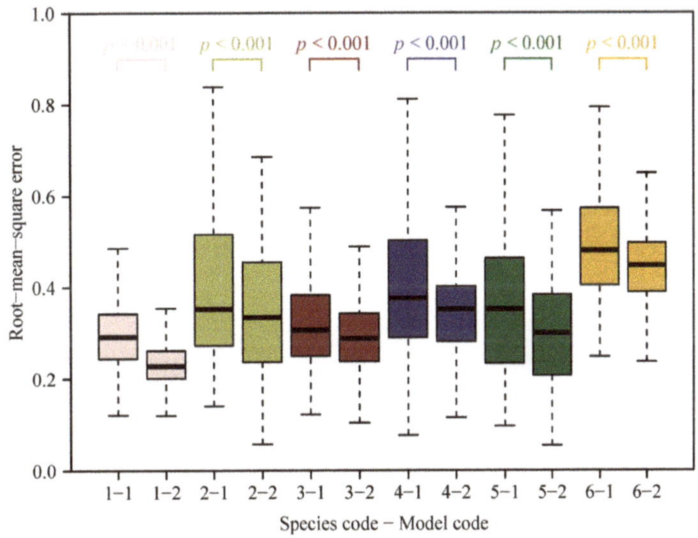

Figure 4. Comparison of the root–mean–square errors (RMSEs) between the two simplified Gielis models (SGE-1 and SGE-2, i.e., Equations (3) and (4)) for the studied six *Pleioblastus* species (Table 1 for species codes). The thick horizontal lines within the boxes represent median values of RMSEs; the box length represents the difference between the 3rd/4th quantile and the 1st/4th quantile; whiskers give 1.5 times the box length or maximum (or minimum) values. The two groups of RMSEs between the two models (1 and 2) for each species (1 to 6) were compared by a paired sample *t*-test.

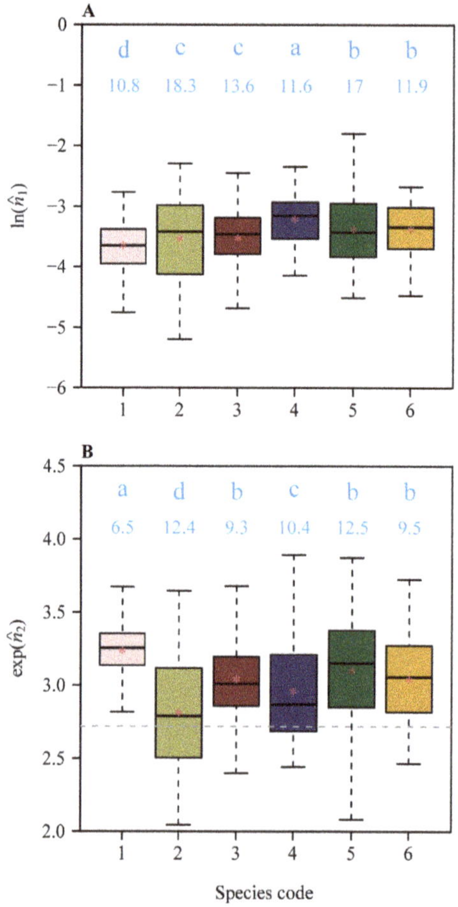

Figure 5. Comparisons of the log-transformed values of the model parameter n_1 (**A**) and the exp-transformed values of the model parameter n_2 (**B**) for the model SGE-2 (Equation (4)) for the six bamboo species (Table 1 for species codes). Different transformations reflect differences in the frequency distributions of the estimated parameter values (right-shewed for n_1 and left-skewed for n_2). In each panel, the lowercase letters show the significance of the differences in the estimated values between any two species at the 0.05 significance level. The numeric values at the top of each box provide the coefficients of variation (%). The horizontal solid line represents the median, and the red asterisk the mean. The whiskers provide the 1.5-fold interquartile range or maximum (or minimum) values. In (**B**), the horizontal gray dashed line shows exp(1).

4. Discussion and Conclusions

In the present study, we found that SEG-2 provided a better goodness of fit than SGE-1 in describing the shape of bamboo leaves. Shi et al. [19] found that SGE-2 also applies to the shape of avian eggs, but SGE-1 cannot reproduce the egg shapes of birds. In Equation (4), let us use an unknown parameter m to replace 1, i.e.,

$$r(\varphi) = a\left(\left|\cos\left(\frac{m}{4}\varphi\right)\right|^{n_2} + \left|\sin\left(\frac{m}{4}\varphi\right)\right|^{n_2}\right)^{-\frac{1}{n_1}} \quad (8)$$

Wang et al. [20] found that Equation (8) can describe the geometries of the outer rims of corolla tubes of *V. major* associated with the flowers that have five or four petals (where $m = 5$ and 4, respectively). Li et al. [26] found that Equation (8) is also applicable to the vertical projection's shape (in the top view) of *Koelreuteria paniculata* fruit when setting $m = 3$. This equation has more applications to other natural geometries owing to its rich symmetrical characteristics (e.g., the profiles of some sea stars) [17,18]. SGE-1 can be regarded as a special case of SGE-2, where SGE-2 is a special case of Equation (8). It will be valuable in future to further examine the validity of Equation (8) for more biological specimens from the same taxon, but with a large variation in morphology (i.e., diatoms, cross-sections of some plant stems that exhibit apparent radial symmetry).

It is necessary to point out that SGE-1 only has two model parameters (a, as the leaf size parameter, and n_1, as the leaf shape parameter), where n_1 is positively correlated with the ratio of leaf width to length [10,21]. That is, n_1 can be used in SGE-1 as a single leaf shape parameter to compare the differences in leaf shape across different bamboo species: a smaller n_1 value corresponds to a narrower leaf with a sharper leaf base, while a greater n_1 value corresponds to a broader leaf with a rounder leaf base [9,11]. Our data also confirm these results (Figure 6). However, there are two leaf shape parameters (i.e., n_1 and n_2) in SGE-2, which means it is not easy to explain the leaf shape variations within a species and across different species if we attempt to use n_1 and n_2 simultaneously for the quantification of leaf shape. In further analyses, we found that the leaf width/length ratio can be expressed as a function of n_1 and n_2 with a higher goodness of fit using the generalized additive models (e.g., models described in Refs. [27,28]). However, the interaction effect between the two parameters on data fitting is still difficult to explain (not shown due to the limitation of space). Thus, we suggest directly using the leaf width/length ratio to reflect or quantify the elongation change of leaf shape rather than using the two parameters in SGE-2. In fact, the leaf width/length ratio has been demonstrated to be closely correlated with the leaf fractal dimension [6]. The main role of SGE-2 in future research should not be to quantify the elongation change (accompanied with the change in tapering and curvature) of leaf shape, but it should be focused on the simulation of the intra- and interspecific variations in leaf shape based on the ranges of the two parameters' empirical estimated values. Another strength of SGE-2 is its ability to simulate a lanceolate leaf whose leaf area can be accurately calculated based on the parameters, and it is valuable in studying the effects of leaf shape and size on leaf structural, chemical and physiological differentiation [29].

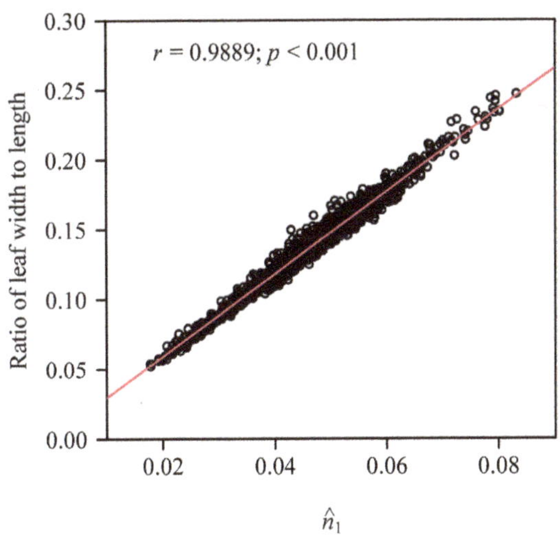

Figure 6. The correlation between the estimated values of the model parameter n_1 for SGE-1 and the ratios of leaf width to length.

In the present study, we compared the validity of two simplified Gielis equations (SGE-1 with two model parameters and SGE-2 with three model parameters) using 1996 leaves from six bamboo species, with more than 300 leaves measured for each species. We found that SGE-2 better characterizes the shape for each of all studied bamboo species. Although SGE-2 is more complex from the viewpoint of the model's structure than SGE-1, the mean percent errors for the six bamboo species were greater than 5%, which indicates that it is worthwhile to include an additional parameter in SGE-2 at the cost of increasing model complexity. Most numerical values of n_2 (1559 out of 1996) were greater than 1.0, further suggesting that an additional parameter needs to be incorporated. This work provides a versatile model tool for the description of the leaf shape of bamboo and other plant species with similar lanceolate leaves.

Supplementary Materials: The following supporting information can be downloaded at: https://www.mdpi.com/article/10.3390/plants11223058/s1, Table S1: The estimated values of parameters and goodness of fit using the SGE-1 to fit empirical leaf boundary data; Table S2: The estimated values of parameters and goodness of fit using the SGE-2 to fit empirical leaf boundary data.

Author Contributions: Methodology, Ü.N., J.G. and P.S.; formal analysis, Ü.N., W.Y. (Wenjing Yao), J.G. and P.S.; investigation, W.Y. (Weihao Yao), Ü.N., W.Y. (Wenjing Yao), J.G., J.S., K.Y. and P.S.; writing—original draft preparation, W.Y. (Weihao Yao) and P.S.; writing—review and editing, Ü.N., W.Y. (Wenjing Yao), J.G. and J.S.; funding acquisition, W.Y. (Wenjing Yao). All authors have read and agreed to the published version of the manuscript.

Funding: This research was funded by Jiangsu Forestry Science and Technology Innovation Project (grant number: LYKJ[2022]02).

Institutional Review Board Statement: Not applicable.

Informed Consent Statement: Not applicable.

Data Availability Statement: The data used in the present work have been listed in the online Supplementary Materials.

Acknowledgments: We thank Yabing Jiao for her valuable help during the preparation of this work.

Conflicts of Interest: The authors declare no conflict of interest.

References

1. Liese, W.; Köhl, M. *Bamboo: The Plant and Its Uses*; Springer: Berlin/Heidelberg, Germany, 2015.
2. Lin, S.; Niklas, K.J.; Wan, Y.; Hölscher, D.; Hui, C.; Ding, Y.; Shi, P. Leaf shape influences the scaling of leaf dry mass vs. area: A test case using bamboos. *Ann. For. Sci.* **2020**, *77*, 11. [CrossRef]
3. Schrader, J.; Shi, P.; Royer, D.L.; Peppe, D.J.; Gallagher, R.V.; Li, Y.; Wang, R.; Wright, I.J. Leaf size estimation based on leaf length, width and shape. *Ann. Bot.* **2021**, *128*, 395–406. [CrossRef] [PubMed]
4. Kincaid, D.T.; Schneider, R.B. Quantification of leaf shape with a microcomputer and Fourier transform. *Can. J. Bot.* **1983**, *61*, 2333–2342. [CrossRef]
5. Baxes, G.A. *Digital Image Processing: Principles and Applications*; John Wiley and Sons, Inc.: New York, NY, USA, 1994.
6. Shi, P.; Yu, K.; Niinemets, Ü.; Gielis, J. Can leaf shape be represented by the ratio of leaf width to length? Evidence from nine species of *Magnolia* and *Michelia* (Magnoliaceae). *Forests* **2021**, *12*, 41. [CrossRef]
7. Li, Y.; Quinn, B.K.; Niinemets, Ü.; Schrader, J.; Gielis, J.; Liu, M.; Shi, P. Ellipticalness index—A simple measure of the complexity of oval leaf shape. *Pak. J. Bot.* **2022**, *54*, 2233–2240. [CrossRef]
8. Dornbusch, T.; Watt, J.; Baccar, R.; Fournier, C.; Andrieu, B. A comparative analysis of leaf shape of wheat, barley and maize using an empirical shape model. *Ann. Bot.* **2011**, *107*, 865–873. [CrossRef]
9. Shi, P.; Xu, Q.; Sandhu, H.S.; Gielis, J.; Ding, Y.; Li, H.; Dong, X. Comparison of dwarf bamboos (*Indocalamus* sp.) leaf parameters to determine relationship between spatial density of plants and total leaf area per plant. *Ecol. Evol.* **2015**, *5*, 4578–4589. [CrossRef]
10. Shi, P.; Ratkowsky, D.A.; Li, Y.; Zhang, L.; Lin, S.; Gielis, J. A general leaf area geometric formula exists for plants—Evidence from the simplified Gielis equation. *Forests* **2018**, *9*, 714. [CrossRef]
11. Lin, S.; Zhang, L.; Reddy, G.V.P.; Hui, C.; Gielis, J.; Ding, Y.; Shi, P. A geometrical model for testing bilateral symmetry of bamboo leaf with a simplified Gielis equation. *Ecol. Evol.* **2016**, *6*, 6798–6806. [CrossRef]
12. Gielis, J.; Shi, P.; Caratelli, D. Universal equations—A fresh perspective. *Growth Form* **2022**. [CrossRef]
13. Gielis, J. A general geometric transformation that unifies a wide range of natural and abstract shapes. *Am. J. Bot.* **2003**, *90*, 333–338. [CrossRef]

14. Lamé, G. *Examen des Différentes Méthodes Employées Pour Résoudre les Problèmes de Géométrie*; V. Courcier: Paris, France, 1818.
15. Shi, P.; Gielis, J.; Quinn, B.K.; Niklas, K.J.; Ratkowsky, D.A.; Schrader, J.; Ruan, H.; Wang, L.; Niinemets, Ü. 'biogeom': An R package for simulating and fitting natural shapes. *Ann. N. Y. Acad. Sci.* **2022**, *1516*, 123–134. [CrossRef] [PubMed]
16. Tian, F.; Wang, Y.; Sandhu, H.S.; Gielis, J.; Shi, P. Comparison of seed morphology of two ginkgo cultivars. *J. For. Res.* **2020**, *31*, 751–758. [CrossRef]
17. Shi, P.; Ratkowsky, D.A.; Gielis, J. The generalized Gielis geometric equation and its application. *Symmetry* **2020**, *12*, 645. [CrossRef]
18. Wang, L.; Ratkowsky, D.A.; Gielis, J.; Ricci, P.E.; Shi, P. Effects of the numerical values of the parameters in the Gielis equation on its geometries. *Symmetry* 2022, *in press*.
19. Shi, P.; Gielis, J.; Niklas, K.J. Comparison of a universal (but complex) model for avian egg shape with a simpler model. *Ann. N. Y. Acad. Sci.* **2022**, *1514*, 34–42. [CrossRef]
20. Wang, L.; Miao, Q.; Niinemets, Ü.; Gielis, J.; Shi, P. Quantifying the variation in the geometries of the outer rims of corolla tubes of *Vinca major* L. *Plants* **2022**, *11*, 1987. [CrossRef]
21. Su, J.; Niklas, K.J.; Huang, W.; Yu, X.; Yang, Y.; Shi, P. Lamina shape does not correlate with lamina surface area: An analysis based on the simplified Gielis equation. *Glob. Ecol. Conserv.* **2019**, *19*, e00666. [CrossRef]
22. R Core Team. *R: A Language and Environment for Statistical Computing*; R Foundation for Statistical Computing: Vienna, Austria, 2022; Available online: https://www.R-project.org/ (accessed on 1 June 2022).
23. Yu, X.; Shi, P.; Schrader, J.; Niklas, K.J. Nondestructive estimation of leaf area for 15 species of vines with different leaf shapes. *Am. J. Bot.* **2020**, *107*, 1481–1490. [CrossRef]
24. Hsu, J.C. *Multiple Comparisons: Theory and Methods*; Chapman and Hall/CRC: New York, NY, USA, 1996.
25. Ratkowsky, D.A. *Handbook of Nonlinear Regression Models*; Marcel Dekker: New York, NY, USA, 1990.
26. Li, Y.; Quinn, B.K.; Gielis, J.; Li, Y.; Shi, P. Evidence that supertriangles exist in nature from the vertical projections of *Koelreuteria paniculata* fruit. *Symmetry* **2022**, *14*, 23. [CrossRef]
27. Hastie, T.; Tibshirani, R. *Generalized Additive Models*; Chapman & Hall: London, UK, 1990.
28. Wood, S.N. *Generalized Additive Models: An Introduction with R*, 2nd ed.; Chapman and Hall/CRC Press: New York, NY, USA, 2017.
29. Niinemets, Ü.; Portsmuth, A.; Tobias, M. Leaf shape and venation pattern alter the support investments within leaf lamina in temperate species, a neglected source of leaf physiological differentiation. *Funct. Ecol.* **2007**, *21*, 28–40. [CrossRef]

plants

Article

Responses of Forest Net Primary Productivity to Climatic Factors in China during 1982–2015

Ziqiang Du [1], Xuejia Liu [2], Zhitao Wu [1], Hong Zhang [3] and Jie Zhao [4,*]

1 Institute of Loess Plateau, Shanxi University, Taiyuan 030006, China
2 Shanxi Academy of Eco-Environmental Planning and Technology, Taiyuan 030000, China
3 College of Environmental & Resource Science, Shanxi University, Taiyuan 030006, China
4 College of Natural Resources & Environment, Northwest A & F University, Xianyang 712100, China
* Correspondence: sxuzhaojie@163.com

Abstract: Forest ecosystems play an important role in the global carbon cycle. Clarifying the large-scale dynamics of net primary productivity (NPP) and its correlation with climatic factors is essential for national forest ecology and management. Hence, this study aimed to explore the effects of major climatic factors on the Carnegie–Ames–Stanford Approach (CASA) model-estimated NPP of the entire forest and all its corresponding vegetation types in China from 1982 to 2015. The spatiotemporal patterns of interannual variability of forest NPP were illustrated using linear regression and geographic information system (GIS) spatial analysis. The correlations between forest NPP and climatic factors were evaluated using partial correlation analysis and sliding correlation analysis. We found that over thirty years, the average annual NPP of the forests was 887×10^{12} g C/a, and the average annual NPP per unit area was 650.73 g C/m^2/a. The interannual NPP of the entire forest and all its corresponding vegetation types significantly increased ($p < 0.01$). The increase in the NPP of evergreen broad-leaved forests was markedly substantial among forest types. From the spatial perspective, the NPP of the entire forest vegetation gradually increased from northwest to southeast. Over the years, the proportions of the entire forest and all its corresponding vegetation types with a considerable increase in NPP were higher than those with a significant decrease, indicating, generally, improvements in forest NPP. We also found climatic factors variably affected the NPP of forests over time considering that the rise in temperature and solar radiation improved the interannual forest NPP, and the decline in precipitation diminished the forest NPP. Such varying strength of the relationship between the interannual forest NPP and climatic factors also varied across many forest types. Understanding the spatiotemporal pattern of forest NPP and its varying responses to climatic change will improve our knowledge to manage forest ecosystems and maintain their sustainability under a changing environment.

Keywords: forest ecology; net primary productivity; varying response; climate change; China

Citation: Du, Z.; Liu, X.; Wu, Z.; Zhang, H.; Zhao, J. Responses of Forest Net Primary Productivity to Climatic Factors in China during 1982–2015. *Plants* **2022**, *11*, 2932. https://doi.org/10.3390/plants11212932

Academic Editors: Jie Gao, Weiwei Huang, Johan Gielis and Peijian Shi

Received: 9 October 2022
Accepted: 26 October 2022
Published: 31 October 2022

1. Introduction

As an essential part of terrestrial ecosystems, vegetation is a natural link between the atmosphere, soil, and water. It not only plays an important role in the global material and energy cycles but also reduces the concentration of greenhouse gases. Net primary productivity (NPP) refers to the rate at which vegetation fixes carbon dioxide from the atmosphere in the ecosystem through photosynthesis minus the rate at which vegetation returns the carbon dioxide to the atmosphere through respiration. NPP stands for the net carbon input from the atmosphere into vegetation. It, therefore, is an important measure of ecosystem carbon sinks, carbon sources, and global carbon balance [1,2]. Increases in temperature, in the frequency, intensity, and distribution of precipitation, and in the spatiotemporal distribution of solar radiation will inevitably have the most important impact on the development, formation, and evolution of terrestrial ecosystems. Changes in NPP can reflect the responses of ecosystems to climatic conditions, so it can serve as an indicator of the responses of ecosystem functions to climate change.

As a dominant vegetation type around most of the world, forests comprise the major component of terrestrial ecosystems [3] and cover 3999 M ha globally in 2015 [4]. They account for about 80% of terrestrial carbon storage on the ground and 40% of it underground [5,6]. Forests can retain existing carbon stocks and effectively increase carbon sinks [7,8]. China's forests represent a significant biomass carbon sink over the past several decades [9]. Because of the vast land area and heterogeneous natural conditions in China, its flora and forest types have diversified. The country harbors nearly all the types of forests in the northern hemisphere [10], ranging from boreal coniferous forests and broad-leaved forests, temperate deciduous broad-leaved forests, warm temperate or subtropical evergreen broad-leaved forests to tropical rainforests [11,12]. Its diverse climatic conditions and forest resources provide a good experimental base for studying the dynamics of the relationship between forest vegetation productivity and climate change [10,13].

Forest NPP is a key parameter that characterizes the functions of forest ecosystems, and it can be estimated to assess the development of forest ecosystems. The large-scale variations in forest NPP and its responses to climatic factors are increasingly gaining concern [2,10,14]. Large scale forest NPP estimates are of increasing interest. For example, Fang et al. [15] explored the spatial pattern of forest NPP in different provinces of China in terms of the third forest survey data. Zhou et al. [10] calculated China's Larix forest NPP based on forest inventory data. Wang et al. [16] measured the NPP in different forests at site level in northern China with the boreal ecosystem productivity simulator. Hasenauer et al. [17] estimated the Australian forest productivity by reconciling satellite with ground data. Cao et al. [14] estimated aboveground NPP in secondary tropical dry forests at the Santa Rosa National Park, Costa Rica using the Carnegie–Ames–Stanford Approach (CASA) model. Tripathi et al. [18] estimated NPP in tropical forest plantations of India during 2009 and 2010 using CASA model. NPP is influenced by different factors, especially climate variables, as the main factors that affect the development and ranges of forest ecosystems. For instance, Peng et al. [19] reported that climate change led to an increase in the NPP of boreal forests in central Canada. Schuur and Matson [20] found that the underground NPP decreased 2.2 times with the increase in average annual precipitation in the mountain forests of Hawaii. Mohamed et al.. [21] stated that the variability in NPP of global ecosystems particularly forests and grasslands was attributed to global anomalies in temperature, precipitation and cloud cover. Cleveland et al. [22] noted that the mean annual temperature could be the strongest predictor of the ground NPP of all tropical forests, but this relationship may be caused by the apparent temperature difference between highland and lowland forests. Reyer et al. [23] suggested that future forest productivity would be subject to climate change, which in turn would largely depend on climate scenarios and the sustainability of CO_2 impacts. Yao et al. [9] quantified Chinese forest biomass carbon sequestration capacity in the near future integrating the effects of stand development, climate change, and increasing CO_2 concentration. Although previous studies have focused mainly on either forest NPP or the impact of climatic factors on forest NPP, most of them were limited at specific regions or within shorter periods [24]. Moreover, previous researchers have paid relatively little attention to the varying responses of vegetation growth (e.g., forest NPP) to climate variables [25].

We aimed to investigate the spatiotemporal variations in China's forest NPP and explored the dynamic responses of forests in terms of their NPP to major climatic factors from 1982 to 2015, which were important for sustainable forest ecology and management.

2. Data and Methods

2.1. Data Sources

2.1.1. Remote Sensing Data

We utilized the Global Inventory Modeling and Mapping Studies (GIMMS) Normalized Difference Vegetation Index-3rd generation (NDVI3g) remote sensing data generated by the National Oceanic and Atmospheric Administration (NOAA) Advanced Very High-resolution Radiometer (AVHRR) and provided by the National Aeronautics and Space Administration (NASA). This data set from 1982 to 2015, with 15-day time resolution and 0.083° spatial resolution, is the longest sequence of NDVI data and has been widely used in the estimation and investigation of large-scale vegetation dynamics, vegetation NPP, and biomass [26]. The effects of cloud and atmospheric interferences were eliminated and the monthly NDVI data were processed through the maximum-value composite (MVC) method [27].

2.1.2. Meteorological Data

The monthly data, such as total solar radiation, average temperature, and precipitation over the same period as the remote sensing data, were obtained from the records of meteorological stations (Figure S1) in the China Meteorological Science Data Sharing Service System (CMSDS). Spatial interpolation was performed through inverse distance weighting (IDW) to generate meteorological grid data with spatial resolution and projection similar to that of the NDVI data [28,29].

2.1.3. Forest-Type Data

Forest data were extracted from China's 1:1,000,000 vegetation map provided by the Environmental and Ecological Science Data Center for West China, National Natural Science Foundation of China (http://westdc.westgis.ac.cn (accessed on 9 October 2022)). Forest vegetation types (Figure S1) included evergreen broad-leaved forest, deciduous broad-leaved forest, evergreen coniferous forest, deciduous coniferous forest, and mixed forest [30]. The distribution of evergreen broad-leaved forests is mostly in subtropical regions; deciduous broad-leaved forests are in temperate regions; evergreen coniferous forests are mostly south of the Qinling–Huaihe line; deciduous coniferous forests are in Northeastern and Northwestern China; and mixed forests are mostly in the Changbai Mountains, Xiaoxinganling Mountains, and subtropical mountains.

2.2. Methods

2.2.1. CASA Model

NPP prediction models have become powerful alternative tools for investigating the NPP scale and geographic distribution of vegetation because the direct measurement of terrestrial vegetation NPP at the regional or global scale has been difficult [31]. The Carnegie–Ames–Stanford Approach (CASA) model is an NPP simulation model considered more realistic than others [1,14,32]. We used it in this study to estimate the NPP of the entire forest and the forest vegetation types and calculated it as follows:

$$NPP(x,t) = APAR(x,t) \times \varepsilon(x,t) \tag{1}$$

where $APAR(x, t)$ represents the photosynthetically active radiation (PAR, in units of MJ/m^2) absorbed at pixel x in month t, and $\varepsilon(x, t)$ represents the actual light energy utilization at pixel x in month t (g C/MJ). APAR was computed as follows:

$$APAR(x,t) = SOL(x,t) \times FPAR(x,t) \times 0.5 \tag{2}$$

where $SOL(x, t)$ represents the total solar radiation (MJ/m^2) at pixel x in month t; $FPAR(x, t)$ represents the fraction of photosynthetically active radiation (FPAR) absorbed by the vegetation layer; and 0.5 represents the proportion of the effective solar radiation relative to the total solar radiation and with a wavelength of 0.4–0.7 μm that can be utilized by vegetation [33]. FPAR is represented by NDVI and vegetation type, and it does not exceed 0.95 in the equation [34]:

$$FPAR = min\left(\frac{SR(x,t) - SR_{min}}{SR_{max} - SR_{min}}, 0.95\right) \tag{3}$$

where $SR(x, t)$ represents the ratio index at pixel x in month t; SR_{min} is 1.08, and SR_{max}, ranging from 4.14 to 6.17, is related to vegetation type. We calculated $SR(x, t)$ using $NDVI(x, t)$ as follows:

$$SR(x,t) = \frac{1 + NDVI(x,t)}{1 - NDVI(x,t)} \tag{4}$$

The light energy utilization rate (ε), subject to temperature and water conditions, refers to the efficiency of the conversion of PAR absorbed by vegetation into organic carbon. We computed it as follows:

$$\varepsilon(x,t) = T_{\varepsilon1}(x,t) \times T_{\varepsilon2}(x,t) \times W_{\varepsilon}(x,t) \times \varepsilon_{max} \tag{5}$$

where $T_{\varepsilon1}(x, t)$ and $T_{\varepsilon2}(x, t)$ represent the effects of temperature on ε; W_{ε} represents the effect of water on ε, and ε_{max} represents the maximum ε under ideal conditions (g C/MJ).

Estimation of temperature stress factors: $T_{\varepsilon1}(x, t)$ and $T_{\varepsilon2}(x, t)$ reflect the effects of temperature on ε. We calculated the former as follows:

$$T_{\varepsilon1}(x,t) = 0.8 + 0.02 \times T_{opt}(x) - 0.0005 \times \left[T_{opt}(x)\right]^2 \tag{6}$$

where $T_{opt}(x)$ represents the average temperature (°C) of the month when the NDVI of a certain area reaches the maximum in a year. If the average temperature of a month is less than or equal to -10 °C, $T_{opt}(x)$ is set to 0.

$$T_{\varepsilon2}(x,t) = \frac{1.184}{\{1 + \exp\left[0.2 \times \left(T_{opt}(x) - 10 - T(x,t)\right)\right]\}} \times \frac{1}{\{1 + \exp\left[0.3 \times \left(-T_{opt}(x) - 10 + T(x,t)\right)\right]\}} \tag{7}$$

If $T(x, t)$, the average temperature of a month is 10 °C higher or 13 °C lower than $T_{opt}(x)$; then, the $T_{\varepsilon2}(x, t)$ value for this month is half the value of $T_{\varepsilon2}(x, t)$ for which $T(x, t)$ is $T_{opt}(x)$.

Estimation of water stress factors: $W_{\varepsilon}(x, t)$ reflects the effect of water on the light utilization efficiency of plants. It gradually increases with the rise in effective water in the environment, and its value ranges from 0.5 (under extreme drought conditions) to 1 (under very wet conditions). We calculated it as follows:

$$W_{\varepsilon}(x,t) = 0.5 + 0.5 \times E(x,t)/E_p(x,t) \tag{8}$$

where $E(x, t)$ represents the actual evapotranspiration (mm) of a region based on the actual evapotranspiration model of Zhou et al. [10], and $E_p(x, t)$ represents the potential evapotranspiration of a region based on the complementary relationship proposed by Boucher [35].

Determination of ε_{max}: We determined the value of ε_{max} through Zhu's method of simulating the ε_{max} of different vegetation types based on the principle of error minimization [36].

2.2.2. Linear Regression Analysis

The annual change rate of NPP based on the total annual NPP of forest vegetation in the same period was calculated through a linear regression model [37,38]. The pixel-by-pixel change rate of NPP was calculated based on the average annual forest NPP for all the pixels from 1982 to 2015, as shown in the equation:

$$y = \alpha t + \beta + \varepsilon \tag{9}$$

where t is a year in the time series; α is the regression coefficient indicating the annual change rate of NPP; β is the regression constant; and ε is the fitted residual. A value of $p < 0.05$ (or 0.01) indicated that a linear regression coefficient was significantly based on the t-test.

2.2.3. Partial Correlation Analysis

The interference of other variables on the impacts of climatic factors on NPP was eliminated through second-order partial correlation analysis [39,40]. The pixel-by-pixel partial correlation coefficient (r) of forest NPP was calculated with total solar radiation, temperature, and precipitation. The partial correlation coefficient between NPP and temperature was calculated while solar radiation and precipitation were fixed, whereas that between NPP and precipitation was determined while solar radiation and temperature were set, and that between NPP and solar radiation was computed while temperature and precipitation were constant [41].

2.2.4. Sliding Partial Correlation Analysis

The sliding correlation coefficient can be used to investigate the variations in the impacts of climatic factors on vegetation NPP [40]. The 10–20a is considered to be the appropriate range of years for a moving window in sliding correlation analysis [42]. Thus, in order to determine the temporal variations in the relationship between vegetation NPP and climate despite the limited study period, we selected 17a as the sliding stride to calculate the second-order partial correlation coefficient of forest NPP with temperature ($R_{NPP\text{-}T}$), precipitation ($R_{NPP\text{-}P}$), and total solar radiation ($R_{NPP\text{-}S}$). A sequence for each correlation coefficient from 1982 to 1998, 1983 to 1999, ... , and 1999 to 2015 was obtained. Thus, the variations in the correlation coefficient between the NPP of China's forest vegetation and the climatic factors with time were analyzed, and statistical significance tests were performed for such changes. Values of $p < 0.05$ indicated that the responses of forest NPP to climatic factors were statistically significant.

3. Results

3.1. Temporal Variations in Forest NPP

The total NPP of China's forest vegetation from 1982 to 2015 was between 770×10^{12} g C/a and 965×10^{12} g C/a, and the average annual NPP was 887×10^{12} g C/a. From the perspective of different vegetation types, the NPP per unit area of evergreen broad-leaved forests was the highest (1323.71 g C/m^2/a), followed by mixed forests (832.06 g C/m^2/a) and deciduous broad-leaved forests (637.21 g C/m^2/a). The NPP per unit area of evergreen coniferous forests and deciduous coniferous forests was 497.59 g C/m^2/a and 442.35 g C/m^2/a, respectively. All five types of forests showed a significantly increasing trend in NPP during the study periods (Figure 1). The NPP growth rate of evergreen broad-leaved forests was the highest (4.78 g C/m^2/a; $p < 0.01$), followed by mixed forests (3.64 g C/m^2/a; $p < 0.01$), deciduous broad-leaved forests (2.42 g C/m^2/a; $p < 0.01$), and evergreen coniferous forests (2.35 g C/m^2/a; $p < 0.01$), whereas that of deciduous coniferous forests was the lowest (1.65 g C/m^2/a; $p < 0.01$). The total NPP of forest vegetation in China extensively increased by 11.67% from 848.86×10^{12} g C in 1982 to 947.89×10^{12} g C in 2015 and at a linear rate of 3.558×10^{12} g C/a per year ($p < 0.01$) over thirty years (Figure 1).

3.2. Spatial Variations in Forest NPP

The annual difference in forest NPP from 1982 to 2015 was between −24.23 g C/m^2/a and 35.59 g C/m^2/a, with an average growth rate of 2.36 g C/m^2/a. The NPP of 89.25% of the total forest area increased, whereas that of only 10.45% decreased. About 51.40% of forest vegetation in the Changbai Mountains, Sichuan–Shaanxi–Gansu bordering region, the Xiaoxinganling Mountains, the northern Daxinganling Mountains, the Han River valley, southern Yunnan, and Eastern Taiwan showed a substantial increase ($p < 0.05$) in their NPP. Only 1.33% of forest vegetation in the Bhareli River and Subansiri River regions in Southeastern Tibet showed a noteworthy decrease ($p < 0.05$) in their NPP. Our results demonstrated that the area of forest vegetation with increasing NPP was greater than that with decreasing NPP and that the NPP of the entire forest increased over thirty years (Figure 2).

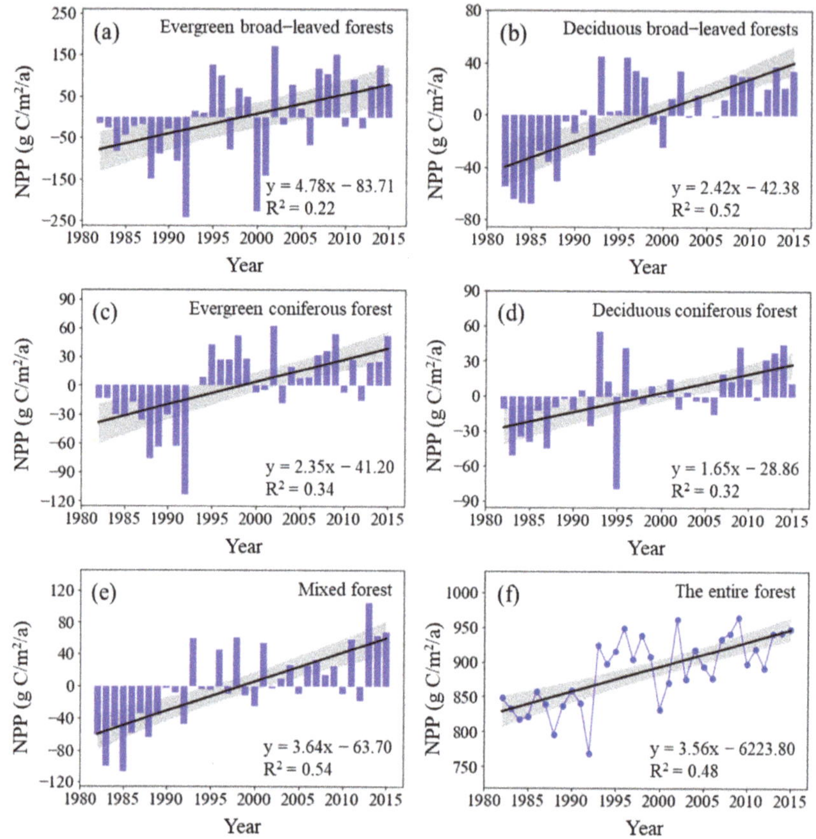

Figure 1. Interannual change in NPP across forest types from 1982 to 2015 (Here, the vertical solid rectangles represent NPP anomalies across forest types; the black lines represent the interannual change trend in forest NPP). Shading denotes 95% prediction intervals.

Based on the annual difference in NPP per unit area of the forest types, the areas of mixed forests with significant increase ($p < 0.05$) comprised the largest proportion (83.33%) of the total area of mixed forests; those of deciduous broad-leaved forests and deciduous coniferous forests with substantial increase both exceeded 60%; whereas those of evergreen broad-leaved forests and evergreen coniferous forests with considerable increase accounted for 44.06% and 33.89%, respectively. On the other hand, the proportions of all forest vegetation types with a notable decrease ($p < 0.05$) in NPP were low. The area of mixed forests with significant decrease in NPP per unit area was almost zero; all those of deciduous broad-leaved forests, evergreen coniferous forests, and deciduous coniferous forests with substantial decrease in NPP per unit area were less than 1%; whereas those of evergreen broad-leaved forests with noteworthy decrease in NPP per unit area comprised the largest proportion that was only 6.67%. Our findings showed that the area of each forest vegetation type with a significant increase ($p < 0.05$) in NPP per unit of area covered a proportion larger than that with significant decrease ($p < 0.05$), and that the NPP of the five forest types generally increased during the study periods.

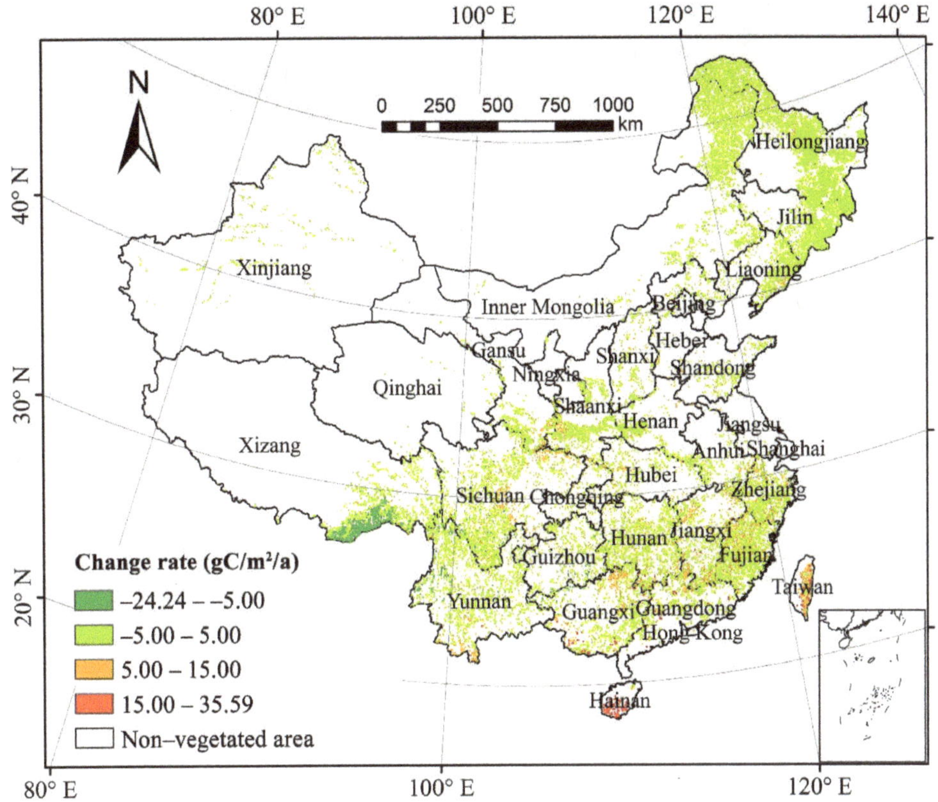

Figure 2. Spatial distribution of China's forest NPP change from 1982 to 2015.

3.3. Interannual Correlation between Forest NPP and Climatic Factors

3.3.1. Total Annual NPP and Temperature

The partial correlation coefficient of the total annual forest NPP and mean annual temperature (MAT) was 0.558 at a 0.01 significant level, suggesting that forest NPP was significantly correlated with MAT in the past three decades (Figure 3). Spatially, the areas with a positive correlation between total forest NPP and MAT accounted for 70.92% of the total forest area (Table 1). Approximately 8.60% of the areas in the northeastern region of the Yunnan–Guizhou Plateau, Chongqing, the southeastern region of Sichuan Basin, and western Hunan passed the significance test ($p < 0.05$). The areas with negative correlation between forest NPP and MAT accounted for 29.08%, and those with significantly negative correlation were mainly distributed in southern Hainan and the Bhareli River as well as in the Subansiri River in Tibet. The areas in southeastern Jilin and southern Hainan that passed the significance test ($p < 0.05$) accounted for only 3.24%.

The pixels of the mixed forests with significant correlation between the NPP of different forest types and MAT ($p < 0.05$) accounted for 36.01%, which is the greatest proportion, of the total pixels of mixed forests; those of green broad-leaved forests ($p < 0.05$) accounted for 20.27%; those of deciduous broad-leaved forests and evergreen coniferous forests ($p < 0.05$) both comprised greater than 10%; whereas those of deciduous broad-leaved forests ($p < 0.05$) accounted for slightly higher than 7%. The pixels of other forest types, except the deciduous broad-leaved forests, with the positive correlation between NPP and MAT, generally constituted proportions greater than those with negative correlation, indicating that temperature evidently altered the NPP across different forest types.

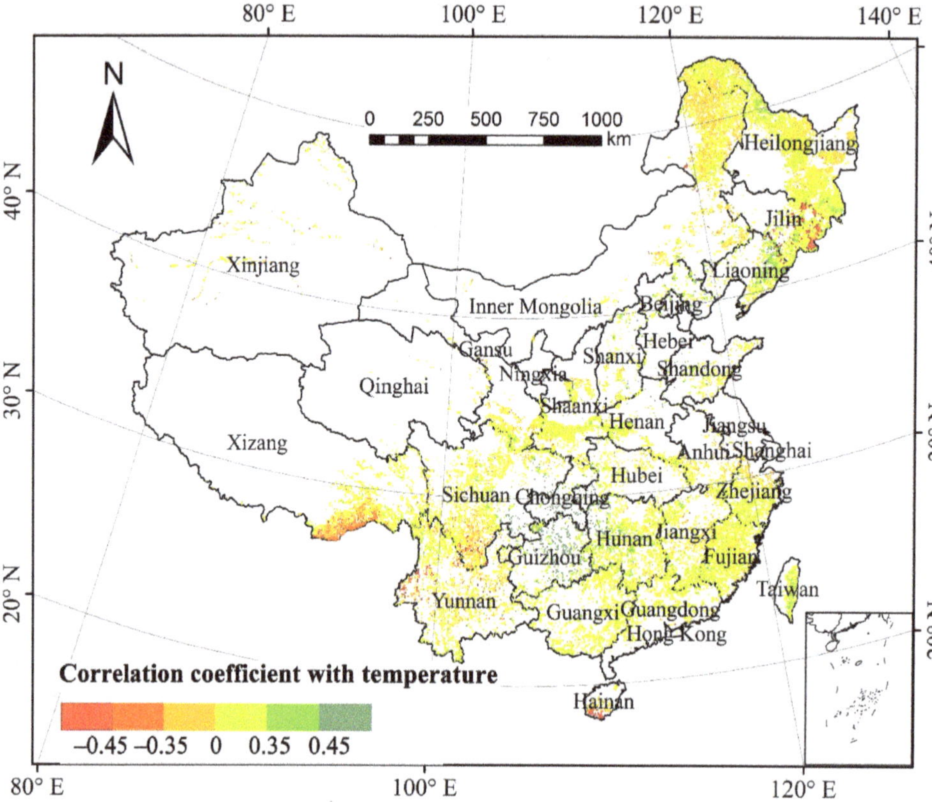

Figure 3. Spatial distribution of partial correlation coefficients between forest NPP and mean annual temperature from 1982 to 2015.

Table 1. Percentages of the area with significant partial correlations ($p < 0.05$) between forest NPP and climatic factors. The numbers in parentheses indicate the percentage of area with significant partial correlations between forest NPP and climatic factors.

Climatic Factors	Positive	Negative
Temperature	70.92% (8.60%)	29.08% (3.24%)
Precipitation	88.10% (31.72%)	11.90% (0.28%)
Solar radiation	98.89% (80.22%)	1.11% (0.03%)

3.3.2. Total Annual NPP and Precipitation

The partial correlation coefficient of the total annual forest NPP and the mean annual precipitation (MAP) was 0.167 ($p > 0.05$), suggesting that there was no significant correlation between the total annual forest NPP and MAP in the past three decades (Figure 4). Spatially, the areas with a positive correlation between forest NPP and MAP accounted for 88.10% of the total forest area (Table 1). About 31.72% of the areas, mainly in the Tibet–Yunnan–Sichuan bordering regions, eastern and western Liaoning, southeastern Guizhou, the Daxinganling Mountains, southeastern Liaoning, the Xiaoxinganling Mountains, Shandong, the Changbai Mountains, Fujian, Yunnan, eastern Guizhou, southern Jiangxi, and eastern Guangdong showed a significantly positive correlation ($p < 0.05$). The areas with negative correlation between forest NPP and MAP accounted for 11.90%. Only 0.28% of the areas mainly in northern Sichuan passed the significance test ($p < 0.05$).

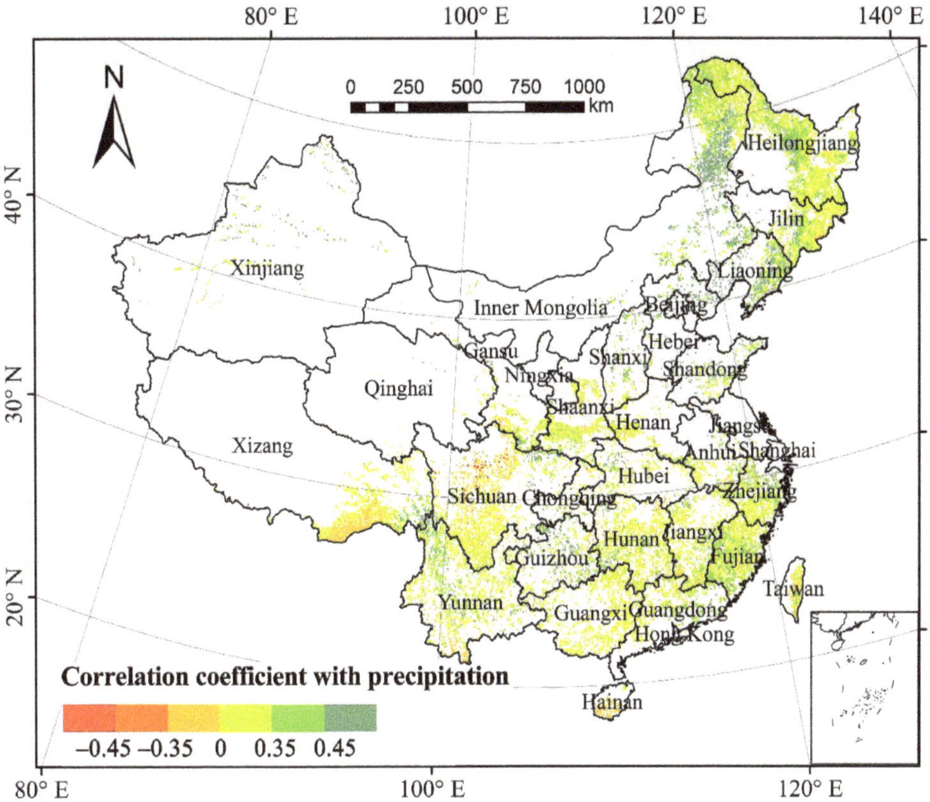

Figure 4. Spatial distribution of partial correlation coefficients between forest NPP and mean annual precipitation from 1982 to 2015.

Based on the proportion of pixels of forest type with correlation between NPP and MAP relative to the total pixels of the same vegetation type, the NPP of 49.47% of deciduous coniferous forests, 38.54% of deciduous broad-leaved forests, 32.96% of mixed forests, 25.19% of evergreen coniferous forests, and 12.66% of evergreen broad-leaved forests were significantly affected by precipitation ($p < 0.05$). All the areas in the five forest vegetation types with a positive correlation between NPP and precipitation comprised a proportion greater than those with a negative correlation. Hence, precipitation minimally affected the forest NPP in the study periods.

3.3.3. Total Annual NPP and Total Solar Radiation

The partial correlation coefficient of the total annual NPP of forest vegetation and total solar radiation (TSR) was 0.476 at 0.05 significant level, suggesting that the total annual forest NPP was significantly correlated with TSR in the past three decades (Figure 5). Spatially, the areas with a positive correlation between forest NPP and TSR accounted for 98.89% of the total forest area (Table 1). About 80.22% of the areas, mainly in the Daxinganling Mountains, eastern Liaoning, Han River valley, Xiaoxinganling Mountains, and south of the Yangtze River, passed the significance test ($p < 0.05$). The areas with the negative correlation between forest NPP and TSR accounted for 1.11%. Those mainly in Sichuan, northwestern Yunnan, and Xinjiang that passed the significance test ($p < 0.05$) only accounted for 0.03% of the total forest area.

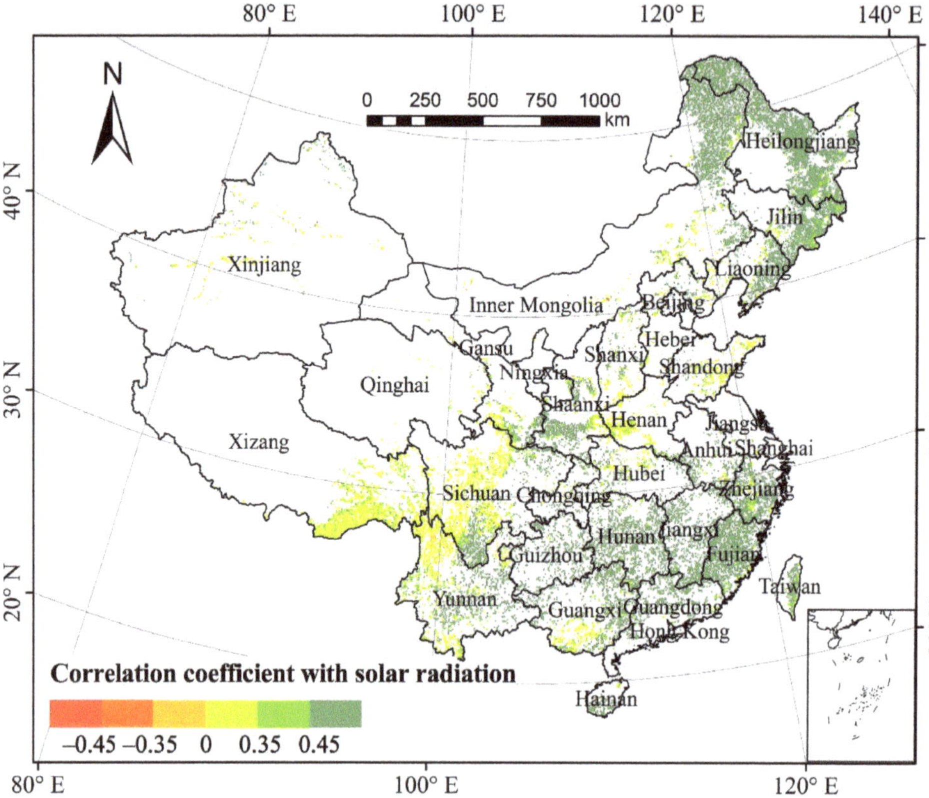

Figure 5. Spatial distribution of partial correlation coefficients between forest NPP and total solar radiation from 1982 to 2015.

Based on the proportion of pixels of each forest type showing correlation between NPP and TSR relative to the total pixels of the corresponding forest type, the proportion of areas across forest types with significantly positive correlation ($p < 0.05$) exceeded 60%, and those with significantly negative correlation ($p < 0.05$) were almost zero. Thus, the increase in solar radiation notably facilitated the growth of forest NPP in the study periods.

3.4. Varying Responses of Forest NPP to Climate Change

3.4.1. Dynamic Relationship between Forest NPP and Temperature

The sliding correlation coefficient R_{NPP-T} between the NPP of the entire forest from 1982 to 2015 and MAT significantly increased ($p < 0.01$) (Figure 6). Similarly, the R_{NPP-T} for deciduous broad-leaved forests and mixed forests also considerably increased ($p < 0.01$). Although the R_{NPP-T} for deciduous coniferous forests increased, it failed the significance test ($p > 0.05$). The R_{NPP-T} for evergreen broad-leaved forests and evergreen deciduous forests decreased, and it failed the significance test ($p > 0.05$).

Considering different time periods, the positive correlation between forest NPP and MAT gradually weakened from the 1982–1998 period to the 1985–2001 period, but it steadily strengthened after 1986–2002 and slowly declined again after 1993–2009 until 1999–2015. The deciduous broad-leaved forests showed an increasingly positive correlation between NPP and MAT, whereas the evergreen coniferous forests and evergreen broad-leaved forests demonstrated no significant change in their corresponding positive correlation. The pertinent correlation in the deciduous coniferous forests and mixed forests changed from negative to positive and continuously increased.

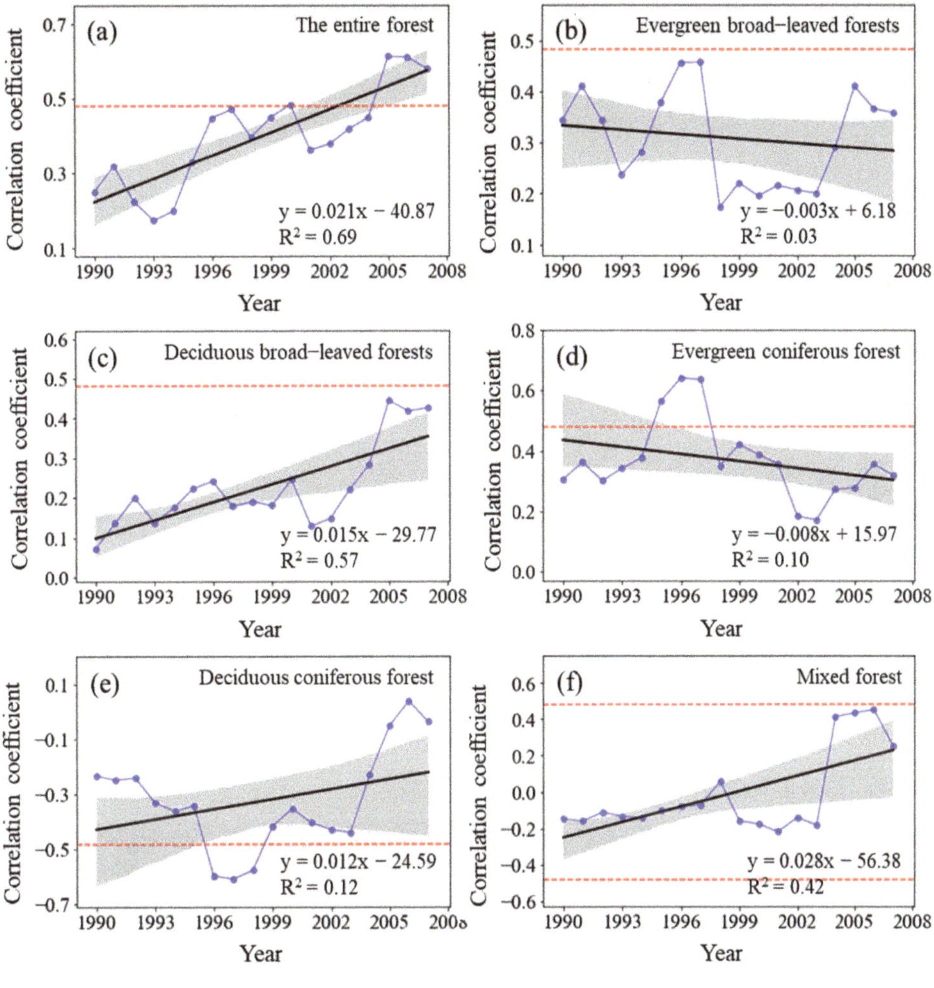

Figure 6. Variations in the sliding correlation coefficients between NPP and mean annual temperature across forest types. Statistically significant partial correlation coefficients are indicated as dotted lines ($p < 0.05$). The black lines represent the interannual change trend in partial correlation coefficients. Shading denotes 95% prediction intervals.

3.4.2. Dynamic Relationship between Forest NPP and Precipitation

The sliding correlation coefficient $R_{NPP\text{-}P}$ between the NPP of the entire forest from 1982 to 2015 and MAP showed a fluctuating downward trend ($p > 0.05$) (Figure 7). The $R_{NPP\text{-}P}$ of evergreen broad-leaved forests significantly decreased ($p < 0.01$), whereas that of deciduous broad-leaved forests substantially increased ($p < 0.01$). The $R_{NPP\text{-}P}$ of evergreen coniferous forests increased ($p > 0.05$), whereas that of deciduous coniferous forests and mixed forests decreased ($p > 0.05$).

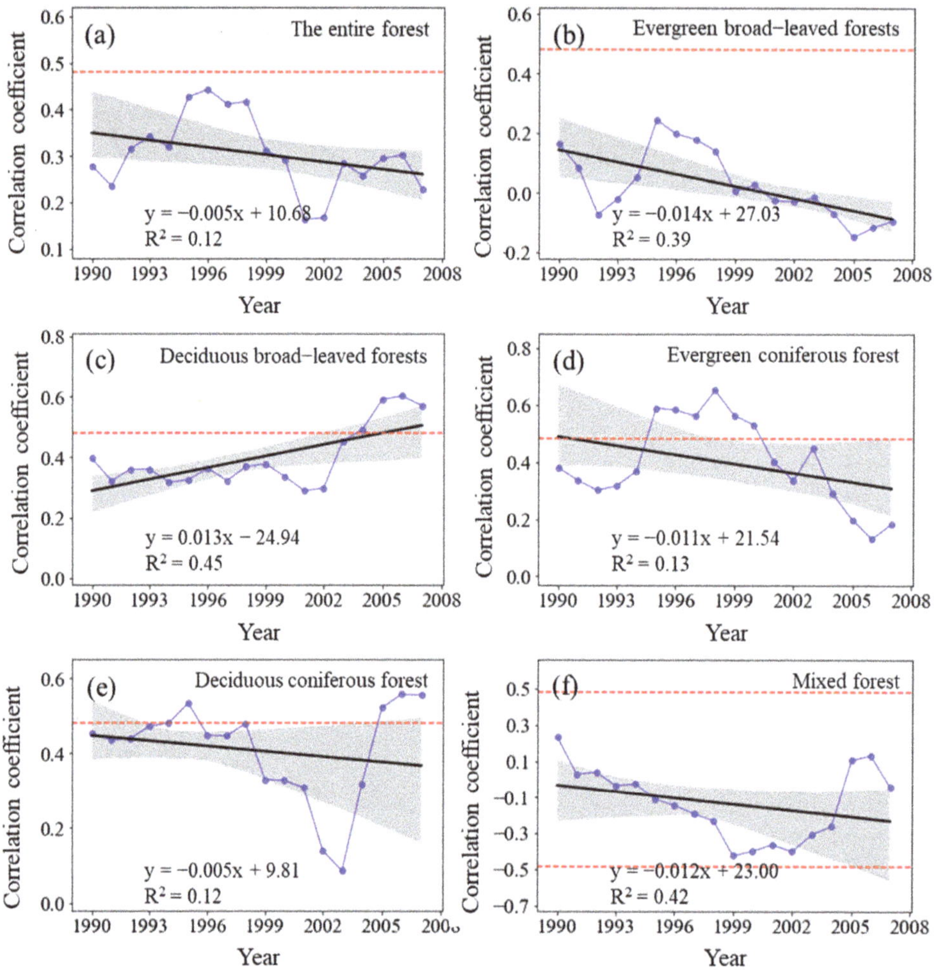

Figure 7. Variations in the sliding correlation coefficients between NPP and mean annual precipitation across forest types. Statistically significant partial correlation coefficients are indicated as dotted lines ($p < 0.05$). The black lines represent the interannual change trend in partial correlation coefficients. Shading denotes 95% prediction intervals.

The $R_{NPP\text{-}P}$ in all the periods showed positive correlation and failed the significance test of $p < 0.05$, but the underlying change was periodic. The $R_{NPP\text{-}P}$ increased from the 1982–1998 period to the 1993–2009 period, gradually decreased from the 1988–2004 period to the 1996–2012 period, and steadily increased again after 1994–2010, but it did not reach a significantly positive correlation. The $R_{NPP\text{-}P}$ of evergreen broad-leaved forests changed from positive to negative, but it was not statistically significant, whereas that of deciduous broad-leaved forests and evergreen coniferous forests was positive in all the study periods. Nevertheless, the changes varied as the $R_{NPP\text{-}P}$ of deciduous broad-leaved forests increased annually, whereas that of evergreen coniferous forests decreased yearly. The $R_{NPP\text{-}P}$ of deciduous coniferous forests similarly changed, and that of the mixed forests changed from positive to negative.

3.4.3. Dynamic Relationship between Forest NPP and Solar Radiation

The sliding correlation coefficient $R_{NPP\text{-}S}$ between the NPP of the entire forest from 1982 to 2015 and TSR did not significantly change ($p > 0.05$) (Figure 8) and the $R_{NPP\text{-}S}$ was positive in each sliding window. The correlation between forest NPP and TSR was significantly positive from 1982 to 1998 ($p < 0.05$), but it gradually weakened from the 1983–1999 period to 1989–2005 period. Its correlation coefficient reached a significantly positive correlation ($p < 0.05$) from the 1990–2006 period to the 1992–2008 period. It sharply declined but remained positively correlated from 1993 to 2009, after which it gradually increased but did not show a significantly positive correlation ($p < 0.05$) until 1998–2014. However, the notable decrease in the $R_{NPP\text{-}S}$ of all the forest types demonstrated the dynamic responses of forest NPP to TSR. The $R_{NPP\text{-}S}$ of evergreen broad-leaved forests and deciduous broad-leaved forests significantly decreased ($p < 0.01$).

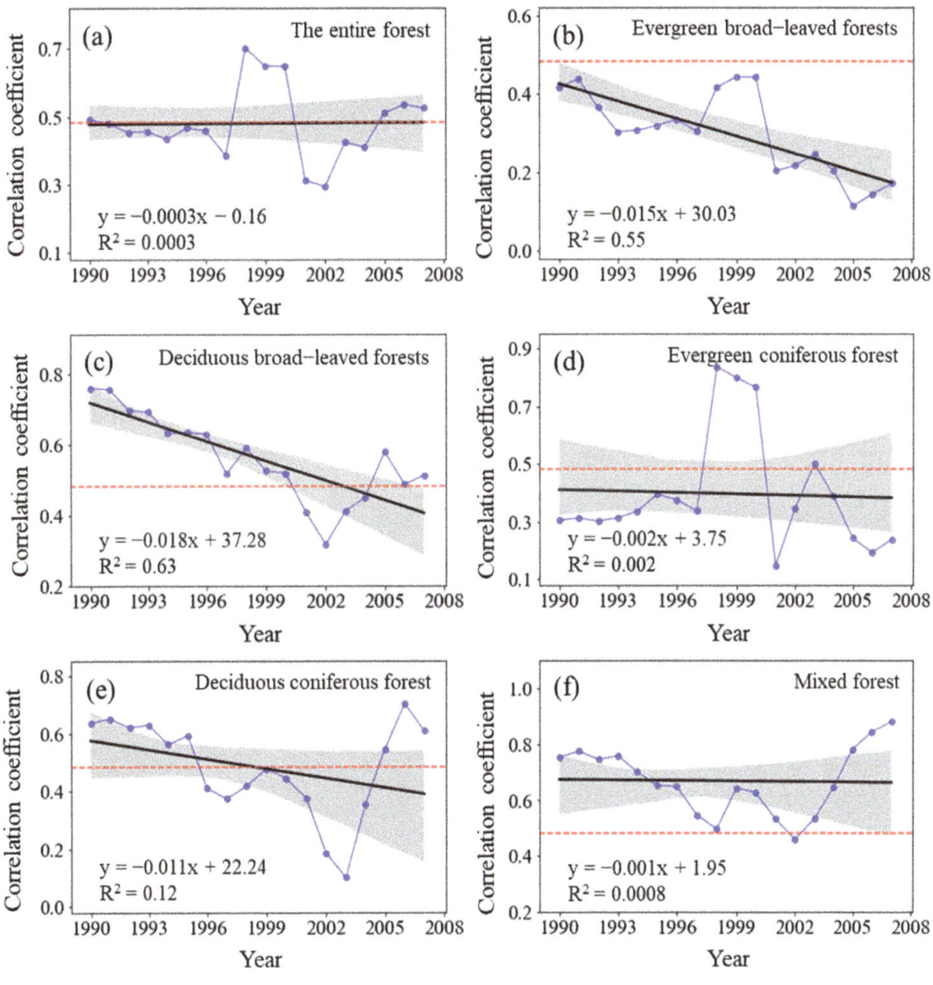

Figure 8. Variations in the sliding correlation coefficients between NPP and total solar radiation across forest types. Statistically significant partial correlation coefficients are indicated as dotted lines ($p < 0.05$). The black lines represent the interannual change trend in partial correlation coefficients. Shading denotes 95% prediction intervals.

4. Discussion

In this study, we simulated China's forest NPP and investigated the spatiotemporal patterns of forest NPP from 1982 to 2015. Even though the CASA model has been widely applied to NPP in China [1,32], the calculation still needs to be verified. Thus, we compared the NPP calculated from the CASA model with other simulation results in the literature. The mean annual NPP estimated for the entire forest (887×10^{12} g C/a) in this study was similar to the forest biomass given by Zhan et al. [43] (approximately 840.3×10^{12} g C/a) and Ni [12] (738.9×10^{12} g C/a). However, it was about twice the estimate of Fang et al. [11] (461.0×10^{12} g C/a) between 1949 and 1998. The average NPP per unit area estimated for the entire forest in this study was slightly lower than that of the MODIS MOD17A3 products (666.19 g C/m^2/a) with a spatial resolution of 1000 m at an annual interval (https://modis.gsfc.nasa.gov/ (accessed on 9 October 2022)), which is commonly used to explore the spatiotemporal patterns of NPP on regional and global scales. Moreover, our estimate of the NPP per unit area of evergreen broad-leaved forests was close to that reported by Wu et al. [44] (1327.22 g C/m^2/a); that of deciduous broad-leaved forests was consistent with those provided by Ni [12] (671.80 g C/m^2/a), Liang et al. [13] (688.50 g C/m^2/a), and Zhu et al. [36] (642.90 g C/m^2/a); that of evergreen coniferous forests was similar to those determined by Wu et al. [44] (515.69 g C/m^2/a), Jiang et al. [45] (519.34 g C/m^2/a), and Liang et al. [13] (542.80 g C/m^2/a); that of deciduous coniferous forests was similar to those determined by Zhu et al. [36] (438.80 g C/m^2/a) and Liang et al. [13] (401.40 g C/m^2/a); and that of the mixed forests was consistent with those reported by MOD17A3 (749.65 g C/m^2/a). Despite the closeness of our pertinent estimates with those in the literature, there were still differences, and they were possible owing to the discrepancies in research periods, data sources, study areas, vegetation types, and classification accuracy. In addition, more detailed field measurements on NPP should be examined in our future studies in order to accurately explore the spatiotemporal pattern of forest NPP in China.

From the spatial perspective, the forest NPP in this study gradually decreased from southeast to northwest, consistent with the spatial distribution of precipitation in China [1,13]. Furthermore, we found that forest NPP was mainly distributed in the humid or very humid monsoon regions of China. The spatial pattern of NPP is generally in line with previous studies [32,46]. For instance, the areas with forest NPP ranging from 1000 to 2100 g C/m^2/a were distributed in Eastern Taiwan, southwestern Yunnan, and southern Hainan, where the main vegetation type is evergreen broad-leaved forests and where productivity was generally high due to abundant precipitation, warm climate, and rich groundwater. Contrarily, annual NPP below 200 g C/m^2/a was mainly in the Qinghai–Tibetan Plateau and Xinjiang Regions, where the climate is characterized by a low precipitation amount. In addition, our findings of the annual NPP at the biome level also agreed well with previous studies [13,46]. For example, evergreen forests exhibited a higher NPP than deciduous forests. The highest NPP witnessed in the evergreen broad-leaved forest, with an average range of 1000–2100 g C/m^2/a. The deciduous coniferous forest has the lowest annual NPP (mostly 200–600 g C/m^2/a) in areas with either dry climate or low temperature.

On average, forest NPP in China significantly increased during the past three decades. Despite its overall trend as a whole, NPP trends in the forest showed an obvious geographical heterogeneity. The overall rising trends in NPP are also consistent among biomes, with the increasing trends being all statistically significant at a significant level of 0.01. Such upward trends in forest NPP were consistent with the increasing NPP of terrestrial vegetation in the northern hemisphere [13], indicating that forest vegetation in China played an increasingly irreplaceable role in carbon sequestration in the past few decades [9,13].

Relative to other areas, those covered with forest vegetation had a significant increase in mean annual temperature and total annual solar radiation but had a decrease in annual precipitation from 1982 to 2015. The annual increase in temperature improved the forest NPP of these areas because it may have prolonged the growth period of their forest vegetation [47] and controlled on plant metabolic activity, water and nutrient availability [48]. The yearly increase in solar radiation also augmented the NPP of the forested areas probably because it increased the rate of photosynthesis among the flora, facilitated the accumulation of organic matter, and intensified the warmth therein. Conversely, the decline in precipitation diminished the forest NPP because it may have reduced the water supply essential for the growth of the forest vegetation. By analyzing the relationship between climatic factors and NPP, we found precipitation had a lower correlation with NPP than temperature and solar radiation, implying that the effect of annual precipitation on forest vegetation growth was smaller than other climatic factors, which was also presented by the previous study on biogeographic patterns of China's forests growth rate and their climatic control [6]. Our findings even showed that forest NPP significantly increased only in eastern Taiwan, southern Yunnan, and south of the Yangtze River, where southeast and southwest monsoon systems were dominant and where rainfall from the Pacific Ocean and the Indian Ocean was abundant during the growing season. As a whole, we showed that ongoing climate change could exert positive impacts on forest growth and carbon sequestration capacity, which was widely acknowledged by previous studies [1,8,9].

Previous researchers have also investigated the climate-driven effect on variations in forests NPP [6,13,22]. However, little is known about the temporal differentiation in the relationships between forests NPP and climate variables, which is essential for future forest ecology and management. Our results of the sliding partial correlation analysis in each 17-year moving window indicated that the interannual partial correlation coefficient between annual NPP and annual TSR ($R_{NPP\text{-}S}$) showed little change, that between annual NPP and MAT ($R_{NPP\text{-}T}$) significantly increased, but that between annual NPP and MAP ($R_{NPP\text{-}P}$) displayed a fluctuating downward trend. Obviously, the correlations exhibited temporally varying strength in our study, which also has been reported by previous studies [29,40,49]. However, unlike previous studies [25,50,51], we did not detect a loss of sensitivity of forest vegetation growth to increasing warming. Instead, we find that the impacts of precipitation on forest NPP were also weakening as the annual precipitation amount decreased, which was similar to the earlier findings that the relationship between NPP and MAP varied with the rising in MAP [52]. Given that the mechanistic explanation for varying forest response to environmental change was still unknown, a complete understanding of the combined environmental effect on forest ecosystem productivity remained a great challenge for further studies.

5. Conclusions

Overall, the interannual NPP of the entire forest and its corresponding vegetation types in China from 1982 to 2015 significantly increased. Our findings demonstrated that climatic factors could variably affect the NPP of forests, as the increase in temperature and solar radiation enhanced the interannual forest NPP, whereas the decrease in precipitation reduced the forest NPP. However, such responses of the forest NPP to climatic factors constantly changed, and they varied with the type of forest vegetation. Studies on the dynamics of the correlation between forest NPP and climatic factors are essential for elucidating the adaptability of forests to global warming.

Supplementary Materials: The following supporting information can be downloaded at: https://www.mdpi.com/article/10.3390/plants11212932/s1, Figure S1: China's forest distribution and meteorological stations.

Author Contributions: Z.D. drafted the article; X.L. performed data analysis and interpreted the results; J.Z. designed the statistical analysis; Z.W. and H.Z. revised the article. All authors contributed to further drafts. All authors have read and agreed to the published version of the manuscript.

Funding: This work was supported by the National Natural Science Foundation of China (U1810101, 41161066, 41977412, and 41871193).

Data Availability Statement: The data used in the present work have been listed in the Data Sources.

Acknowledgments: We sincerely appreciate the editors and anonymous reviewers for their valuable comments.

Conflicts of Interest: The authors declare no conflict of interest.

References

1. Fang, J.; Piao, S.; Field, C.B.; Pan, Y.; Guo, Q.; Zhou, L.; Peng, C.; Tao, S. Increasing net primary production in China from 1982 to 1999. *Front. Ecol. Environ.* **2003**, *1*, 293–297. [CrossRef]
2. Jiang, H.; Apps, M.J.; Zhang, Y.; Peng, C.; Woodard, P.M. Modelling the spatial pattern of net primary productivity in Chinese forests. *Ecol. Model.* **1999**, *122*, 275–288. [CrossRef]
3. Binkley, D.; Campoe, O.C.; Alvares, C.; Carneiro, R.L.; Cegatta, Í.; Stape, J.L. The interactions of climate, spacing and genetics on clonal Eucalyptus plantations across Brazil and Uruguay. *For. Ecol. Manag.* **2017**, *405*, 271–283. [CrossRef]
4. Keenan, R.J.; Reams, G.A.; Achard, F.; de Freitas, J.V.; Grainger, A.; Lindquist, E. Dynamics of global forest area: Results from the FAO Global Forest Resources Assessment 2015. *For. Ecol. Manag.* **2015**, *352*, 9–20. [CrossRef]
5. Dixon, R.K.; Solomon, A.; Brown, S.; Houghton, R.; Trexier, M.; Wisniewski, J. Carbon pools and flux of global forest ecosystems. *Science* **1994**, *263*, 185–190. [CrossRef]
6. Zhang, H.; Wang, K.; Zeng, Z.; Du, H.; Zou, Z.; Xu, Y.; Zeng, F. Large-scale patterns in forest growth rates are mainly driven by climatic variables and stand characteristics. *For. Ecol. Manag.* **2019**, *435*, 120–127. [CrossRef]
7. Magney, T.S.; Bowling, D.R.; Logan, B.A.; Grossmann, K.; Stutz, J.; Blanken, P.D.; Burns, S.P.; Cheng, R.; Garcia, M.A.; Khler, P.; et al. Mechanistic evidence for tracking the seasonality of photosynthesis with solar-induced fluorescence. *Proc. Natl. Acad. Sci. USA* **2019**, *116*, 11640–11645. [CrossRef]
8. McMahon, S.M.; Parker, G.G.; Miller, D.R. Evidence for a recent increase in forest growth. *Proc. Natl. Acad. Sci. USA* **2010**, *107*, 3611–3615. [CrossRef]
9. Yao, Y.; Piao, S.; Wang, T. Future biomass carbon sequestration capacity of Chinese forests. *Sci. Bull.* **2018**, *63*, 1108–1117. [CrossRef]
10. Zhou, G.; Wang, Y.; Jiang, Y.; Yang, Z. Estimating biomass and net primary production from forest inventory data: A case study of China's Larix forests. *For. Ecol. Manag.* **2002**, *169*, 149–157. [CrossRef]
11. Fang, J.; Chen, A.; Peng, C.; Zhao, S.; Ci, L. Changes in forest biomass carbon storage in China between 1949 and 1998. *Science* **2001**, *292*, 2320–2322. [CrossRef]
12. Ni, J. Net primary productivity in forests of China: Scaling-up of national inventory data and comparison with model predictions. *For. Ecol. Manag.* **2003**, *176*, 485–495. [CrossRef]
13. Liang, W.; Yang, Y.; Fan, D.; Guan, H.; Zhang, T.; Long, D.; Zhou, Y.; Bai, D. Analysis of spatial and temporal patterns of net primary production and their climate controls in China from 1982 to 2010. *Agric. For. Meteorol.* **2015**, *204*, 22–36. [CrossRef]
14. Cao, S.; Sanchez-Azofeifa, G.A.; Duran, S.M.; Calvo-Rodriguez, S. Estimation of aboveground net primary productivity in secondary tropical dry forests using the Carnegie–Ames–Stanford approach (CASA) model. *Environ. Res. Lett.* **2016**, *11*, 075004. [CrossRef]
15. Fang, J.; Liu, G.; Xu, S. Biomass and net production of forest vegetation in China. *Acta Ecol. Sin.* **1996**, *5*, 497–508.
16. Wang, P.; Sun, R.; Hu, J.; Zhu, Q.; Zhou, Y.; Li, L.; Chen, J.M. Measurements and simulation of forest leaf area index and net primary productivity in Northern China. *J. Environ. Manag.* **2007**, *85*, 607–615. [CrossRef] [PubMed]
17. Hasenauer, H.; Petritsch, R.; Zhao, M.; Boisvenue, C.; Running, S.W. Reconciling satellite with ground data to estimate forest productivity at national scales. *For. Ecol. Manag.* **2012**, *276*, 196–208. [CrossRef]
18. Tripathi, P.; Patel, N.R.; Kushwaha, S.P.S. Estimating net primary productivity in tropical forest plantations in India using satellite-driven ecosystem model. *Geocarto Int.* **2018**, *33*, 988–999. [CrossRef]
19. Peng, C.; Apps, M.J. Modelling the response of net primary productivity (NPP) of boreal forest ecosystems to changes in climate and fire disturbance regimes. *Ecol. Model.* **1999**, *122*, 175–193. [CrossRef]
20. Schuur, E.A.; Matson, P.A. Net primary productivity and nutrient cycling across a mesic to wet precipitation gradient in Hawaiian montane forest. *Oecologia* **2001**, *128*, 431–442. [CrossRef] [PubMed]
21. Mohamed, M.A.A.; Babiker, I.S.; Chen, Z.M.; Ikeda, K.; Ohta, K.; Kato, K. The role of climate variability in the inter-annual variation of terrestrial net primary production (NPP). *Sci. Total Environ.* **2004**, *332*, 123–137. [CrossRef] [PubMed]
22. Cleveland, C.C.; Townsend, A.R.; Taylor, P.; Alvarez-Clare, S.; Bustamante, M.M.; Chuyong, G.; Marklein, A. Relationships among net primary productivity, nutrients and climate in tropical rain forest: A pan-tropical analysis. *Ecol. Lett.* **2011**, *14*, 939–947. [CrossRef] [PubMed]
23. Reyer, C.; Lasch-Born, P.; Suckow, F.; Gutsch, M.; Murawski, A.; Pilz, T. Projections of regional changes in forest net primary productivity for different tree species in Europe driven by climate change and carbon dioxide. *Ann. For. Sci.* **2013**, *71*, 211–225. [CrossRef]
24. Yao, J.; He, X.Y.; Li, X.Y.; Chen, W.; Tao, D.L. Monitoring responses of forest to climate variations by MODIS NDVI: A case study of Hun River upstream, northeastern China. *Eur. J. For. Res.* **2011**, *131*, 705–716. [CrossRef]
25. Chen, A.; He, B.; Wang, H.; Huang, L.; Zhu, Y.; Lv, A. Notable shifting in the responses of vegetation activity to climate change in China. *Phys. Chem. Earth Parts A/B/C* **2015**, *87–88*, 60–66. [CrossRef]

26. Tei, S.; Sugimoto, A. Time lag and negative responses of forest greenness and tree growth to warming over circumboreal forests. *Glob. Chang. Biol.* **2018**, *24*, 4225–4237. [CrossRef] [PubMed]
27. Holben, B.N. Characteristics of maximum-value composite images from temporal AVHRR Data. *Int. J. Remote Sens.* **1986**, *7*, 1417–1434. [CrossRef]
28. Dong, L.; Zhang, M.; Wang, S.; Qiang, F.; Zhu, X.; Ren, Z. The freezing level height in the Qilian Mountains, northeast Tibetan Plateau based on reanalysis data and observations, 1979–2012. *Quat. Int.* **2015**, *380*, 60–67. [CrossRef]
29. He, B.; Chen, A.; Jiang, W.; Chen, Z. The response of vegetation growth to shifts in trend of temperature in China. *J. Geogr. Sci.* **2017**, *27*, 801–816. [CrossRef]
30. Wang, Q.; Zhao, P.; Ren, H.; Kakubari, Y. Spatiotemporal dynamics of forest net primary production in China over the past two decades. *Glob. Planet. Chang.* **2008**, *61*, 267–274. [CrossRef]
31. Cramer, W.; Kicklighter, D.W.; Bondeau, A.; Moore, B., III; Churkinea, G.; Nemry, B.; Ruimy, A. Comparing golbal models of terrestrial net primary productivity (NPP): Overview and key results. *Glob. Chang. Biol.* **1999**, *5*, 1–15. [CrossRef]
32. Li, J.; Wang, Z.; Lai, C.; Wu, X.; Zeng, Z.; Chen, X.; Lian, Y. Response of net primary production to land use and land cover change in mainland China since the late 1980s. *Sci. Total Environ.* **2018**, *639*, 237–247. [CrossRef]
33. McCree, K.J. Photosynthetically active radiation. In *Physiological Plant Ecology I*; Springer: Berlin/Heidelberg, Germany, 1981; pp. 41–55.
34. Demetriades-Shah, T.; Kanemasu, E.; Flitcroft, I.; Su, H. Comparison of ground-and satellite-based measurements of the fraction of photosynthetically active radiation intercepted by tallgrass prairie. *J. Geophys. Res. Atmos.* **1992**, *97*, 18947–18950. [CrossRef]
35. Zhou, G.; Zhang, X. A natural vegetation NPP model. *Acta Phytoecol. Sin.* **1995**, *19*, 193–200.
36. Zhu, W.; Pan, Y.; Zhang, J. Estimation of Net Primary Productivity of Chinese Terrestrial Vegetation Based on Remote Sensing. *J. Plant Ecol.* **2007**, *31*, 413–424. [CrossRef]
37. Du, Z.; Zhao, J.; Pan, H.; Wu, Z.; Zhang, H. Responses of vegetation activity to the daytime and nighttime warming in Northwest China. *Environ. Monit. Assess.* **2019**, *191*, 721. [CrossRef]
38. Piao, S.; Wang, X.; Ciais, P.; Zhu, B.; Wang, T.A.O.; Liu, J.I.E. Changes in satellite-derived vegetation growth trend in temperate boreal Eurasia from 1982 to 2006. *Glob. Chang. Biol.* **2011**, *17*, 3228–3239. [CrossRef]
39. Peng, S.; Piao, S.; Ciais, P.; Myneni, R.B.; Chen, A.; Chevallier, F.; Dolman, A.J.; Janssens, I.A.; Peñuelas, J.; Zhang, G.; et al. Asymmetric effects of daytime and night-time warming on Northern Hemisphere vegetation. *Nature* **2013**, *501*, 88–92. [CrossRef]
40. Piao, S.; Nan, H.; Huntingford, C.; Ciais, P.; Friedlingstein, P.; Sitch, S.; Peng, S.; Ahlstrom, A.; Canadell, J.G.; Cong, N.; et al. Evidence for a weakening relationship between interannual temperature variability and northern vegetation activity. *Nat. Commun.* **2014**, *5*, 5018. [CrossRef]
41. Du, Z.; Zhao, J.; Liu, X.; Wu, Z.; Zhang, H. Recent asymmetric warming trends of daytime versus nighttime and their linkages with vegetation greenness in temperate China. *Environ. Sci. Pollut. Res.* **2019**, *26*, 35717–35727. [CrossRef]
42. Lin, X. Unstability of correlations in statistical weather forecasting. *Sci. Atmos. Sin.* **1978**, *2*, 55–63.
43. Zhan, X.; Guo, M.; Zhang, T. Joint Control of Net Primary Productivity by Climate and Soil Nitrogen in the Forests of Eastern China. *Forests* **2018**, *9*, 322. [CrossRef]
44. Wu, Y.; Luo, Z.; Wu, Z. Net Primary Productivity Dynamics and Driving Forces in Guangzhou City, China. *Appl. Ecol. Environ. Res.* **2018**, *16*, 6667–6690. [CrossRef]
45. Jiang, C.; Wu, Z.F.; Cheng, J.; Yu, Q.; Rao, X.Q. Impacts of urbanization on net primary productivity in the Pearl River Delta, China. *Int. J. Plant Prod.* **2015**, *9*, 581–598.
46. Shang, E.; Xu, E.; Zhang, H.; Liu, F. Analysis of Spatiotemporal Dynamics of the Chinese Vegetation Net Primary Productivity from the 1960s to the 2000s. *Remote Sens.* **2018**, *10*, 860. [CrossRef]
47. Ju, W.M.; Chen, J.M.; Harvey, D.; Wang, S. Future carbon balance of China's forests under climate change and increasing CO_2. *J. Environ. Manag.* **2007**, *85*, 538–562. [CrossRef]
48. Kimball, J.S.; McDonald, K.C.; Running, S.W.; Frolking, S.E. Satellite radar remote sensing of seasonal growing seasons for boreal and subalpine evergreen forests. *Remote Sens. Environ.* **2004**, *90*, 243–258. [CrossRef]
49. Cong, N.; Shen, M.; Yang, W.; Yang, Z.; Zhang, G.; Piao, S. Varying responses of vegetation activity to climate changes on the Tibetan Plateau grassland. *Int. J. Biometeorol.* **2017**, *61*, 1433–1444. [CrossRef]
50. Andreu-Hayles, L.; Arrigo, R.D.; Anchukaitis, K.J.; Beck, P.S.A. Varying boreal forest response to Arctic environmental change at the Firth River, Alaska. *Environ. Res. Lett.* **2011**, *6*, 045503. [CrossRef]
51. Briffa, K.R.; Schweingruber, F.H.; Jones, P.D.; Osborn, T.J.; Shiyatov, S.G.; Vaganov, E.A. Reduced sensitivity of recent northern tree-growth to temperature at northern high latitudes. *Nature* **1998**, *391*, 678–682. [CrossRef]
52. Zhu, L.; Southworth, J. Disentangling the Relationships between Net Primary Production and Precipitation in Southern Africa Savannas Using Satellite Observations from 1982 to 2010. *Remote Sens.* **2013**, *5*, 3803–3825. [CrossRef]

plants

Article

Does the Spatial Pattern of Plants and Green Space Affect Air Pollutant Concentrations? Evidence from 37 Garden Cities in China

Chengkang Wang [1,*], Mengyue Guo [1], Jun Jin [2], Yifan Yang [1], Yujie Ren [3], Yang Wang [4] and Jiajie Cao [1,*]

1 College of Landscape Architecture, Nanjing Forestry University, Nanjing 210037, China
2 Research Institute of Architecture, Southeast University, Nanjing 210096, China
3 Graduate School of Human-Environment Studies, Kyushu University, Fukuoka 819-0395, Japan
4 State Key Laboratory of Vegetation and Environmental Change, Institute of Botany, Chinese Academy of Sciences, Beijing 100093, China
* Correspondence: chengkang.wang@njfu.edu.cn (C.W.); caojiajie@njfu.edu.cn (J.C.)

Abstract: Relevant studies have demonstrated that urban green spaces composed of various types of plants are able to alleviate the morbidity and mortality of respiratory diseases, by reducing air pollution levels. In order to explore the relationship between the spatial pattern of urban green spaces and air pollutant concentrations, this study takes 37 garden cities with subtropical monsoon climate in China as the research object and selects the urban air quality monitoring data and land use type data in 2019 to analyze the relationship between the spatial pattern and the air pollutant concentration through the landscape metrics model and spatial regression model. Moreover, the threshold effect of the impact of green space on air pollutant concentrations is estimated, as well. The results showed that the spatial pattern of urban green space was significantly correlated with the concentrations of $PM_{2.5}$ (PM with aerodynamic diameters of 2.5 mmor less), NO_2 (Nitrogen Dioxide), and SO_2 (Sulfur dioxide) pollutants in the air, while the concentrations of PM_{10} (PM with aerodynamic diameters of 10 mmor less) pollutants were not significantly affected by the green space pattern. Among them, the patch shape index (LSI), patch density (PD) and patch proportion in landscape area (PLAND) of forest land can affect the concentration of $PM_{2.5}$, NO_2, and SO_2, respectively. The PLAND, PD, and LSI of grassland and farmland can also have an additional impact on the concentration of SO_2 pollutants. The study also found that there was a significant threshold effect within the impact mechanism of urban green space landscape pattern indicators (LSI, PD, PLAND) on the concentrations of $PM_{2.5}$, NO_2, and SO_2 air pollutants. The results of this study not only clarified the impact mechanism of the spatial pattern of urban green space on air pollutant concentrations but also provided quantitative reference and scientific basis for the optimization and updating of urban green space to promote public health.

Keywords: public health; urban green spaces; landscape pattern; air pollution; quantitative analysis; threshold effect

Citation: Wang, C.; Guo, M.; Jin, J.; Yang, Y.; Ren, Y.; Wang, Y.; Cao, J. Does the Spatial Pattern of Plants and Green Space Affect Air Pollutant Concentrations? Evidence from 37 Garden Cities in China. *Plants* **2022**, *11*, 2847. https://doi.org/10.3390/plants11212847

Academic Editors: Jie Gao, Weiwei Huang, Johan Gielis, Peijian Shi and Aneta Baczewska-Dąbrowska

Received: 15 September 2022
Accepted: 24 October 2022
Published: 26 October 2022

1. Introduction

The emergence and development of modern urban green spaces are closely related to the improvement of human living environments and public health [1]. As an important discipline for the development of human living environments, landscape architecture provides important support in the establishment of public safety systems in urban and rural spaces and promotes the health and wellness of residents [2]. With the accelerated global urbanization and rapid industrial development, including the frequent occurrence of public health incidents such as the new coronavirus pneumonia, the public health of residents is under unprecedented threat [3]. The severe challenges posed by the pandemic have prompted the entire society to deeply recognize the importance of public health and have triggered unprecedented attention to the urban built environment and the construction of healthy urban green spaces.

As a global environmental and public health problem, air pollution has severe adverse effects on human health. The pollutants can enter the body through the respiratory system and affect the lungs and heart, causing cardiovascular and respiratory diseases [4]. The World Health Organization (WHO) report demonstrated that respiratory diseases caused by air pollution in 2019 ranked fourth in the top 10 causes of death worldwide (https://www.who.int/zh/ (accessed on 15 July 2021)). Numerous studies have verified that short-term or long-term exposure to air pollutants, including $PM_{2.5}$, PM_{10}, NO_2, and SO_2, increased the risks of mortality and morbidity, thereby posing a severe threat to the public health of the residents [5]. As an important part of the urban ecosystem, in addition to providing residents with green spaces for recreation and entertainment, urban green spaces also improve air quality. The ability of different plant species, plant communities, and green spaces along roads to retain dust and mitigate air pollution has been widely demonstrated [6]. In addition, by optimizing the structure of urban green spaces, increasing the areas of green spaces and green coverage could effectively reduce the concentration of particulate matter and gas pollutants in the air, which effectively improves the urban environment, and ultimately play an important role in promoting public health [7].

Domestic and international research on the mitigation of air pollutants in green spaces at the micro-level has primarily focused on the ability of urban garden green spaces to reduce air pollutant concentrations, using individual plant microstructure processes such as sedimentation, retardation, adsorption, and absorption. For example, Latha and Highwood demonstrated that changes in structures, such as the roughness of plant leaf surfaces, affect the sedimentation pattern of dust particles [8]. Beckett et al. deduced that plants maintained a higher humidity in a certain range during transpiration, and dust sedimentation is more likely to occur when it increases in weight after absorbing moisture, while the ability of the leaves to adsorb dust increases with an increase in their humidity [9]. In addition, numerous studies have demonstrated the variability in the absorption of different gaseous pollutants by different landscape plants [10–12].

The meso-level was adopted to investigate the reduction effect of the size, shape, plant configuration, vertical structure, etc., of small- and medium-scale urban green spaces on different pollutants. In his study on the ecological mechanism of urban open space planning, Wang Shaozeng pointed out that the more practical the mix of green space levels, the better the filtration effect on the atmospheric particulate matter [13]. By analyzing the relationship between the three-dimensional green volume of green spaces along roads and $PM_{2.5}$ concentration, Sheng found that high 3D green volume did not indicate low $PM_{2.5}$ concentration, and such green spaces with the uniform vertical distribution of biomass and diverse vegetation were more effective in reducing $PM_{2.5}$ concentration [14]. Fan analyzed the correlation between the daily PM_{10} and $PM_{2.5}$ concentrations and particulate matter concentrations of seven typical land cover types and different scales of land cover patterns and determined that pavement-type and low-to-medium canopy density vegetation exerted a more significant effect on PM_{10} levels, while $PM_{2.5}$ concentrations were more sensitive to the response of building-type and low-to-medium canopy density vegetation [15]. In several ways, the aforementioned studies verified that green spaces, as living media, are one of the most important vehicles for mitigating air pollution and play a significant role in the public health of residents and the urban environment.

Recently, studies on how to optimize the landscape pattern of urban green spaces to reduce air pollution concentrations at the macro-level have increasingly garnered considerable attention, and the current research results primarily focused on exploring the correlation between the concentrations of $PM_{2.5}$, PM_{10}, and other pollutants, including land use or the landscape patterns of land cover. Ye et al. [16] explored the relationship between $PM_{2.5}$ growth and land use changes in China from 1998–2015 and inferred that $PM_{2.5}$ concentrations were higher in the eastern plains and Taklamakan Desert in China, and higher $PM_{2.5}$ concentrations existed on artificial land surfaces, croplands, and deserts, while forests, grasslands, and unused land usually contained lower $PM_{2.5}$ concentrations. Simultaneously, the average annual increase in $PM_{2.5}$ concentrations in a desert land and artificial land surfaces was higher than that of other land types. Yue et al. [17] studied the quantitative relationship between vegetation coverage and atmospheric particulate matter based on remote sensing inversion and deduced that vegetation coverages of $\leq$10% and >45% exhibited a significant effect on mitigating atmospheric particulate matter pollution. Lei et al. [18] explored the effect of green spaces on particulate matter pollution by studying the landscape patterns of urban green spaces at multiple spatial scales and determined that increasing the biodiversity of green spaces and increasing the number of large green spaces significantly reduced PM_{10} concentrations at almost all scales. Zhao et al. [19] conducted a land use regression (LUR) analysis on the green spaces of lakes and wetlands, including the surrounding 500-m built environment in Wuhan City, and their results demonstrated that the lakes, wetlands, and nearby greenery exerted a positive and significant effect on $PM_{2.5}$ concentrations within a buffer zone of 300 m or closer. Via a multivariate linear regression modeling of the daily average PM_{10} concentration data with land use pattern information for the cities of Vienna and Dublin, McNabola et al. [20] determined that adding transboundary air pollution and traffic activity representations to the predictor variables significantly improved the accuracy of LUR-based methods. Zhang et al. [21] applied the LUR model to analyze the correlation between air pollution levels and childhood asthma hospitalization rates, by establishing a spatial distribution LUR model of the daily pollutant concentration data of PM_{10} and SO_2 with associated influencing factors in Shenyang, China and deduced that the number of childhood asthma hospitalizations was highly correlated with PM_{10} and SO_2 pollution levels. Lee and Koutrakis adopted satellite ozone monitoring instruments together with land use parameters to develop a mixed-effects model through which they estimated the daily NO_2 concentrations in New England, and explored the source areas of emissions (e.g., high population or traffic areas) in the study area and elucidated the seasonal characteristics of NO_2, based on NO_2 spatial distribution patterns [22]. Lu et al. [23] analyzed the landscape pattern indices and $PM_{2.5}$ data in China and demonstrated that differences exist in the significant impact indicators of different land use types on $PM_{2.5}$ concentration and that the landscape pattern indices exhibit a significant effect on $PM_{2.5}$ concentrations. De Jalón et al. [24] conducted a comparative analysis on the effect of the dry deposition of trees on air pollution reduction for different land use types in the Basque Country, and finally deduced that coniferous forests are the most effective in eliminating air pollution. These research results mainly elucidated the effect of urban land use changes on univariate air pollutants, and the pollutants were mainly $PM_{2.5}$, PM_{10}, and other fine particulate matter. A fewer number of studies have been conducted on NO_2, SO_2, and other gaseous pollutants, and these studies rarely involved any comprehensive air pollution research on the quantification and regulation strategies of the concentration of mixed pollutants of gases and particulate matter, including the spatial distribution of urban landscape patterns.

On this basis, this study takes 37 garden cities with subtropical monsoon climate in China as the research object, selects the urban air quality monitoring data and land use remote sensing data in 2019 to analyze the relationship between the landscape pattern index and the air pollutant concentration through the spatial regression model method. Also, a new threshold effect estimation method based on a polynomial model is designed to explore the threshold effect of green space landscape patterns on air quality [25,26].

2. Results

In this paper, regression modeling tools in GeoDa software are used to conduct SEM regression analysis, and the results obtained are shown in Table 1.

Table 1. Conclusion of thresholds for landscape pattern indices.

	Variable	p	Threshold	p	Threshold	p	Threshold	p	Threshold
		$PM_{2.5}$		PM_{10}		NO_2		SO_2	
Forest land	PLAND	0.569		0.161		0.774		**0.09** *	**50**
	PD	0.41		0.513		**0.091** *	**0.0718**	**0.003** ***	**0.038**
	LSI	**0.059** *	**18.018**	0.323		0.204		0.493	
Grassland	PLAND	0.847		0.764		0.816		**0.031** **	**3.33**
	PD	0.571		0.751		0.9902		**0.000** ***	**0.121**
	LSI	0.182		0.499		0.818		**0.000** ***	**14.13**
Farm land	PLAND	0.247		0.537		0.592		**0.001** ***	**32.56**
	PD	0.652		0.921		0.624		0.925	
	LSI	0.309		0.592		0.774		0.139	

*** $p < 0.01$, ** $p < 0.05$, * $p < 0.1$. Correlations significant at the level of 0.1 are marked in bold.

2.1. Regression Analysis of Landscape Pattern Indices and Air Pollutants

$PM_{2.5}$ concentrations were significantly and negatively correlated with the LSI of forestlands; each unit increase in the landscape shape index of the forestlands was followed by a 0.68 unit decrease in the $PM_{2.5}$ pollutant concentration. NO_2 concentration was significantly and negatively correlated with the PD of forestlands. With each unit of increase in the PD of forestlands, the concentration of the NO_2 pollutant was subsequently reduced by 258.409 units. However, Grasslands and farmlands had no significant correlation with $PM_{2.5}$ and NO_2 pollutant concentrations. SO_2 concentrations were significantly and positively correlated with the PD of forestlands, PLAND, and LSI of grasslands, where the relationship between SO_2 pollutant concentrations and the PD of forestlands and LSI of grasslands was highly significant ($p < 0.01$). SO_2 concentration was significantly and negatively correlated with the PLAND of forestlands, PD of grasslands, and PLAND of farmlands, where the PD of grasslands and PLAND of farmlands were highly significantly correlated with the SO_2 pollutant concentration ($p < 0.01$). When the PD of forestlands, PLAND, and LSI of grasslands increased by one unit, the SO_2 concentration increased by 149.939, 0.752, and 0.429 units, respectively. When each unit of PLAND of forestlands, PD of grasslands, and PLAND of farmlands increased, the SO_2 concentration was reduced by 0.073, 214.564, and 0.172 units, respectively. In contrast, SO_2 concentrations were not significantly affected by the LSI of forestlands, and the PD and LSI of farmlands.

From the results of the spatial correlation analysis between the landscape pattern indices of green spaces and air pollutants in the 37 cities, it was determined that the landscape pattern of urban green spaces was significantly associated with the concentrations of $PM_{2.5}$, NO_2, and SO_2 pollutants in the air, while the concentrations of PM_{10} pollutants were not significantly affected by the pattern of green spaces. Meanwhile, LSI, PD, and PLAND of forestlands influenced the concentration of the three types of air pollutants, respectively. In addition, the PLAND, PD, and LSI of grasslands and PLAND of farmlands exert an additional influence on the concentration of SO_2 pollutants.

2.2. Threshold Effect of the Impact of Landscape Pattern Indices on Air Quality

To further validate the relationship between the landscape pattern indices and the concentrations of the four pollutants, this study conducted regression analysis and linear fitting on the four pollutant concentrations and their landscape pattern indices, with EXCEL revealing significant effects on them. After comparing the R^2 of the function's trend line, it was deduced that the overall effect of the polynomial fit was better than several other types of functions in the experiment; hence, the polynomial was chosen for linear fitting in this study.

2.2.1. Threshold Effect of the Impact of Landscape Pattern Indices on $PM_{2.5}$

Table 1 indicated that the LSI of forest land has a significant negative effect on the concentration of $PM_{2.5}$ ($p < 0.1$). Therefore, this study analyzes the influence trend of $PM_{2.5}$

concentration change to determine the threshold range of forest land LSI. It is found that the concentration of $PM_{2.5}$ will first decrease and then increase with the increase of forest land LSI. According to the fitting calculation of the polynomial function, the coordinate value of the polynomial vertex of the forest landscape shape index is [18.02, 37.94]. Therefore, when the landscape shape index of the forest is 18.02, the minimum concentration of $PM_{2.5}$ reaches 37.94 $\mu g/m^3$. Therefore, this study found that the LSI value of 18.02 is the threshold of forest land (Figure 1).

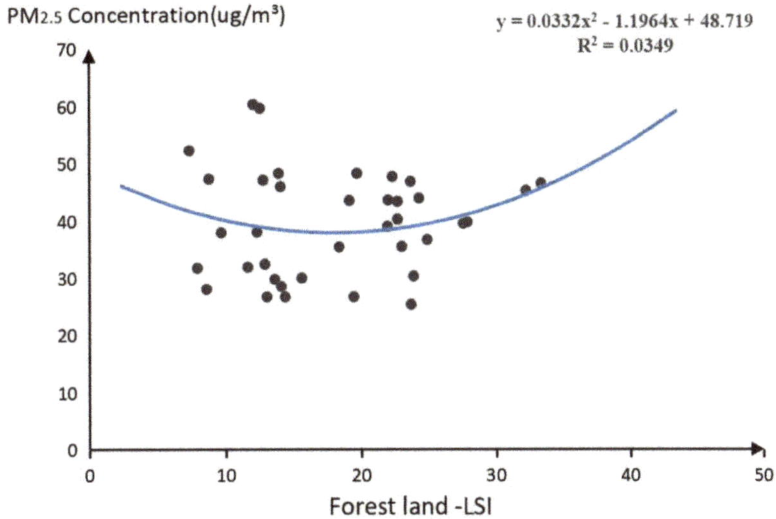

Figure 1. Threshold of significant landscape pattern indices for $PM_{2.5}$ pollutant.

2.2.2. Threshold Effect of the Impact of Landscape Pattern Indices on NO_2

From Table 1, it was clear that the PD of forestland exerted a significant effect on NO_2 concentration ($p < 0.1$); therefore, the study determined the range of both thresholds by analyzing the trend of the effect of NO_2 concentration changes. From the equation curve in Figure 2, it can be observed that the NO_2 concentration exhibited a decreasing trend, and then started increasing with the increase in the PD of the forestland. From the function calculation, the value of the polynomial vertex coordinates of the PD of forestland was [0.072, 29.161]; hence, when the forestland patch density was 0.072, the NO_2 concentration reached the minimum of 29.161 $\mu g/m^3$. Therefore, 0.072 was deduced to be the threshold value for the PD of forestland in this study (Figure 2).

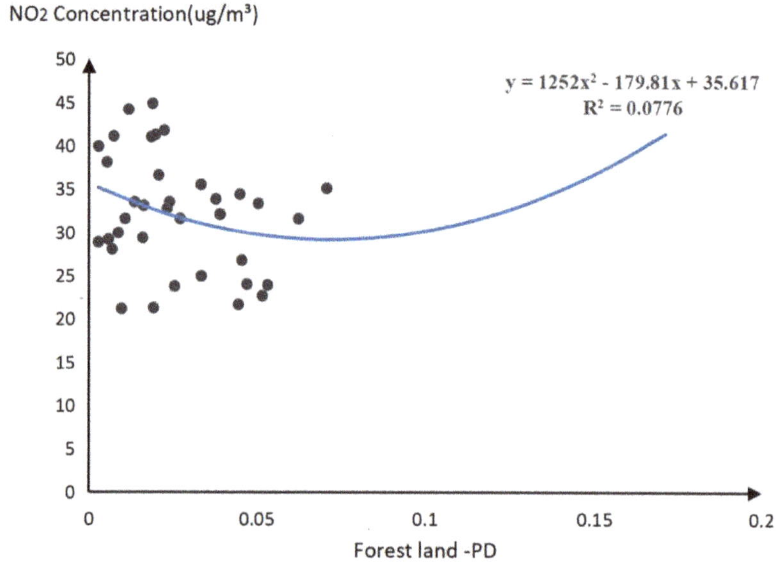

Figure 2. Threshold of significant landscape pattern indices for NO_2 pollutant.

2.2.3. Threshold Effect of the Impact of Landscape Pattern Indices on SO_2

PD and PLAND of the forest, PD, and LSI of grassland, and PLAND of farmland will affect the concentration of SO_2 pollutants, significantly. Through linear analysis of SO_2 concentration change influence trend, the threshold results of 6 indicators are determined as follows.

For forest land, the concentration of SO_2 will increase first and then decrease in the forest. Through polynomial fitting, it is found that when the PLAND of forest value is 50%, the SO_2 concentration reaches the maximum value of 8.7 μg/m^3. When the PD value of forest land is 0.038, the SO_2 concentration reaches the maximum value of 8.72 μg/m^3. In consideration of air quality and public health of residents, the optimal range of PD and PLAND of forest land is greater than or less than 50% and 0.038.

For grassland, the concentration of SO_2 will increase first and then decrease with the increase of PD and PLAND of grassland. When the PLAND value of grassland is 3.33%, the SO_2 concentration reaches the maximum value of 8.72 μg/m^3. When the PD value of grassland is 0.121, the SO_2 concentration reaches the maximum value of 9.545 μg/m^3. Also, the SO_2 concentration will decrease first and then increase with the increase of LSI of grassland, when the grassland LSI value is 14.13, the SO_2 concentration reaches the minimum value of 7.84 μg/m^3.

For farmland, the concentration of SO_2 will increase first and then decrease with the increase of PLAND of farmland. When the PLAND value of farmland is 32.56, the PLAND value of 8.78 μg/m^3. Therefore, this study found that 32.56 is the threshold of the PLAND value of farmland (Figure 3).

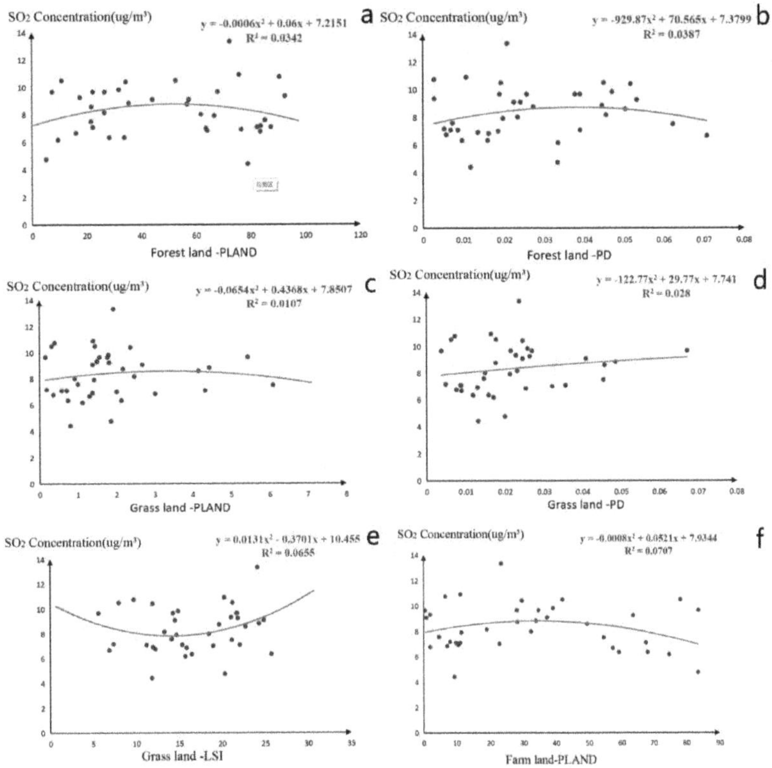

Figure 3. Threshold of significant landscape pattern indices for SO_2 pollutant ((**a–f**) represents the threshold effect of SO_2 and forest land PLAND, forest land PD, grassland PLAND, grassland PD, grassland LSI, and farmland PLAND landscape pattern index).

3. Discussions

3.1. Impact Mechanism of Landscape Pattern Index on Air Pollutant Concentration

3.1.1. Impact Mechanism of Landscape Pattern Index on $PM_{2.5}$

The results of spatial regression analysis showed that the LSI of forest land had a significant negative effect on the concentration of $PM_{2.5}$, which was consistent with the results obtained by previous studies and indicated that increasing the contact area between the edges of the green space patches and surrounding urban areas at large spatial scales significantly reduces $PM_{2.5}$ concentrations [18]. $PM_{2.5}$ primarily originates from industrial emissions, traffic emissions, and the burning of biomass, and is mainly present on roads, factories, and surrounding farmland in built-up areas [27]. When the LSI of forestland was elevated, the complexity of its patch edge also increased, and the contact area between the forestland patch edge and urban construction land was subsequently elevated. As the contact area between vegetation and $PM_{2.5}$ particulate matter increased, the effect of dust reduction via vegetation leaf surface villi retardation, stem adsorption, and stomatal absorption by plants was enhanced, while the concentration of $PM_{2.5}$ pollutants reduced [28,29].

3.1.2. Impact Mechanism of Landscape Pattern Index on NO_2

The results of spatial regression analysis showed that the PD of forest land had a significant negative effect on the concentration of NO_2, which indicated that increasing the contact area of forestland with NO_2 pollutants was beneficial to the reduction of NO_2 pollutant concentration.

The main sources of NO_2 in the atmosphere are industrial and motor vehicle exhaust emissions [30]. Therefore, NO_2 air pollution mainly exists in urban industrial areas with large traffic emissions. Studies have demonstrated that plants can absorb NO_2 through their leaves [31]. The forestland patches outside the urban constructed land have a large area; however, NO_2 was mainly generated in the built-up area and blocked by high-density buildings; hence, the contact area between urban forestland and NO_2 pollutant gas was small, and increasing the forestland area exhibited no significant effect on the weakening effect of the overall urban NO_2 pollutant concentration. Increasing the density of forestland patches in the built-up area enhanced the number of public green spaces in the city, and measures were taken to insert greenery in the edges, especially the greening of industrial areas, and street greenery was appropriately sufficient in moving plant leaves to come into contact with more polluting gases while playing a more significant role in the improvement of urban NO_2 pollutant gases.

3.1.3. Impact Mechanism of Landscape Pattern Index on SO_2

The results of spatial regression analysis showed that the PLAND of forest land, PD of grassland, and PLAND of farmland had a significant negative effect on the concentration of SO_2. And the PD of forest land, PLAND of grassland, and LSI of grassland had a significant positive effect on the concentration of SO_2.

The production of SO_2 in urban air pollution mainly originates from coal combustion and industrial emissions, and studies have demonstrated that SO_2 concentrations were lower in areas with rich vegetation under different spaces in cities [32]; hence, the increase in the area share of forestland patches effectively enhanced the efficiency of SO_2 absorption by urban plants and reduced the SO_2 air content in cities. Simultaneously, as a gas, SO_2 was more susceptible to wind speed, temperature, and humidity; hence, the single vertical structure of grass greenery was more conducive to the circulation of urban winds and the transportation of SO_2 pollutants from the inner city to the outer city, to achieve a lower average SO_2 concentration [33]. The diffusion of SO_2 is mainly influenced by wind speed and direction, and in areas with high building density due to the blockage of tall buildings resulting in low internal wind speed, SO_2 concentration was not easily diffused. Therefore, with the increase in the proportion of farmland area, the degree of SO_2 diffusion was accelerated and the concentration was reduced [34]. Meanwhile, cities with relatively large areas of farmland generally have industries positioned as agricultural cities, such as Henan Province, which is a largely agricultural province in China, including Xinyang and Nanyang, and these areas have low average SO_2 concentrations.

As the density of forestland patches increased, SO_2 concentration increased, and related studies have demonstrated that the larger the average area of green space patches in urban landscapes, the lower the fragmentation index, and the greater the role of green spaces in air pollution purification. In addition, when the degree of fragmentation of forestland was more severe, it could not function as an effective urban green heart, and the fragility of the landscape structure reduced the absorption capacity of SO_2 [35]. As the ratio of the area occupied by grass increased, SO_2 concentration increased. It was speculated that the reason for this result might be that the decrease in the number of plant leaves (the main organ of SO_2 absorption by plants) in grass patches resulted in the subsequent decrease in the efficiency of SO_2 absorption by green spaces [36]. Therefore, it was inferred that in addition to improving the air purification capacity, increasing the density of grass patches with small areas and low edge complexity also reduces economic costs.

3.2. Threshold Mechanism of Landscape Pattern Index on Air Pollutant Concentration

3.2.1. Threshold Effect of $PM_{2.5}$

The LSI of forest land is related to the edge complexity and patch density of green space patches. The increase in LSI value can enhance the energy flow and exchange between green patches and surrounding patches and create more interaction opportunities for source and sink landscapes, so as to absorb and settle more pollution particles and reduce the concentration of $PM_{2.5}$ [37]. However, as the LSI of forest land continues to increase, the patch density then increases the degree of green space fragmentation increases, and the abatement between forest land edges and air pollutants cannot offset the increasing amount of air pollution due to urban green space fragmentation, and $PM_{2.5}$ concentrations then continue to rise to show a trend of first decreasing and then increasing [38].

3.2.2. Threshold Effect of NO_2

As the PD of forest land keeps increasing, NO_2 concentration shows a trend of first decreasing and then increasing. When the density of forested land patches increases continuously from 0, the intra-urban green space gradually transitions from single-core large green space to multi-core green space. With the connection of streets, rivers, and other green channels, the multi-core urban green space plays a greater ecosystem service function, more contact area of air pollutants, and higher efficiency of material exchange and energy circulation, which is conducive to the reduction of NO_2 pollutant concentration [39]. However, with the increasing density of forest land patches, the urban green space system transitions from multi-core green space to green space patch fragmentation, and the concentration of NO_2 pollutants then increases [40].

3.2.3. Threshold Effect of SO_2

Except for the LSI of grassland, with the increasing values of PLAND and PD of forest land, PLAND and PD of grassland, and PLAND of agricultural land, the trend between the landscape pattern index and SO_2 concentrations of different green space types was first increasing and then decreasing.

Green space is a sink landscape for mitigating urban SO_2 pollution, which can adsorb and absorb and deter SO_2 through plant leaves and branches, while urban construction land is a source landscape for SO_2 pollutants [41]. The decrease in the area share of forest land patches is often associated with cities with high levels of development. According to relevant studies showing more, developed cities with relatively well-developed environmental measures and more rational urban master plans, as well as industrial structures with the upgrading and transformation of heavy industries to light industries and high-tech industries, tend to have lower pollution levels. The opposite is often true in medium-sized developing cities. Such cities are still at the stage of economic development, with more types of industries and insufficient attention to the environment, resulting in the reduction of SO_2 air pollution by vegetation in such cities being much lower than the SO_2 emissions, and therefore the pollution concentration is higher. With the increasing area of forest land emissions and reductions gradually reaching a balance or even vegetation reduction exceeding the local pollution emissions air pollution levels are reduced again [42,43].

3.3. *Implications for Urban Planning and Management Policies*

In the context of optimizing and renewing the landscape patterns of urban green spaces, several studies have demonstrated that regulating land use patterns by carefully planning the morphology and layout of green space networks effectively improves air quality and enhances the public health of residents. Based on the aforementioned findings, we made the following recommendations for the improvement of air pollution in cities with subtropical monsoon climates.

Forestland, grassland, and farmland can inhibit the concentration of $PM_{2.5}$, NO_2, and SO_2. The grassland area in the urban built-up area should be properly controlled, the forest coverage should be gradually increased, the restoration and reconstruction of damaged forest land should be accelerated, the integrity and stability of the ecosystem should be improved, the urban green space landscape pattern should be reasonably controlled through scientific basis, and the production and living ecological space layout should be coordinated.

For cities with NO_2 and SO_2 as the main pollutants, the pollutants can be reduced systematically by reasonably arranging the land use pattern of forest land, grassland, farmland, and other green areas, and cooperating with each other. When the PD value of forest land is about 0.072, the overall value of urban NO_2 pollutant concentration reaches the optimum. When the land and PD values of forest land are away from 50% and 0.038, the land and PD values of grassland are away from 3.33% and 0.121, and the LSI of grassland reaches 14.13, the urban SO_2 pollutant concentration reduction effect is the best. Therefore, reasonable urban green space planning should be carried out to optimize the density of green space patches and evenly distribute the types of green space patches, to achieve a close relationship between various land uses and better alleviate air pollution. Through the way of inserting green in the gap, the coverage of urban green space can be improved economically and efficiently, and the contact area between green space and air pollutants can be increased to better improve the air quality level.

$PM_{2.5}$ mainly comes from industrial waste gas emission, traffic emission, and biomass combustion, and mainly exists in roads and factories in built-up areas. China's subtropical monsoon climate zone is in the stage of rapid urban development, and it is difficult to reduce traffic emissions and industrial emissions. However, $PM_{2.5}$ pollution can be improved by optimizing the landscape pattern of urban green space. When the LSI of forest land reaches 18.02, the overall value of $PM_{2.5}$ pollution concentration is the best. It is suggested to reasonably arrange the urban green space area. For the streets and factories with large traffic and industrial emissions, we will focus on improving the greening level of the streets, appropriately increasing the contact area between the street green space and $PM_{2.5}$, improving the vertical structure of the urban green belt, block, and absorb $PM_{2.5}$ pollutants to the greatest extent, and reduce the transmission route. In addition, the treatment efficiency of polluted gas in the factory shall be strictly controlled to reduce air pollution.

The comprehensive management of unused land and inefficient land should be promoted to form a reasonable and efficient urban green landscape pattern. At the same time, it is suggested to promote the development of residents' lifestyle and consumption concepts towards green, healthy, and low-carbon. The scientific achievements in the prevention and control of $PM_{2.5}$, NO_2, and SO_2 pollution should be actively disseminated to the residents so that the relevant departments can implement the relevant pollution control measures more smoothly, and the residents can support the air pollution control.

In addition, the results of the spatial regression model in this study also show that air pollution has obvious spatial effects, so it is not feasible to carry out internal air pollution control for a single city. From the perspective of urban agglomerations, we should coordinate and plan the air pollution control policies among cities, jointly improve the regional air quality level and reduce the threat of air pollution to residents' public health. At the same time, our results show that the careful planning of urban green space landscape patterns has brought some positive benefits to air pollution, but compared with traffic emissions, industrial emissions, and biomass combustion, the role of pollutant gas emissions is still limited. Therefore, improving the rationality of urban green space landscape patterns is on the one hand. On the other hand, we should also attach great importance to energy efficiency and traffic management, further promote the reform of industrial structure, and eliminate backward industries with high pollution and high consumption. Finally, it aims to increase the proportion of high-tech industries, promote China's transformation from incremental expansion to stock revitalization, improve development quality and resource utilization efficiency, and reduce air pollutant emissions from the source.

3.4. Research Innovations and Limitations

As early as the end of the last century Wickham et al. [44] proposed an integrated assessment of the environmental condition of a large region, by combining data on land cover, population, roads, rivers, air pollution, and topography. In 2009, Rafiee et al. [45] applied a combination of remote sensing image classification, landscape indicator assessment, and vegetation indices, to explore changes in urban landscape patterns and provide an assessment of changes and trends in urban living environments. Recently, there has been an increasing interest in applying remote sensing images to investigate the role of changes in urban land use patterns on air pollution concentrations. In 2015, Wu et al. [46] adopted landscape indicators such as PLAND, PD, ED, SHEI, and CONTAG to explore the effect of urban landscape patterns on $PM_{2.5}$ pollution and determined that vegetation and water bodies were significant landscape components that reduced $PM_{2.5}$ concentrations. In 2016, Xu et al. [47] explored the quantitative relationship between land use and air quality (SO_2, NO_2, and PM_{10}) through binary correlation analysis, and the results indicated that land use had a significant effect on air quality. For each standard deviation increase in construction land, NO_2 concentrations increased by 2%. For one standard deviation increase in water bodies, SO_2 or PM_{10} concentrations decreased by 3–6%. Although this study quantified the role of land use type on air pollution levels, it did not explore the quantitative relationship between land use effects on $PM_{2.5}$. In 2020, Li et al. [48] studied the non-linear effects of land use distribution on $PM_{2.5}$, using the boosted regression tree method to capture the effects of land use scale on $PM_{2.5}$ in different seasons. It was inferred that when the grassland and forestland areas were below 8% and 20%, respectively, the air quality improved significantly with the increase in grassland and forestland areas. When the distribution of construction land was greater than approximately 10%, $PM_{2.5}$ pollution increased significantly with the increase in the construction land area. This study was the first to identify the threshold for the effect of land use type on $PM_{2.5}$ concentrations. However, the role of the effect on the three air pollutants such as PM_{10}, NO_2, and SO_2 is yet to be addressed. Most of the aforementioned studies adopted linear regression models to explore the role of land use types on air pollutants, which could not comprehensively incorporate the spatial effects between cities into the study, and also failed to demonstrate how the effects of landscape patterns on multiple air pollutants varied with changes in landscape pattern indices. On this premise, this study employed a spatial regression model approach to regress air pollutant concentrations and landscape pattern indices in 37 subtropical monsoon climate garden cities of China in 2019, to explore the core landscape pattern indices that exhibited significant effects on air pollution concentrations and the optimal core indicator thresholds for mitigating urban air pollution, to quantify the spatial effects of the landscape pattern indices on air pollutants, using urban spatial effects as a starting point, and then a general framework that differed from existing studies was proposed. In addition to focusing on quantifying the negative or positive effects of landscape pattern indices on air pollutant concentrations, to a certain extent, this study also reflected the appropriate threshold values of landscape pattern indices for reducing air pollution concentrations, which provides quantitative reference and technical support for urban planning in a more targeted manner.

However, in general, this study had some limitations. Firstly, the spatial resolution of satellite remote sensing images in this study is low, and there may be a certain misclassification of patch types, which may cause deviation to the impact indicators of air pollution. At the same time, because the research object is a large city with a subtropical monsoon climate, it is not universal for other climate belt cities. In addition, the image data used in our study provide a reference for the overall land use pattern of the whole city. The conclusion is that it provides a reference for the overall planning of urban green space landscape patterns from a macro perspective. The role of small-scale urban green space in reducing pollutants needs to be further discussed. At the same time, in order to ensure the accuracy of the urban green space landscape pattern indicators, the image data selected in this study are all summer image data, but the air pollution data is the annual average value of pollutants. Due to the seasonal changes, the green space coverage of a few cities in winter is reduced, so the impact analysis of this study on air pollutants is still insufficient. Therefore, based on the above shortcomings, in future research, we will expand the sample number of research objects, improve the accuracy of remote sensing data, and analyze the urban landscape pattern and air pollutant concentration in different seasons and climate zones respectively, so as to reduce variables, improve universality and research accuracy, and put forward more targeted urban green space landscape pattern planning strategies.

4. Materials and Methods

In this study, firstly, we take 37 garden cities with subtropical monsoon climate as the research unit and take the annual average concentration of $PM_{2.5}$, PM_{10}, NO_2, and SO_2 and the landscape pattern index of these 37 cities in 2019 as the dependent variable and independent variable, respectively, to carry out the spatial regression model. Based on the output of the model, the impact mechanism between landscape pattern index and $PM_{2.5}$, PM_{10}, NO_2, and SO_2 pollution was explored. Finally, a new threshold effect estimation method based on a polynomial model is designed to explore the threshold effect of green space landscape patterns on air quality (Figure 4).

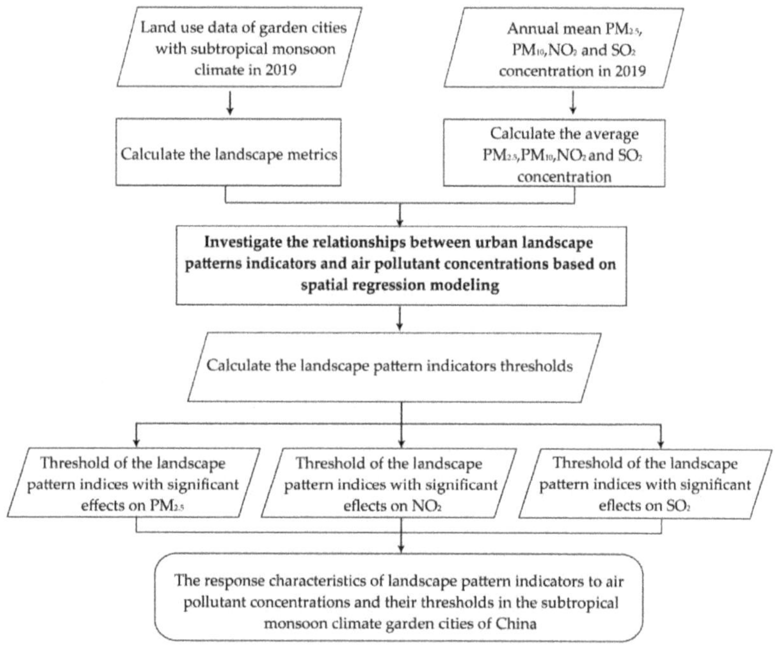

Figure 4. Flow chart Schemes follow the same formatting.

4.1. Study Region

The subtropical monsoon climate zone is one of the four major climate zones in China, covering approximately a quarter of China's land area. It supports almost 600 million people in China, is the most densely populated region in China, and mainly includes the Yangtze River Delta region, Pearl River Delta region, and middle reaches of the Yangtze River, including other areas of China's economic center, where the level of economic development and urban landscaping are at a relatively high level in the country. With the advancement of urbanization, the original urban ecological network structure has been destroyed, owing to the drastic changes in urban land use patterns caused by the population concentration and rapid industrial development; in addition, the subtropical monsoon climate region has also become the region most threatened by air pollution in China, which exposes the public health of residents to tremendous pressure [49,50]. Considering the substantial influence of natural climate on urban air quality, to improve the accuracy of this study, 37 "national garden cities" with a population size of more than 4 million people and complete air quality data for 2019 were selected from the subtropical monsoon climate zone of China as the research objects of this study. According to the "National Garden City Standards" adopted by the Ministry of Housing and Urban-Rural Development of the People's Republic of China, garden cities are cities with balanced distribution, practical structure, perfect functions, beautiful landscapes, fresh and comfortable human living, and ecological environments, safe, and pleasant. In addition, they were adopted as examples to investigate the effect of landscape patterns of urban green spaces on air pollution levels (see Figure 5, Appendix A).

Figure 5. Distribution map of the selected 37 cities.

4.2. Data Sources

Air quality index data were obtained from the "2019 National Air Quality Monthly Report" published by the Ministry of Ecology and Environment of the People's Republic of China (http://www.mee.gov.cn/ (accessed on 13 September 2020)). In addition, the specific distributions of the air pollution index data ($PM_{2.5}$, PM_{10}, SO_2, and NO_2) for the 37 cities are presented in Figure 6, while the population data were obtained from the "2020 City Yearbook" for each city provided by the National Bureau of Statistics of China (http://www.stats.gov.cn/ (accessed on 1 September 2020)), and the number of permanent residents in 2019 was selected as the population index (Figure 6). The 37 urban land use classification maps in this study were obtained from the MCD12Q1.006 land use type product of the 2019 MODIS/Terra, provided by the official USGS online platform with a resolution of 500 m (https://www.usgs.gov/ (accessed on 20 September 2020)). The urban land use map was converted to a tiff format by ArcGIS10.3 and then imported into FRAGSTATS for calculations, to obtain the data for patch-level landscape pattern indices (Appendix B).

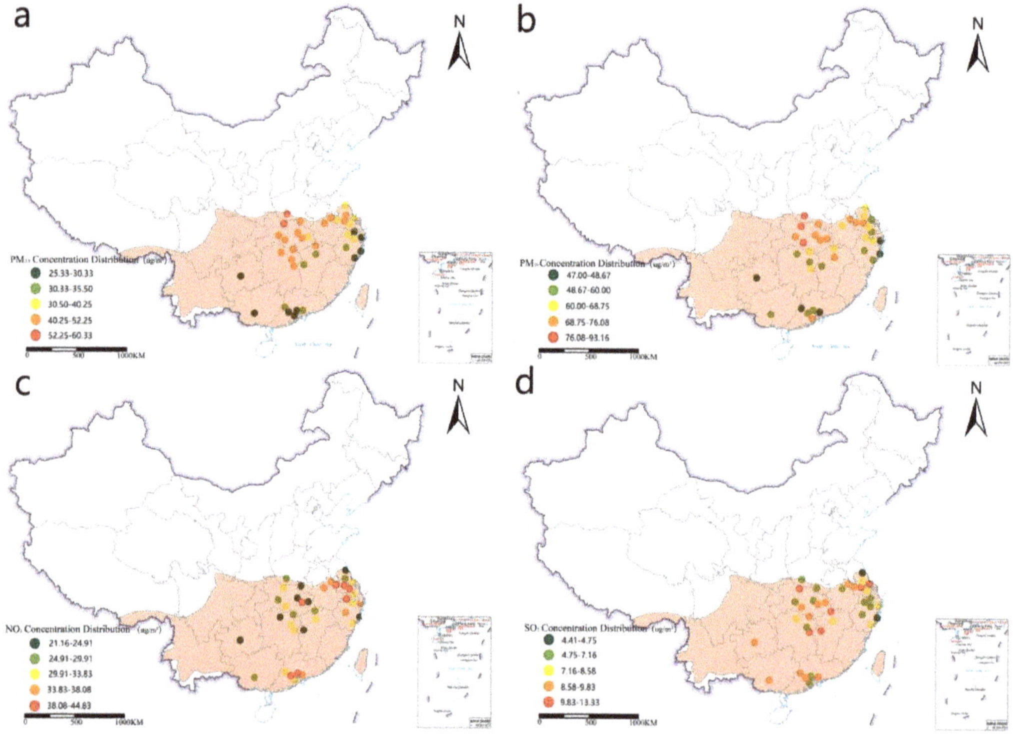

Figure 6. Distribution map of $PM_{2.5}$, PM_{10}, NO_2, and SO_2 concentration. (**a**) Distribution map of $PM_{2.5}$ concentration in the study city. (**b**) Distribution map of PM_{10} concentration in the study city. (**c**) Distribution map of NO_2 concentration in the study city. (**d**) Distribution map of SO_2 concentration in the study city.

4.3. Landscape Pattern Indices

To some extent, the landscape pattern indices capture the intrinsic spatial structure of the environment and enhance the interpretation of spatial patterns and characteristics of the landscape and are now widely used to measure landscape patterns [51]. Based on the results obtained from previous related studies and validation experiments, we selected three landscape indicators to measure the urban landscape pattern of the study region, these indexes are patch proportion in landscape area (PLAND), patch density (PD), and landscape shape index (LSI). The reason why these three indicators are selected is that they are not only widely used to describe the fragmentation, irregularity, and complexity of urban landscape patterns, but also have a more direct macro control effect on the adjustment and optimization of urban green space landscape pattern in the later stage, which is conducive to clarifying and standardizing the follow-up policymaking. Patch proportion of landscape area (PLAND) is the proportion of different landscape patch types in the overall land area percentage measurement, which can help us judge the proportion of this patch type in the overall spatial pattern The patch density (PD) represents the number of patches of a certain type within 100 hectares and can reflect the spatial pattern of the landscape. Its value has a positive correlation with the fragmentation of patch types. The greater the density index, the higher the fragmentation of the patch type. The landscape shape index (LSI) is a robust index used to describe the complexity of urban morphology through the ratio of urban patch perimeter. The larger the LSI value, the more irregular the patch shape, the higher the landscape complexity, and the lower the stability. The above indicators are calculated using FRAGSTATS 4.2 software (Table 2).

Table 2. Landscape pattern indices.

Metrics (Abbreviation)	Calculation Formula	Description
Percentage of landscape types (PLAND)	PLAND = $\frac{1}{A}\sum_{j=1}^{n} aij$	PLAND quantifies the proportional abundance of each patch type in the landscape (percent)
Patch density (PD)	PD = $\frac{1}{A}\sum_{j=1}^{M} Nj$	PD expresses number of patches on a per unit area for considered class (number per 100 hectares)
Landscape shape index (LSI)	LSI = $\frac{1}{A}\sum_{j=1}^{n} E$	LSI expresses the larger LSI value is, the more complex landscape shape is.

4.4. Spatial Regression Modeling Methods

In this study, to investigate the quantitative influence of the landscape patterns of urban green spaces on air pollutants, the PLAND, PD, and LSI of three land use types (forestland, grassland, and farmland) were selected as the independent variables of air pollutant concentration that influence factors for spatial regression analysis.

First, in the selection of the model, after the regression analysis. The calculation of Moran's I index for air pollutants using the GeoDa software revealed that the Moran's I index was 0.57, and the model exhibited a significant spatial correlation. This indicated that air pollutants exhibit spatial interactions, and their effects could spread through adjacent regions, which confirmed that local and regional landscape pattern changes directly or indirectly affected their surrounding air pollutant concentrations. Therefore, the spatial regression model can better reveal the relationship between air pollutant concentration and landscape green space pattern.

Secondly, the regression results of the spatial autoregression model (SAR) and spatial error model (SEM) were compared in Table 3. The evaluation indicators for the goodness of fit of the spatial regression model included the coefficient of determination R2, the natural logarithm of the likelihood function (log-likelihood, logL), Akaike information criterion (AIC), and Schwarz Criterion (SC). Among them, the value range of R2 is (0,1), and the closer R2 is to 1, the better the regression fit of the model. In addition, the higher the logL value, and the smaller the AIC and SC values, the better the regression effect of the spatial regression model. For the selection of the optimal spatial regression model, the R2 and logL values of the SEM model were higher than those of the SAR model in the regression model information of the four pollutants, while the AIC and SC values of the SEM model were smaller for the remaining three pollutants, except for PM10 of the SEM model, which was 0.04 greater than the SAR model, thus implying that the regression results of the SEM model simulating air pollutants were better. thus, implying that the regression results of the SEM model simulating air pollutants were better. After comprehensive analysis and comparison, the SEM was selected for the spatial regression of the four atmospheric pollutants in this study. Therefore, this research established an SEM using the air pollutant concentrations of 37 Chinese cities as the dependent variables and the landscape pattern indices of the green spaces of each city as the independent variables [52,53].

$$Y = \beta_1 X_1 + \beta_2 X_2 + \beta_3 X_3 + \ldots \ldots + \beta_9 X_9 + \varepsilon \quad (1)$$

where Y denotes the dependent variable, i.e., pollutant concentration; Wy is the spatial weight matrix; ρ denotes the spatial regression coefficient of the spatial weight matrix WY, which is adapted to indicate the spatial interaction of air pollution; X_1–X_9 represents the impact factors of landscape pattern indexes (PLAND, PD, LSI) for three types of urban green space: forest land, grassland, and farmland, respectively (The independent variables); β_1–β_9 denote the regression coefficients of independent factors such as landscape pattern indices; ε denotes the random error. This was ultimately utilized to establish the SEM spatial regression model between pollutant concentrations and landscape patterns in 2019.

Table 3. Information comparison of SEM and SAR models.

	Model	R^2	LogL	AIC	SC
$PM_{2.5}$	SEM	0.747	−116.730	265.46	291.662
	SAR	0.716	−116.772	267.543	295.382
PM_{10}	SEM	0.744	−123.79	279.581	305.782
	SAR	0.73	−122.789	279.577	307.416
NO_2	SEM	0.549	−108.694	249.39	275.591
	SAR	0.525	−111.464	256.928	284.767
SO_2	SEM	0.6339	−60.004	152.01	178.211
	SAR	0.41	−67.446	168.894	196.733

5. Conclusions

Urban green space is an effective tool to improve air quality. Quantifying the relationship between urban green space landscape patterns and air pollution concentration is of great significance to promote the public health level and sustainable development of residents in high-density population areas. This study takes 37 garden cities with subtropical monsoon climate in China as the research object, selects the urban air quality monitoring data and land use remote sensing data in 2019, carries out regression analysis on urban air pollutant concentration and landscape pattern index through spatial regression model method, and explores the relationship between landscape pattern index and air pollutant concentration, According to the above regression analysis results, the landscape pattern index threshold with significant correlation with air pollutant concentration was explored. The specific conclusions are as follows:

(1) The landscape pattern of urban green space was significantly correlated with the concentrations of $PM_{2.5}$, NO_2, and SO_2 pollutants in the air, while the concentrations of PM_{10} pollutants were not significantly affected by the green space pattern.

(2) Among them, the patch shape index (LSI), patch density (PD), and patch proportion in landscape area (PLAND) of forest land can affect the concentration of $PM_{2.5}$, NO_2 and SO_2, respectively. The PLAND, PD, and LSI of grassland and farmland can also have an additional impact on the concentration of SO_2 pollutants.

(3) The study also found that there was a significant threshold effect on the impact mechanism of urban green space landscape pattern indicators (LSI, PD, PLAND) on the concentrations of $PM_{2.5}$, NO_2, and SO_2 air pollutants. When the PD value of forest land is about 0.072, the overall value of urban NO_2 pollutant concentration reaches the optimum. When the land and PD values of forest land are away from 50% and 0.038, the land and PD values of grassland are away from 3.33% and 0.121, and the LSI of grassland reaches 14.13, the urban SO_2 pollutant concentration reduction effect is the best.

The results of this study not only clarify the impact mechanism of the landscape pattern of urban green space on air quality but also propose a polynomial-based threshold effect estimation method, which provides quantitative reference and scientific basis for the optimization and updating of urban green space landscape patterns to promote public health.

Author Contributions: C.W.: Conceptualization, Methodology, Software; M.G.: Writing—Reviewing and Editing; J.J.: Data curation, Writing—Original draft preparation; Y.Y.: Visualization, Investigation; Y.R.: Software, Methodology; Y.W.: Validation; J.C.: Supervision. All authors have read and agreed to the published version of the manuscript.

Funding: This research was funded by the National Key R&D Program of China (Funder: Jun Jin; Funding No.: 2020YFD1100405); the National Natural Science Foundation of China (Funder: Jiajie Cao; Funding No.: 32071832); the Humanity and Social Sciences Youth Foundation of the Ministry of

Education of China (Funder: Chengkang Wang; Funding No.: 18YJCZH167) and the Natural Science Foundation of Jiangsu Province (Funder: Chengkang Wang; Funding No.: BK20190751).

Data Availability Statement: Not applicable.

Conflicts of Interest: The authors declare no conflict of interest.

Appendix A

Table A1. The effect of landscape patterns of urban green spaces on air pollution levels.

		Greening Indicators	Air Quality Indicators				Social Indicators
Serial Number	Name	Greening Rate	$PM_{2.5}$ (μg/m^3)	PM_{10} (μg/m^3)	NO_2 (μg/m^3)	SO_2 (μg/m^3)	Population Ten Thousand
1	Guiyang	38.20	26.66	47.00	21.16	9.66	497.14
2	Taizhou	38.20	26.66	48.67	21.66	4.41	614.00
3	Nanning	34.08	30.33	52.75	29.33	9.08	734.48
4	Ningbo	37.87	28.50	47.33	35.50	7.91	854.20
5	Wenzhou	34.49	28.00	52.92	33.83	7.58	930.00
6	Jiangmen	41.62	26.66	73.83	32.08	7.00	463.03
7	Zhaoqing	37.99	31.75	48.00	33.50	9.33	415.17
8	Jinhua	38.15	32.41	54.4	34.41	7.16	562.40
9	Yichun	44.65	35.42	59.41	23.75	13.33	558.26
10	Yancheng	39.10	39.50	66.08	24.00	4.75	720.89
11	Jiaxing	36.15	25.33	56.25	32.75	6.66	480.00
12	Nantong	40.00	36.66	55.00	31.58	10.50	731.80
13	Dongguan	37.50	31.91	48.42	36.58	9.66	846.45
14	Shaoxing	37.19	38.00	61.58	31.58	6.91	505.70
15	Foshan	42.50	29.75	56.16	41.08	9.08	815.86
16	Nanchang	38.51	35.5	69.33	33.50	8.58	560.05
17	Guangzhou	39.91	30.00	52.66	44.83	6.83	1530.59
18	Changde	39.31	47.66	60.00	22.66	8.00	577.13
19	Xiaogan	unavailable	43.33	72.25	21.25	7.08	492.10
20	Huanggang	unavailable	40.25	73.08	24.91	9.66	633.30
21	Jiujiang	44.75	45.92	63.50	29.91	10.91	492.03
22	Yueyang	39.41	43.5	67.83	26.75	8.75	577.13
23	Taizhou	38.6	43.92	67.08	28.00	7.50	463.61
24	Changsha	35.25	47.08	57.41	33.08	7.08	839.45
25	Xinyang	38.00	48.25	76.08	23.91	6.33	646.00
26	Hangzhou	37.23	37.91	66.41	41.25	6.75	1036.00
27	Zhuzhou	34.91	47.25	65.25	33.33	10.75	402.85
28	Yichang	36.10	52.25	72.83	29.16	7.08	413.79
29	Wuxi	39.84	39.00	68.75	39.91	8.16	659.15
30	Hefei	39.75	43.58	65.75	38.08	6.16	818.90
31	Nannjing	41.00	39.75	69.16	41.75	9.83	850.00
32	Jingzhou	33.10	46.50	83.00	32.08	9.25	557.01
33	Chuzhou	42.86	48.25	72.08	35.08	9.66	414.70
34	Changzhou	39.20	46.83	71.00	41.00	10.41	473.60
35	WUhan	34.47	45.25	70.75	44.16	8.83	1121.20
36	Xiangyang	33.34	60.33	84.58	31.58	10.50	568.00
37	Nanyang	38.10	59.66	93.16	28.83	6.33	1003.00

Appendix B

Table A2. The data for patch-level landscape pattern indices.

Number	City	Forest-PL	Forest-PD	Forest-LSI	Grass-PL	Grass-PD	Grass-LSI	Water-PL	Water-PD	Water-LSI	Farm-PL	Farm-PD	Farm-LSI	Construction-PL	Construction-PD	Construction-LSI
1	Guiyang	67.91	0.02	14.36	0.15	0.00	5.62	0.45	0.00	3.86	27.90	0.03	18.40	3.57	0.00	7.29
2	Taizhou	78.94	0.01	13.00	0.80	0.01	11.81	2.54	0.01	10.71	9.30	0.04	23.85	8.36	0.02	13.31
3	Nanning	57.40	0.02	23.86	1.39	0.02	24.82	1.11	0.01	11.88	37.16	0.02	27.25	2.94	0.01	15.06
4	Ningbo	66.70	0.02	14.07	1.42	0.02	14.65	3.21	0.01	12.45	11.36	0.038	20.95	17.22	0.02	12.41
5	Wenzhou	85.19	0.01	8.59	1.00	0.01	14.13	1.60	0.01	10.37	4.62	0.02	18.27	7.50	0.01	10.79
6	Jiangmen	63.61	0.02	19.43	2.01	0.03	18.94	5.43	0.02	14	22.74	0.02	22.09	6.04	0.01	10.31
7	Zhaoqing	92.56	0.00	7.93	1.50	0.02	21.04	2.17	0.01	12.42	1.98	0.01	13.48	1.77	0.01	11.32
8	Jinhua	83.64	0.01	12.86	0.18	0.01	7.35	0.32	0.00	7.18	8.01	0.03	21.79	7.83	0.02	17.34
9	Yichun	72.37	0.02	18.35	1.93	0.02	24.09	0.83	0.01	12.24	23.26	0.02	24.02	1.60	0.01	11.61
10	Yancheng	5.12	0.03	27.57	1.86	0.02	20.31	5.84	0.01	10.80	83.39	0.00	8.80	3.53	0.01	11.79
11	Jiaxing	16.13	0.07	23.68	1.31	0.01	6.85	4.45	0.01	9.24	57.23	0.01	11.89	20.87	0.03	12.25
12	Nantong	10.85	0.04	24.88	0.31	0.01	7.95	1.83	0.00	5.64	78.03	0.01	9.15	8.92	0.01	10.51
13	Dongguan	26.50	0.04	11.61	5.43	0.07	14.27	4.42	0.03	9.57	0.38	0.01	4.45	62.62	0.01	6.71
14	Shaoxing	76.63	0.01	12.29	1.38	0.01	11.92	0.89	0.01	8.48	10.43	0.03	18.52	10.66	0.02	12.65
15	Foshan	44.00	0.02	13.59	2.68	0.04	14.49	8.07	0.03	14.02	0.83	0.01	5.38	42.85	0.02	9.32
16	Nanchang	21.78	0.05	22.99	4.14	0.05	22.72	17.85	0.02	11.95	49.34	0.01	16.50	6.79	0.01	7.06
17	Guangzhou	64.12	0.02	15.59	3.01	0.03	15.85	2.08	0.02	13.69	7.11	0.03	17.65	23.47	0.01	11.66
18	Changde	61.78	0.02	22.29	0.92	0.01	18.44	4.00	0.01	17.97	32.21	0.01	24.54	1.08	0.00	8.46
19	Xiaogan	22.24	0.04	22.67	4.32	0.04	22.05	3.58	0.01	8.97	67.43	0.01	14.38	2.43	0.01	8.29
20	Huanggang	22.24	0.04	22.67	1.56	0.02	21.73	4.92	0.01	16.62	34.45	0.02	28.23	2.35	0.01	11.59
21	Jiujiang	75.83	0.01	14.01	1.40	0.02	20.23	9.90	0.00	11.38	11.13	0.02	21.65	1.66	0.01	10.75
22	Yueyang	56.73	0.03	19.11	2.17	0.02	19.63	11.53	0.01	13.25	28.05	0.017	23.50	1.48	0.00	7.71
23	Taizhou	21.62	0.06	24.27	6.09	0.05	21.13	7.02	0.01	10.55	54.43	0.018	13.40	10.76	0.015	10.47
24	Changsha	82.30	0.01	12.73	0.71	0.01	11.18	0.63	0.00	7.58	11.05	0.03	23.63	5.29	0.01	8.13
25	Xinyang	28.20	0.02	13.88	0.74	0.01	16.47	1.54	0.01	10.94	68.01	0.01	9.65	1.50	0.01	9.47
26	Hangzhou	83.47	0.01	9.68	0.36	0.01	12.21	3.81	0.00	11.94	1.97	0.01	16.25	10.36	0.01	11.57
27	Zhuzhou	90.53	0.00	8.78	0.39	0.01	9.68	0.66	0.00	8.50	6.51	0.02	18.69	1.90	0.00	7.06
28	Yichang	87.35	0.01	7.36	0.57	0.01	15.34	1.21	0.00	12.80	9.81	0.00	12.80	1.03	0.00	8.33
29	Wuxi	26.43	0.05	21.93	2.47	0.02	13.26	19.74	0.01	5.33	18.94	0.03	16.54	32.39	0.02	10.37
30	Hefei	9.52	0.03	21.98	1.12	0.02	15.70	8.41	0.01	5.91	74.66	0.01	10.69	6.28	0.01	7.56
31	Nannjing	31.67	0.05	27.87	1.79	0.03	14.853	10.58	0.01	8.36	38.99	0.03	21.35	16.85	0.02	9.40
32	Jingzhou	17.56	0.05	33.33	1.81	0.03	21.83	14.92	0.01	16.36	63.47	0.01	16.36	2.21	0.01	9.52
33	Chuzhou	7.35	0.03	19.66	1.77	0.03	21.67	4.80	0.01	13.71	83.47	0.00	11.46	2.59	0.01	10.91
34	Changzhou	34.23	0.05	23.64	2.37	0.02	11.87	9.09	0.01	6.11	29.42	0.03	18.12	24.88	0.02	8.55
35	Wuhan	35.31	0.04	32.20	4.43	0.05	24.25	15.11	0.02	16.60	33.69	0.032	24.62	11.29	0.01	10.32
36	Xiangyang	52.55	0.02	12.00	1.45	0.02	21.20	1.23	0.01	12.96	41.86	0.01	14.72	2.90	0.01	17.15
37	Nanyang	33.64	0.01	12.49	2.13	0.02	25.68	1.52	0.00	6.34	59.12	0.01	15.14	3.57	0.025	28.08

References

1. Gianfredi, V.; Buffoli, M.; Rebecchi, A.; Croci, R.; Oradini-Alacreu, A.; Stirparo, G.; Marino, A.; Odone, A.; Capolongo, S.; Signorelli, C. Association between Urban Greenspace and Health: A Systematic Review of Literature. *Int. J. Environ. Res. Public Health* **2021**, *18*, 5137. [CrossRef] [PubMed]
2. Reyes-Riveros, R.; Altamirano, A.; De La Barrera, F.; Rozas-Vásquez, D.; Vieli, L.; Meli, P. Linking public urban green spaces and human well-being: A systematic review. *Urban For. Urban Green.* **2021**, *61*, 127105. [CrossRef]
3. Barrett, E.; Apodaca, K. Response to "Alcohol and COVID-19: How Do We Respond to This Growing Public Health Crisis?". *J. Gen. Intern. Med.* **2021**, *36*, 2476. [CrossRef] [PubMed]
4. Jaafari, S.; Shabani, A.A.; Moeinaddini, M.; Danehkar, A.; Sakieh, Y. Applying landscape metrics and structural equation modeling to predict the effect of urban green space on air pollution and respiratory mortality in Tehran. *Environ. Monit. Assess.* **2020**, *192*, 412. [CrossRef] [PubMed]
5. Kim, S.Y.; Kim, S.H.; Wee, J.H.; Min, C.; Han, S.M.; Kim, S.; Choi, H.G. Short and long term exposure to air pollution increases the risk of ischemic heart disease. *Sci. Rep.* **2021**, *11*, 51085.
6. Luo, J.; Niu, Y.; Zhang, Y.; Zhang, M.; Tian, Y.; Zhou, X. Dynamic analysis of retention $PM_{2.5}$ by plant leaves in rainfall weather conditions of six tree species. *Energy Sources Part A Recovery Util. Environ. Eff.* **2020**, *42*, 1014–1025. [CrossRef]
7. Cai, L.; Zhuang, M.; Ren, Y. A landscape scale study in Southeast China investigating the effects of varied green space types on atmospheric PM2.5 in mid-winter. *Urban For. Urban Green.* **2020**, *49*, 126607. [CrossRef]
8. Latha, K.M.; Highwood, E. Studies on particulate matter (PM10) and its precursors over urban environment of Reading, UK. *J. Quant. Spectrosc. Radiat. Transf.* **2006**, *101*, 367–379. [CrossRef]
9. Beckett, K.P.; Freer-Smith, P.H.; Taylor, G. Urban woodlands: Their role in reducing the effects of particulate pollution. *Environ. Pollut.* **1998**, *99*, 347–360. [CrossRef]
10. Sheng, Q.; Zhu, Z. Physiological Response of European Hornbeam Leaves to Nitrogen Dioxide Stress and Self-recovery. *J. Am. Soc. Hortic. Sci.* **2019**, *144*, 23–30. [CrossRef]
11. Thomas, M.D.; Hill, G.R., Jr. Absorption of sulphur dioxide by alfalfa and its relation to leaf injury. *Plant Physiol.* **1935**, *10*, 291. [CrossRef] [PubMed]
12. Wang, H.; Sun, Y.-Q.; Zhang, M.; Liu, W.J.; Yang, L.; Zhou, X.W. Adsorption ability of air pollutants by indigenous tree species in ta-pieh mountains. *Fresenius Environ. Bull.* **2019**, *28*, 2908–2915.
13. Wang, S.; Li, M. Study on the Principle of Urban Open Space Ecological Planning. *Chin. Landsc. Arch.* **2001**, *5*, 33–37.
14. Sheng, Q.; Zhang, Y.; Zhu, Z.; Li, W.; Xu, J.; Tang, R. An experimental study to quantify road greenbelts and their association with PM2.5 concentration along city main roads in Nanjing, China. *Sci. Total Environ.* **2019**, *667*, 710–717. [CrossRef] [PubMed]
15. Fan, S.; Li, X.; Dong, L. Field assessment of the effects of land-cover type and pattern on PM_{10} and $PM_{2.5}$ concentrations in a microscale environment. *Environ. Sci. Pollut. Res.* **2019**, *26*, 2314–2327. [CrossRef] [PubMed]
16. Ye, L.; Fang, L.; Tan, W.; Wang, Y.; Huang, Y. Exploring the effects of landscape structure on aerosol optical depth (AOD) patterns using GIS and HJ-1B images. *Environ. Sci. Process. Impacts* **2016**, *18*, 265–276. [CrossRef]
17. Yue, F.; Dai, F.; Guo, X. Study on Correlation Between Air Pollutants and Vegetation Coverage in Wuhan Based on Remote Sensing Inversion. *Landsc. Arch.* **2019**, *26*, 76–81.
18. Lei, Y.; Davies, G.M.; Jin, H.; Tian, G.; Kim, G. Scale-dependent effects of urban greenspace on particulate matter air pollution. *Urban For. Urban Green.* **2021**, *61*, 127089. [CrossRef]
19. Zhao, L.; Li, T.; Przybysz, A.; Guan, Y.; Ji, P.; Ren, B.; Zhu, C. Effect of urban lake wetlands and neighboring urban greenery on air PM_{10} and $PM_{2.5}$ mitigation. *Build. Environ.* **2021**, *206*, 108291. [CrossRef]
20. Alam, M.S.; McNabola, A. Exploring the modeling of spatiotemporal variations in ambient air pollution within the land use regression framework: Estimation of PM_{10} concentrations on a daily basis. *J. Air Waste Manag. Assoc.* **2015**, *65*, 628–640. [CrossRef]
21. Zhang, H.; Zhao, Y. Land use regression for spatial distribution of urban particulate matter (PM10) and sulfur dioxide (SO2) in a heavily polluted city in Northeast China. *Environ. Monit. Assess.* **2019**, *191*, 712. [CrossRef] [PubMed]
22. Lee, H.J.; Koutrakis, P. Daily ambient NO2 concentration predictions using satellite ozone monitoring instrument NO2 data and land use regression. *Environ. Sci. Technol.* **2014**, *48*, 2305–2311. [PubMed]
23. Lu, D.; Xu, J.; Yue, W.; Mao, W.; Yang, D.; Wang, J. Response of $PM_{2.5}$ pollution to land use in China. *J. Clean. Prod.* **2019**, *244*, 118741. [CrossRef]
24. de Jalón, S.G.; Burgess, P.; Yuste, J.C.; Moreno, G.; Graves, A.; Palma, J.; Crous-Duran, J.; Kay, S.; Chiabai, A. Dry deposition of air pollutants on trees at regional scale: A case study in the Basque Country. *Agric. For. Meteorol.* **2019**, *278*, 107648. [CrossRef]
25. Anselin, L. Local Indicators of Spatial Association—LISA. *Geogr. Anal.* **1995**, *27*, 93–115. [CrossRef]
26. Getis, A.; Ord, J.K. The Analysis of Spatial Association by Use of Distance Statistics. *Geogr. Anal.* **2010**, *24*, 189–206. [CrossRef]
27. Alastuey, A.; Querol, X.; Plana, F.; Viana, M.; Ruiz, C.R.; De La Campa, A.S.; De La Rosa, J.; Mantilla, E.; Dos Santos, S.G. Identification and chemical characterization of industrial particulate matter sources in southwest Spain. *J. Air Waste Manag. Assoc.* **2006**, *56*, 993–1006. [CrossRef]
28. Chen, L.; Liu, C.; Zhang, L.; Zou, R.; Zhang, Z. Variation in Tree Species Ability to Capture and Retain Airborne Fine Particulate Matter ($PM_{2.5}$). *Sci. Rep.* **2017**, *7*, 3206. [CrossRef]
29. Liang, D.; Ma, C.; Wang, Y.-Q.; Wang, Y.-J.; Chen-Xi, Z. Quantifying $PM_{2.5}$ capture capability of greening trees based on leaf factors analyzing. *Environ. Sci. Pollut. Res.* **2016**, *23*, 21176–21186. [CrossRef]

30. Gasmi, K.; Aljalal, A.; Al-basheer, W.; Abdulahi, M. Analysis of NO_x, NO and NO_2 ambient levels as a function of meteorological parameters in Dhahran, Saudi Arabia. *WIT Trans. Ecol. Environ.* **2017**, *211*, 77–86.
31. Morikawa, H.; Erkin, Ö.C. Basic processes in phytoremediation and some applications to air pollution control. *Chemosphere* **2003**, *52*, 1553–1558. [CrossRef]
32. Guo, Z.; Jiang, H.; Chen, J.; Cheng, M.; Wang, B.; Jiang, Z. The relationship between atmospheric SO_2 column density and land use in Zhejiang, China. In Proceedings of the 2009 Joint Urban Remote Sensing Event, Shanghai, China, 20–22 May 2009; pp. 1–6. [CrossRef]
33. Li, C.; Xu, Y.; Liu, M.; Hu, Y.; Huang, N.; Wu, W. Modeling the Impact of Urban Three-Dimensional Expansion on Atmospheric Environmental Conditions in an Old Industrial District: A Case Study in Shenyang, China. *Pol. J. Environ. Stud.* **2020**, *29*, 3171–3181. [CrossRef]
34. Li, Y.; Shi, J. Characteristics of SO2 Dispersion in Urban of Taiyuan, Shanxi. *J. Shanxi Univ.* **2011**, *34*, 153–157.
35. Shao, T.; Zhou, Z.; Wang, P.; Tang, W.; Liu, X.; Hu, X. Relationship between urban green-land landscape patterns and air pollution in the central district of Yichang city. *Ying Yong Sheng Tai Xue Bao J. Appl. Ecol.* **2004**, *15*, 691–696.
36. Spedding, D.J. Uptake of Sulphur Dioxide by Barley Leaves at Low Sulphur Dioxide Concentrations. *Nature* **1969**, *224*, 1229–1231. [CrossRef]
37. Li, K.; Li, C.; Liu, M.; Hu, Y.; Wang, H.; Wu, W. Multiscale analysis of the effects of urban green infrastructure landscape patterns on $PM_{2.5}$ concentrations in an area of rapid urbanization. *J. Clean. Prod.* **2021**, *325*, 129324. [CrossRef]
38. Yang, Y. The effect of urban form on $PM_{2.5}$ concentration: Evidence from china's 340 prefecture-level cities. *Remote Sens.* **2021**, *14*, 7.
39. Rao, M.; George, L.A.; Rosenstiel, T.N.; Shandas, V.; Dinno, A. Assessing the relationship among urban trees, nitrogen dioxide, and respiratory health. *Environ. Pollut.* **2014**, *194*, 96–104. [CrossRef]
40. Klingberg, J.; Broberg, M.; Strandberg, B.; Thorsson, P.; Pleijel, H. Influence of urban vegetation on air pollution and noise exposure—A case study in Gothenburg, Sweden. *Sci. Total Environ.* **2017**, *599*, 1728–1739. [CrossRef]
41. Geng, Y.C.; Li, J. The value evaluation on the forest resource of forest industry zone of heilongjiang province in acid rain prevention. In Proceedings of the Asia-Pacific Power and Energy Engineering Conference, Wuhan, China, 25–28 March 2011; IEEE Computer Society: Washington, DC, USA, 2011.
42. He, Z.; Shi, X.; Wang, X.; Xu, Y. Urbanisation and the geographic concentration of industrial SO_2 emissions in China. *Urban Stud.* **2017**, *54*, 3579–3596. [CrossRef]
43. Wickham, J.D.; Jones, K.B.; Riitters, K.H.; O'Neill, R.V.; Tankersley, R.D.; Smith, E.R.; Chaloud, D.J. An integrated environmental assessment of the us mid—Atlantic region. *Environ. Manag.* **1999**, *24*, 553–560. [CrossRef] [PubMed]
44. Rafiee, R.; Mahiny, A.S.; Khorasani, N. Assessment of changes in urban green spaces of Mashad city using satellite data. *Int. J. Appl. Earth Obs. Geoinf.* **2009**, *11*, 431–438. [CrossRef]
45. Wu, J.; Xie, W.; Li, W.; Li, J. Effects of Urban Landscape Pattern on PM2.5 Pollution—A Beijing Case Study. *PLoS ONE* **2015**, *10*, e0142449. [CrossRef] [PubMed]
46. Xu, G.; Jiao, L.; Zhao, S.; Yuan, M.; Li, X.; Han, Y.; Zhang, B.; Dong, T. Examining the Impacts of Land Use on Air Quality from a Spatio-Temporal Perspective in Wuhan, China. *Atmosphere* **2016**, *7*, 62. [CrossRef]
47. Li, C.; Zhang, K.; Dai, Z.; Ma, Z.; Liu, X. Investigation of the impact of land-use distribution on $PM_{2.5}$ in Weifang: Seasonal variations. *Int. J. Environ. Res. Public Health* **2020**, *17*, 5135. [CrossRef]
48. Garcia-Menendez, F.; Monier, E.; Selin, N.E. The role of natural variability in projections of climate change impacts on U.S. ozone pollution. *Geophys. Res. Lett.* **2017**, *44*, 2911–2921. [CrossRef]
49. Lu, D.; Mao, W.; Yang, D.; Zhao, J.; Xu, J. Effects of land use and landscape pattern on $PM_{2.5}$ in Yangtze River Delta, China. *Atmos. Pollut. Res.* **2018**, *9*, 705–713. [CrossRef]
50. Xiao, H.; Huang, Z.; Zhang, J.; Zhang, H.; Chen, J.; Zhang, H.; Tong, L. Identifying the impacts of climate on the regional transport of haze pollution and inter-cities correspondence within the Yangtze River Delta. *Environ. Pollut.* **2017**, *228*, 26–34. [CrossRef]
51. Parker, D.C.; Evans, T.; Meretsky, V. *Measuring Emergent Properties of Agent-Based Landcover/Landuse Models Using Spatial Metrics*; Society for Computational Economics: Dallas, TX, USA, 2001.
52. Lee, W.J.; Park, C. Prediction of apartment prices per unit in Daegu-Gyeongbuk areas by spatial regression models. *J. Korean Data Inf. Sci. Soc.* **2015**, *26*, 561–568. [CrossRef]
53. Tu, M.; Liu, Z.; He, C.; Fang, Z.; Lu, W. The relationships between urban landscape patterns and fine particulate pollution in China: A multiscale investigation using a geographically weighted regression model. *J. Clean. Prod.* **2019**, *237*, 117744. [CrossRef]

plants

MDPI

Article

Varying Responses of Vegetation Greenness to the Diurnal Warming across the Global

Jie Zhao [1,2], Kunlun Xiang [3], Zhitao Wu [1] and Ziqiang Du [1,*]

[1] Institute of Loess Plateau, Shanxi University, Taiyuan 030006, China
[2] College of Natural Resources and Environment, Northwest A & F University, Xianyang 712100, China
[3] Guangdong Ecological Meteorology Center, Guangzhou 510275, China
[*] Correspondence: duzq@sxu.edu.cn; Tel.: +86-25-8542-7231

Abstract: The distribution of global warming has been varying both diurnally and seasonally. Little is known about the spatiotemporal variations in the relationships between vegetation greenness and day- and night-time warming during the last decades. We investigated the global inter- and intra-annual responses of vegetation greenness to the diurnal asymmetric warming during the period of 1982–2015, using the normalized different vegetation index (NDVI, a robust proxy for vegetation greenness) obtained from the NOAA/AVHRR NDVI GIMMS3g dataset and the monthly average daily maximum (T_{max}) and minimum temperature (T_{min}) obtained from the gridded Climate Research Unit, University of East Anglia. Several findings were obtained: (1) The strength of the relationship between vegetation greenness and the diurnal temperature varied on inter-annual and seasonal timescales, indicating generally weakening warming effects on the vegetation activity across the global. (2) The decline in vegetation response to T_{max} occurred mainly in the mid-latitudes of the world and in the high latitudes of the northern hemisphere, whereas the decline in the vegetation response to T_{min} primarily concentrated in low latitudes. The percentage of areas with a significantly negative trend in the partial correlation coefficient between vegetation greenness and diurnal temperature was greater than that of the areas showing the significant positive trend. (3) The trends in the correlation between vegetation greenness and diurnal warming showed a complex spatial pattern: the majority of the study areas had undergone a significant declining strength in the vegetation greenness response to T_{max} in all seasons and to T_{min} in seasons except autumn. These findings are expected to have important implications for studying the diurnal asymmetry warming and its effect on the terrestrial ecosystem.

Keywords: varying response; diurnal warming; vegetation activity; NDVI

Citation: Zhao, J.; Xiang, K.; Wu, Z.; Du, Z. Varying Responses of Vegetation Greenness to the Diurnal Warming across the Global. *Plants* **2022**, *11*, 2648. https://doi.org/10.3390/plants11192648

Academic Editors: Jie Gao, Weiwei Huang, Johan Gielis and Peijian Shi

Received: 13 September 2022
Accepted: 5 October 2022
Published: 8 October 2022

1. Introduction

Surface vegetation cover is a core component of terrestrial ecosystems, as it plays an important role in connecting material circulation and energy flow in the atmosphere, hydrosphere, and soil circles [1–4]. Current satellite observations have revealed long-term greening and browning changes in the vegetation greenness (i.e., the normalized difference vegetation index, NDVI) across the global from the 1980s until the present time [2,5–8]. Such variations in the vegetation activity are closely related to climate change, particularly the increasing global temperatures over the last several decades [1,8–11], which can be characterized by the increase in both daytime and night-time temperature [12,13]. The link between global warming and vegetation activity is overwhelming. Recent studies have investigated the responses of satellite-derived vegetation productivity to temperatures during daytime and nighttime between regions and ecosystem [12,14–20]. However, most of the measurements were static and insufficient to clarify the varying responses of vegetation dynamics to climate warming [21].

The relationship between vegetation activity and temperature variations is not fixed; it constantly changes over time because of the interventions of other environmental and anthropogenic factors, in the face of persistent global warming [22,23]. Therefore, the current focus is on the dynamic responses of vegetation greenness to temperature change over time. For instance, Andreu-Hayles et al. [24] explored the varying boreal forest responses to arctic environmental change near the Firth River, Alaska. Piao et al. [23] found a weakening relationship between inter-annual temperature variability and northern vegetation activity. Fu et al. [25] reported declining global warming effects on the phenology of spring leaf unfolding. Cong et al. [26] examined the varied responses of vegetation activity to climate change on the grasslands of the Tibetan Plateau. He et al. [21] identified obvious shifts in the relationship between vegetation growth and temperature in China's temperate desert and rainforest areas. Although these findings confirm the dynamic relationship between vegetation greenness and temperature change, the link between diurnal warming and vegetation activity, particularly the variations in their correlation over time, remains unclear.

The main objective of this study was to explore the varying relationships between vegetation greenness and day- and night-time warming on temporal and spatial scale, by simultaneously employing time-series satellite-derived vegetation NDVI data and gridded meteorological data. The understandings drawn from the findings are expected to have important implications for studying the diurnal asymmetry warming and its effect on the terrestrial ecosystem.

2. Results

2.1. Trends of Correlations between Vegetation Greenness and Diurnal Warming

2.1.1. Inter-Annual Changes in $R_{NDVI\text{-}Tmax}$ and $R_{NDVI\text{-}Tmin}$

Recent changes in satellite-based vegetation greenness across different areas were closely related to global warming. Figure 1 shows the inter-annual variation in the partial correlation coefficients of the NDVI and diurnal temperature for the period of 1982–2015. In the mid-latitudes of the southern hemisphere, $R_{NDVI\text{-}Tmax}$ is approximately 0.323 for the first window of 1982–1998 and decreased to −0.436 for the last window of 1999–2015. In the high latitudes of the northern hemisphere, $R_{NDVI\text{-}Tmin}$ shows a significant increasing trend ($\beta = 0.098$, $p < 0.01$) in the first six windows and then a significant decreasing trend ($\beta = -0.056$, $p < 0.01$) in the remaining windows. Similarly, in the low latitudes of the southern hemisphere, $R_{NDVI\text{-}Tmin}$ shows a significant increasing trend ($\beta = 0.039$, $p < 0.01$) in the first seven windows, followed by a significant decreasing trend ($\beta = -0.055$, $p < 0.01$) in the remaining windows. In contrast, in the mid-latitudes of the southern hemisphere, $R_{NDVI\text{-}Tmin}$ is generally on an upward trend, increasing from 0.022 in the window of 1982–1998 to 0.737 in the window of 1998–2014. These findings indicate a notable shift in the responses of vegetation activity to diurnal temperatures over the last three decades.

Overall, $R_{NDVI\text{-}Tmax}$ exhibits only a significant downward trend in the mid-latitudes of the southern hemisphere ($\beta = -0.051$, $p < 0.05$). In contrast, $R_{NDVI\text{-}Tmin}$ shows a significant upward trend in the mid-latitudes of the southern hemisphere ($\beta = 0.032$, $p < 0.05$), but a significant downward trend in the low latitudes of the southern hemisphere ($\beta = -0.021$, $p < 0.05$). $R_{NDVI\text{-}Tmax}$ does not exhibit a significant trend in the northern hemisphere ($p > 0.05$), and $R_{NDVI\text{-}Tmin}$ exhibits a significant downward trend only in the high latitudes of the northern hemisphere ($\beta = -0.021$, $p < 0.05$).

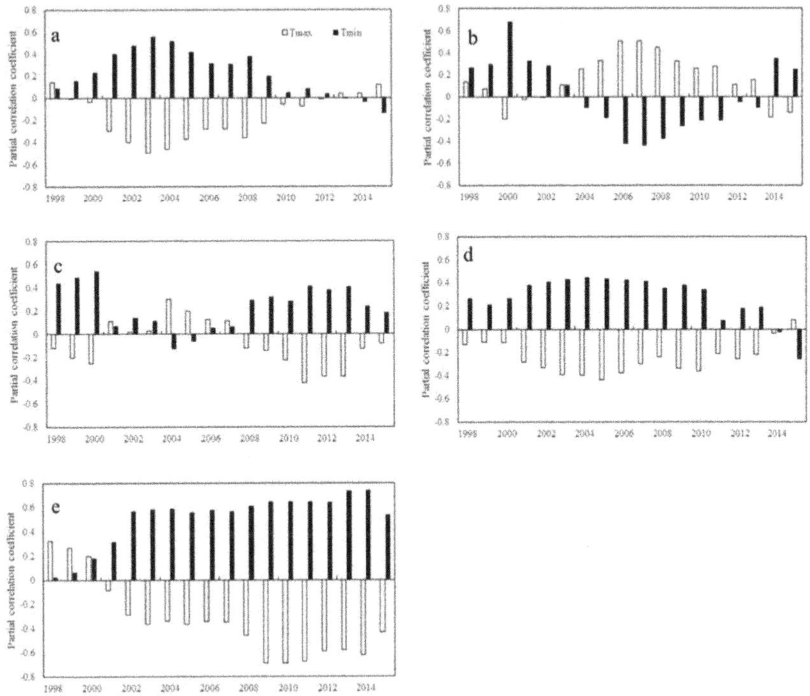

Figure 1. Temporal variations in the partial correlation coefficients between mean annual NDVI and the diurnal temperature (T_{max} and T_{min}) for each 17-year moving window across latitudes intervals. (**a–e** represents latitudes intervals at 60°~90° N, 30°~60° N, 0°~30° N, −30°~0° S, and −60°~0° S, respectively (N and S indicates the Northern and southern hemisphere, respectively). The x axis is the last year of the 17-year moving-window (for example, 1998 stands for a moving-window from 1982 to 1998, . . . , 2015 stands for a moving-window from 1999 to 2015). The Y axis is the partial correlation coefficients).

2.1.2. Inter-Annual Changes in $R_{NDVI\text{-}Tmax}$ and $R_{NDVI\text{-}Tmin}$

Variations in the temperature on a seasonal scale may not correspond to similar changes on inter-annual scale. The photosynthesis of vegetation in different growth stages depends on the seasonal cycle of the temperature. Therefore, the sensitivity of vegetation greenness to day- and nighttime warming and variations in seasonality is likely to be highly complex.

During spring, $R_{NDVI\text{-}Tmax}$ shows a significant downward trend at the mid-latitudes in both the northern hemisphere ($\beta = -0.031$, $p < 0.05$) and the southern hemisphere ($\beta = -0.055$, $p < 0.05$) (Figure 2). $R_{NDVI\text{-}Tmin}$ shows a significant downward trend ($\beta = -0.040$, $p < 0.05$) at the low latitudes in the northern hemisphere while a significant upward trend ($\beta = 0.035$, $p < 0.05$) at the mid-latitudes in the southern hemisphere (Figure 2).

During summer, $R_{NDVI\text{-}Tmax}$ shows a significant downward trend not only at the low latitudes in the northern hemisphere ($\beta = -0.020$, $p < 0.05$), but also at the low and mid-latitudes in the southern hemisphere ($\beta = -0.029$, $p < 0.05$) (Figure 3). $R_{NDVI\text{-}Tmin}$ shows a significant downward trend ($\beta = -0.022$, $p < 0.05$) at the high latitudes in the northern hemisphere, but a significant upward trend ($\beta = 0.023$, $p < 0.05$) at the mid-latitudes in the southern hemisphere (Figure 3).

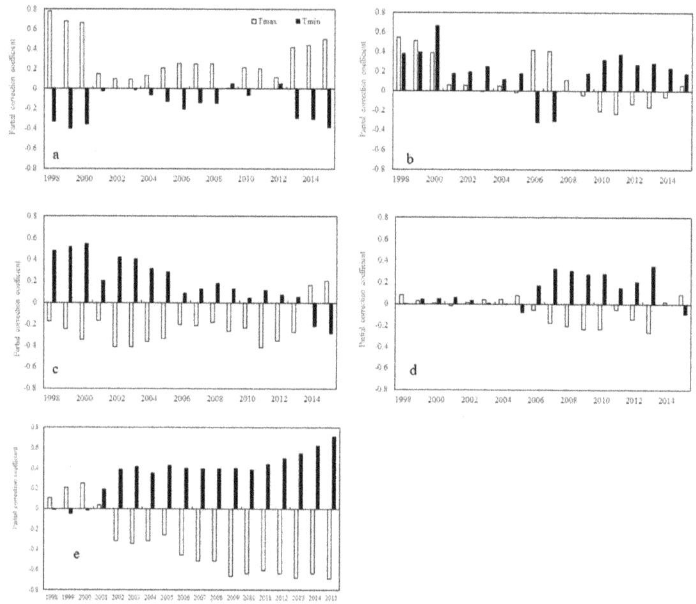

Figure 2. Temporal variations in the partial correlation coefficients between mean NDVI and the diurnal temperature (T_{max} and T_{min}) in spring for each 17-year moving window across latitudes intervals. (**a–e** represents latitudes intervals at 60°~90° N, 30°~60° N, 0°~30° N, −30°~0° S, and −60°~0° S, respectively (N and S indicates the Northern and southern hemisphere, respectively). The x axis is the last year of the 17-year moving-window (for example, 1998 stands for a moving-window from 1982 to 1998, . . . , 2015 stands for a moving-window from 1999 to 2015). The Y axis is the partial correlation coefficients).

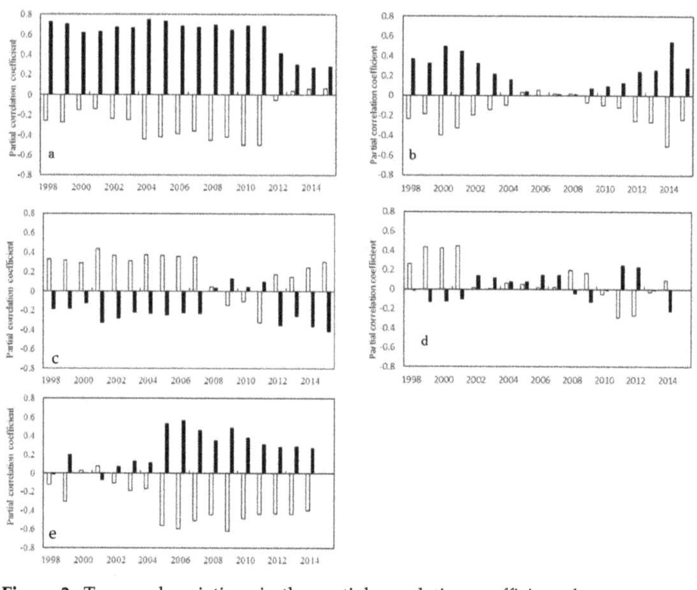

Figure 3. Temporal variations in the partial correlation coefficients between mean NDVI and the diurnal temperature (T_{max} and T_{min}) in summer for each 17-year moving window across latitudes in-

tervals. (**a–e** represents latitudes intervals at 60°~90° N, 30°~60° N, 0°~30° N, −30°~0° S, and −60°~0° S, respectively (N and S indicates the Northern and southern hemisphere, respectively). The x axis is the last year of the 17-year moving-window (for example, 1998 stands for a moving-window from 1982 to 1998, . . . , 2015 stands for a moving-window from 1999 to 2015). The Y axis is the partial correlation coefficients).

During autumn, $R_{NDVI\text{-}Tmax}$ shows a significant upward trend at the low-latitudes ($\beta = 0.057$, $p < 0.01$) and the mid-latitudes ($\beta = 0.021$, $p < 0.01$) in the northern hemisphere. (Figure 4) $R_{NDVI\text{-}Tmin}$ shows a significant downward trend at the mid-latitudes ($\beta = -0.017$, $p < 0.05$) and the low latitudes ($\beta = -0.048$, $p < 0.01$) in the northern hemisphere (Figure 4). However, in the southern hemisphere, we did not find any statistically significant variations in $R_{NDVI\text{-}Tmax}$ and $R_{NDVI\text{-}Tmin}$.

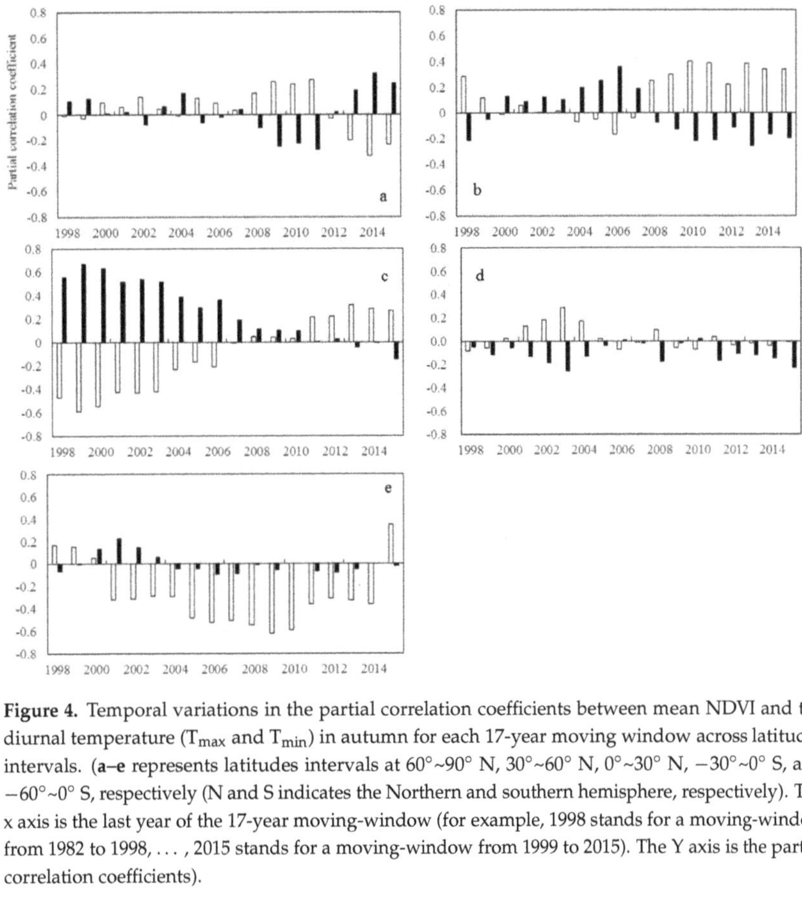

Figure 4. Temporal variations in the partial correlation coefficients between mean NDVI and the diurnal temperature (T_{max} and T_{min}) in autumn for each 17-year moving window across latitudes intervals. (**a–e** represents latitudes intervals at 60°~90° N, 30°~60° N, 0°~30° N, −30°~0° S, and −60°~0° S, respectively (N and S indicates the Northern and southern hemisphere, respectively). The x axis is the last year of the 17-year moving-window (for example, 1998 stands for a moving-window from 1982 to 1998, . . . , 2015 stands for a moving-window from 1999 to 2015). The Y axis is the partial correlation coefficients).

During winter, $R_{NDVI\text{-}Tmax}$ shows a significant upward trend at the mid-latitudes in the northern hemisphere ($\beta = 0.041$, $p < 0.01$) and the low latitudes in the southern hemisphere ($\beta = 0.016$, $p < 0.05$), but a significant downward trend ($\beta = -0.025$, $p < 0.01$) at the mid-latitudes in the southern hemisphere (Figure 5). $R_{NDVI\text{-}Tmin}$ shows a significant downward trend at the low ($\beta = -0.010$, $p < 0.05$) and the mid-latitudes ($\beta = -0.036$, $p < 0.01$) in the northern hemisphere and at the low latitudes ($\beta = -0.025$, $p < 0.01$) in the southern hemisphere, but a significant upward trend at the mid-latitudes ($\beta = 0.027$, $p < 0.01$) in the southern hemisphere (Figure 5). Although we tried to characterize the sensitivity of vegetation greenness for each season, the possible mechanism leading to the seasonal divergence in the response of vegetation greenness to diurnal warming needs to be studied further.

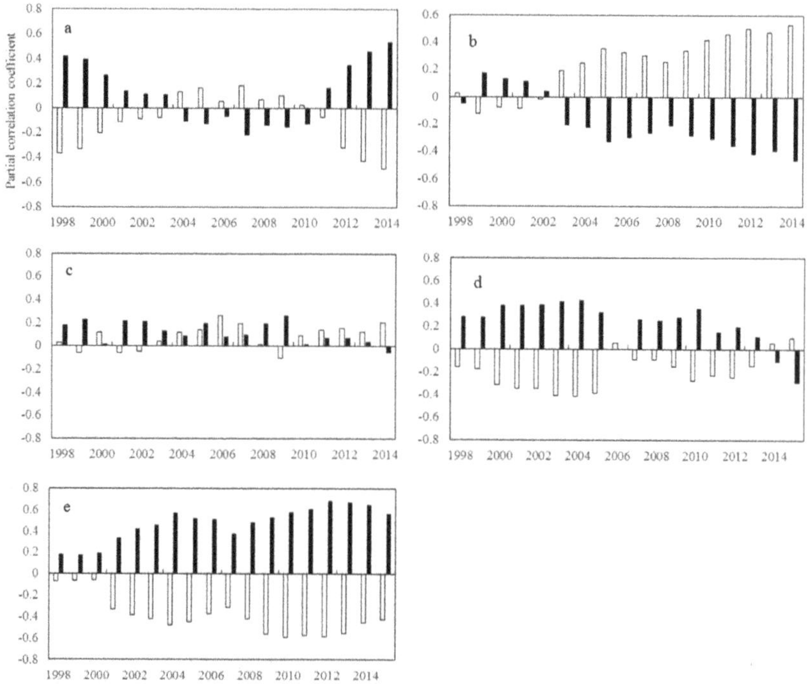

Figure 5. Temporal variations in the partial correlation coefficients between mean NDVI and the diurnal temperature (T_{max} and T_{min}) in winter for each 17-year moving window across latitudes intervals. (**a–e** represents latitudes intervals at 60°~90° N, 30°~60° N, 0°~30° N, −30°~0° S, and −60°~0° S, respectively (N and S indicates the Northern and southern hemisphere, respectively). The x axis is the last year of the 17-year moving-window (for example, 1998 stands for a moving-window from 1982 to 1998, ... , 2015 stands for a moving-window from 1999 to 2015). The Y axis is the partial correlation coefficients).

2.2. *Spatial Patterns of the Trends in the Correlations between Vegetation Greenness and Diurnal Temperatures*

2.2.1. Inter-Annual Patterns of $R_{NDVI\text{-}Tmax}$ and $R_{NDVI\text{-}Tmin}$

Figure 6 shows a summary of the spatial variations in the correlations between vegetation activity and diurnal temperatures over 30 years. The areas with a significant increase in $R_{NDVI\text{-}Tmax}$ account for 26.01% of the total vegetated areas, mainly located in the southeastern part of Oceania, the Ganges Plain, the western plains of Eastern Europe, and the Amazonian Plain. The areas with a significant decrease in $R_{NDVI\text{-}Tmax}$ account for 35.65% of the total vegetated areas, mainly distributed in the northwestern part of Oceania, southern South America, eastern Africa, and the cold temperate zones of the northern hemisphere. About 28.62% of the total vegetated areas show a significant upward trend in $R_{NDVI\text{-}Tmin}$, these areas are mainly located in southern South America, southern Africa, Oceania, and the northwestern boreal regions of North America. Over 32.49% of the total vegetated areas shows a significant downward trend in $R_{NDVI\text{-}Tmin}$, these areas are mainly distributed in eastern Oceania, central Africa, and the tropical regions in the northern hemisphere. In particular, the number of pixels with a significant decrease in $R_{NDVI\text{-}Tmax}$ were greater than those with a significant increase in $R_{NDVI\text{-}Tmax}$ in each latitude interval.

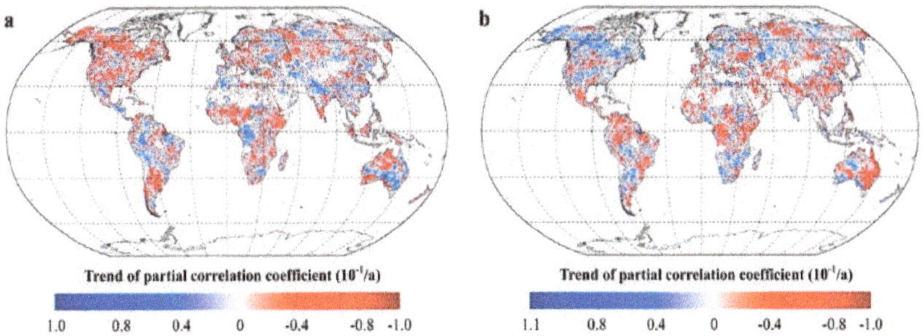

Figure 6. The response of vegetation greenness to the diurnal temperature. (**a**. spatial distribution of the temporal trend of the partial coefficients between mean annual NDVI and T_{max}. **b**. spatial distribution of the temporal trend of the partial coefficients between mean annual NDVI and T_{min}. Supplementary Table S1).

This difference in the number of pixels between $R_{NDVI\text{-}Tmax}$ and $R_{NDVI\text{-}Tmin}$ was most pronounced at the mid-latitudes in the southern hemisphere and the mid- and high latitudes in the northern hemisphere. At the mid- and high latitudes in the northern hemisphere and at the mid-latitudes in the southern hemisphere, the proportion of $R_{NDVI\text{-}Tmin}$ showing a significant upward trend is slightly higher than that of $R_{NDVI\text{-}Tmin}$ showing a significant downward trend. At low latitudes, however, the percentage of areas where $R_{NDVI\text{-}Tmin}$ shows a significant downward trend is much higher than that of areas where $R_{NDVI\text{-}Tmin}$ shows the opposite trend.

Overall, regions with a declining correlation between vegetation greenness and diurnal temperatures were more than those with an increasing correlation. This implies that worldwide vegetation activity on a larger scale became less sensitive to variations in the day- and night temperature with global warming during the last several decades.

2.2.2. Intra-Annual Patterns of $R_{NDVI\text{-}Tmax}$ and $R_{NDVI\text{-}Tmin}$

The seasonal variation in the vegetation greenness response to global warming may increase the complexity of the intra-annual patterns of $R_{NDVI\text{-}Tmax}$ and $R_{NDVI\text{-}Tmin}$. For spring, the areas with positive trends in $R_{NDVI\text{-}Tmax}$ accounts for 25.66% of the total vegetated areas, mostly in the northeastern part of South America, the eastern plains of Eastern Europe, the Iberian Peninsula, the southeastern parts of Australia, and the Ganges Plain (Figure 7). The areas with negative trends in $R_{NDVI\text{-}Tmax}$ accounts for 32.43% of the total vegetated areas, mainly distributed in Western Australia, the Katanga Plateau, and the cold temperate zones in the northern hemisphere. Similarly, about 24.09% of the vegetated areas show positive trends in $R_{NDVI\text{-}Tmin}$, these areas are mainly located in northern North America, southern South America, southern Europe, the Indian Peninsula, and southwestern Australia. Over 32.11% of the vegetated areas show negative trends in $R_{NDVI\text{-}Tmin}$, these areas are mainly distributed in northern South America, central Africa, southeastern Australia, and the temperate regions of the northern hemisphere.

For summer, $R_{NDVI\text{-}Tmax}$ shows significant positive trends in 29.80% of the vegetated areas, located mainly in the northern and eastern North America, southern South America, Armenian plateau, Tulan lowland, Kazakh hills, and Ganges plain (Figure 8). $R_{NDVI\text{-}Tmax}$ shows significant negative trends in 31.78% of the vegetated areas, located mainly in southern North America, Central European Plains, Scandinavia, Northeast Asia, India Peninsula, and Indochina. Similarly, $R_{NDVI\text{-}Tmin}$ shows significant positive trends in 26.68% of the vegetated areas, mainly in the southern Patagonia Plateau of South America, the Yukon Plateau in North America, the south-central United States, the northwestern part of Asia, the Indo-China Peninsula, and Southern Australia. $R_{NDVI\text{-}Tmin}$ shows significant negative trends in 34.71% of the vegetated areas, located mainly in northern South America, southern Africa, northwestern Australia, and the cold temperate zones of the northern hemisphere (Figure 8).

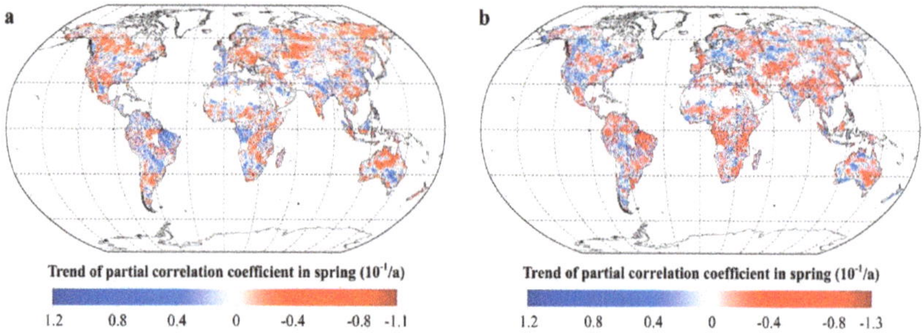

Figure 7. The response of vegetation greenness to the diurnal temperature in spring. (**a**. spatial distribution of the temporal trend of the partial coefficients between spring NDVI and T_{max}. **b**. spatial distribution of the temporal trend of the partial coefficients between spring NDVI and T_{min}. Supplementary Table S2).

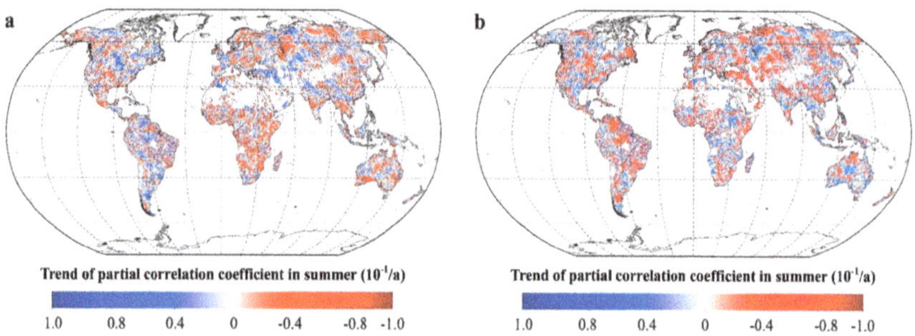

Figure 8. The response of vegetation greenness to the diurnal temperature in summer. (**a**. spatial distribution of the temporal trend of the partial coefficients between summer NDVI and T_{max}. **b**. spatial distribution of the temporal trend of the partial coefficients between summer NDVI and T_{min}. Supplementary Table S3).

For autumn, about 27.43% of the total vegetated areas show significant positive trends in $R_{NDVI\text{-}Tmax}$, these areas are mainly distributed in central Eurasia, central and southern North America, northern Africa, and central Australia (Figure 9). Over 34.07% of the total vegetated areas show significant negative trends in $R_{NDVI\text{-}Tmax}$, these areas are mainly distributed in northern Eurasia, central North America, southern Africa, western Africa, southern South America, and northeastern Australia. In contrast, over 32.67% of the total vegetated areas show significant positive trends in $R_{NDVI\text{-}Tmin}$, these areas are widely distributed in southern Africa, northern Australia, and the cold temperate zones of the northern hemisphere. About 28.76% of the total vegetated areas show significant negative trends in $R_{NDVI\text{-}Tmin}$, these areas are mainly distributed in southern North America, northwestern South America, central Asia, southern Europe, central Africa, and southern Europe (Figure 9).

For winter, the positive trends in $R_{NDVI\text{-}Tmax}$ are found in 25.06% of the total vegetated areas, located mainly in central Asia, eastern Australia, Eastern Europe, and southern China (Figure 10). The negative trends in $R_{NDVI\text{-}Tmax}$ are found in 25.74% of the total vegetated areas, mainly distributed in Western Europe, central and southern North America, central South America, eastern Africa, Western Australia, Indian Peninsula, and Indochina. Similarly, the positive trends in $R_{NDVI\text{-}Tmin}$ are found in 23.32% of the total vegetated areas, located mainly in central North America, Western Europe, southwestern Australia, and the north-central La Plata Plain in South America. The negative trends in $R_{NDVI\text{-}Tmin}$ are found in 26.71% of the total vegetated areas, mainly distributed in central Asia, central and eastern Australia, and Eastern Europe (Figure 10).

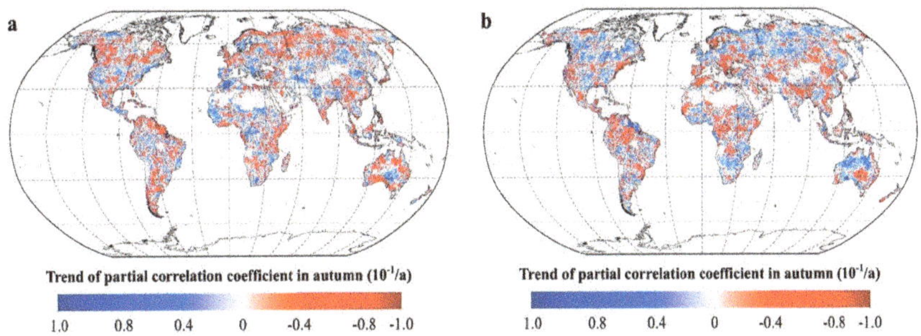

Figure 9. The response of vegetation greenness to the diurnal temperature in autumn. (**a.** spatial distribution of the temporal trend of the partial coefficients between autumn NDVI and T_{max}. **b.** spatial distribution of the temporal trend of the partial coefficients between autumn NDVI and T_{min}. Supplementary Table S4).

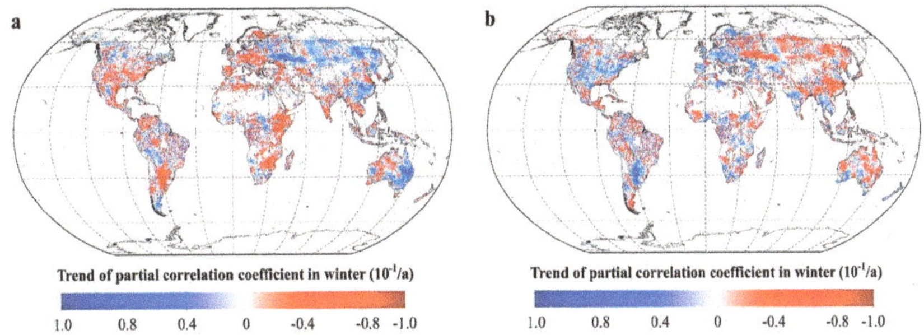

Figure 10. The response of vegetation greenness to the diurnal temperature in winter. (**a.** spatial distribution of the temporal trend of the partial coefficients between winter NDVI and T_{max}. **b.** spatial distribution of the temporal trend of the partial coefficients between winter NDVI and T_{min}. Supplementary Table S5).

3. Discussion

The global climate has undergone continuous warming since the 19th century. Such warming has been characterized by a faster change in the global land surface during the nighttime than during the daytime [12,27,28]. However, the current understanding of the sensitivity of vegetation activity to daytime and nighttime temperatures (not to mean temperature) during the last decades is still poor. Therefore, we applied the moving-window-based partial correlation analysis and ordinary least squares linear regression method within a moving 17-year window to the GIMMS NDVI3g dataset and gridded meteorological data covering the global vegetated areas to reveal the variations in the response of vegetation greenness to the day- and nighttime warming during the period of 1982–2015.

We reported a wide reduction in the vegetation NDVI and diurnal temperature sensitivity since the early 1980s across the global. We also showed that the strength of the relationship between vegetation greenness and diurnal temperature varied among seasons and not just at inter-annual intervals. We further confirmed declining warming effects on vegetation activity, in line with some earlier studies conducted from local to global scale [21–26].

We tried to reveal the dynamic responses of vegetation greenness to day and night warming and explore the difference in the responses of vegetation to such diurnal warming. However, some uncertainties in this study still constrained our understanding of these responses. For example, the changing response could be related to any one, or a combination, of various factors, such as increasing drought stress [29–31], rising CO2 fertilization [22,23], fire disturbances [32,33], global dimming [29], and human-associated activities [21]. These factors were not considered in our analyses because of the lack of sufficient records. Regardless of the cause, our findings on the change in the response of vegetation greenness to diurnal warming have important implications for studies on climate change related to the past as well as the future. Further studies are needed, combining process studies (e.g., ecosystem warming experiments) with multi-scale data, to clarify the mechanisms of the observed decline in the correlation between the diurnal warming and vegetation activity. Our study mainly examined the immediate responses of terrestrial vegetation activity to asymmetric warming. In fact, such asymmetric warming has a non-uniform time-lag impact on terrestrial vegetation growth [34]. Previous studies have also found different responses of vegetation activity to warming across terrestrial ecosystems [12,23]. We did not pay enough attention to this difference in the biome involved. Although the combined effects of hydrothermal regime, ecosystem types, and the time-lag effects increase the uncertainties associated with the weakening relationship between vegetation greenness and current asymmetric warming during day- and night-time, our findings demonstrate the need to consider asymmetric warming and its uneven effect rather than variations in the annual mean temperature when modeling the future behavior of vegetation activity. Hence, significant efforts are required to examine the varying responses of vegetation activity to climate change in such a dynamic climate system in light of the substantial spatial and temporal variations in the diurnal warming and its effects.

4. Materials and Methods

4.1. Datasets

The third-generation NDVI data for the period of 1982–2015 was produced by the Global Inventory Modelling and Mapping Studies group (GIMMS NDVI3g) from the NOAA/AVHRR series satellite imageries, with a spatial resolution of ~0.083° and fortnightly interval. To further minimize the atmospheric noise, the value composites method was used to reconstruct the monthly and yearly NDVI datasets [35]. These NDVI data were aggregated to $0.5° \times 0.5°$ to match the resolution of Meteorological data. Pixels with the mean NDVI (annual and monthly) below 0.1 served as non-vegetation areas and were employed as a mask [5,26]. The GIMMS NDVI3g can be used as a proxy for vegetation greenness and has been widely applied to explore the diverse spatiotemporal-scale vegetation activity [5,12,23,26,34].

Climatological data for the same period, including the monthly average daily maximum (T_{max}) and minimum temperature (T_{min}) and precipitation, with a spatial resolution of ~ $0.5° \times 0.5°$ were obtained from the gridded Climate Research Unit, University of East Anglia (CRU TS 4.01). The CRU TS 4.01 data were produced using angular-distance weighting interpolation. These meteorological materials have been well used in studies on the relationship between global and regional vegetation cover and climate change [7,12,23,34,36].

4.2. Methods

4.2.1. Moving-Window-Based Partial Correlation Analysis

The relationships between vegetation NDVI and temperature were examined using partial correlation analysis (the partial correlation coefficient of NDVI and day- and nighttime temperature were denoted by $R_{NDVI\text{-}Tmax}$ and $R_{NDVI\text{-}Tmin}$, respectively), to eliminate the covariate effects of other factors [12,37–39]. In this case, for instance, the partial correlation coefficient between NDVI and T_{max} was calculated when conditioning the precipitation and T_{min}; and that between NDVI and T_{min} was calculated when conditioning the precipitation and T_{max}. The partial correlations of the NDVI with temperature were calculated in the first 17-year window (1982–1998). The window was then moved forward by 1 year to represent the second 17-year window (1983–1999), and so on until the last 17-year window (1999–2015). Thus, we obtained the partial correlation coefficients within all 17-year windows, which were applied to describe the responses of vegetation greenness to day- and nighttime warming. The NDVI and relevant variables were detrended by a linear detrending method prior to the partial correlation analysis [12,17].

4.2.2. Ordinary Least Squares Linear Regression

To explore the dynamics in the correlation between global vegetation greenness and day- and night temperature, the Ordinary least-squares linear regression analysis method was used to calculate the slope of the partial correlation coefficients of NDVI and T_{max} and T_{min} within each 17-year window over the entire temporal domain. The partial correlation coefficient and P-value were used to determine the relationship and its significance. The slope of the linear regression represents the changing trend in the correlations between the vegetation greenness and diurnal warming during the period 1982–2015. A positive slope indicates a positive growth response to temperature, while a negative slope indicates a weakening relationship with diurnal warming.

$$y_{(t)} = \alpha + \beta t + \varepsilon \quad (1)$$

Here, α and β were the fitted intercept and slope, respectively. A two-tailed *t*-test was used to examine the significance level of the trend. Thus, we described the change in vegetation greenness responses to the diurnal warming from 1982 to 2015.

5. Conclusions

In the most recent period (i.e., 1982–2015), most parts of the world exhibited an overall significant change in the response of vegetation greenness to diurnal warming. On an interannual timescale, the percentage of areas showing a significantly negative trend in the partial correlation coefficient between vegetation greenness and day- and nighttime temperatures was greater than that of areas showing a significant positive trend. For daytime temperatures, this decline in the vegetation response occurred mainly at the mid-latitudes of the world and at the high latitudes of the northern hemisphere. For nighttime temperatures, the decline was primarily concentrated at the low-latitudes across the global. Our study also showed that the strength of the associations between vegetation greenness and diurnal warming varied among seasons. Although the trends in the correlation between vegetation greenness and diurnal warming showed a complex spatial pattern, most of the study areas had undergone a significant declining strength in the vegetation greenness response to daytime temperatures in all seasons and to nighttime temperatures in seasons except autumn. This further implies that the recent reduction in the strength of the response of vegetation greenness to diurnal temperature fluctuations became widespread throughout the world as such as the inter-annual tendency. These findings shew new light on our understanding of the response of vegetation greenness to global warming. We demonstrate the need to carefully consider the asymmetric diurnal warming when unraveling the influence of global warming on terrestrial ecosystem.

Supplementary Materials: The following supporting information can be downloaded at: https://www.mdpi.com/article/10.3390/plants11192648/s1, Table S1: Slopes of the partial correlation coefficients between mean annual NDVI and the duirnal temperature (T_{max} and T_{min}) for each 17-year moving window across latitudes intervals; Table S2: Slopes of the partial correlation coefficients between mean NDVI and the duirnal temperature (T_{max} and T_{min}) in spring for each 17-year moving window across latitudes intervals; Table S3: Slopes of the partial correlation coefficients between mean NDVI and the duirnal temperature (T_{max} and T_{min}) in summer for each 17-year moving window across latitudes intervals; Table S4: Slopes of the partial correlation coefficients between mean NDVI and the duirnal temperature (T_{max} and T_{min}) in autumn for each 17-year moving window across latitudes intervals; Table S5: Slopes of the partial correlation coefficients between mean NDVI and the duirnal temperature (T_{max} and T_{min}) in winter for each 17-year moving window across latitudes intervals.

Author Contributions: Methodology, J.Z., and Z.D.; formal analysis, J.Z. and Z.W.; writing—original draft preparation, Z.D. and K.X.; writing—review and editing, Z.D. and Z.W.; funding acquisition, Z.D. and Z.W. All authors have read and agreed to the published version of the manuscript.

Funding: Z.D. and Z.W. were funded by the National Natural Science Foundation of China (grant number: U1810101, 41161066, 41977412, and 418711934), and Shanxi Province Disciplines Group Construction Project of Service to Industrial Innovation: Ecological Remediation of Soil Pollution Disciplines Group (grant number: 20181401).

Institutional Review Board Statement: Not applicable.

Informed Consent Statement: Not applicable.

Data Availability Statement: The data used in the present work have been listed in the Supplementary Materials.

Acknowledgments: The authors thank the editor, the associate editor, and anonymous reviewers for their helpful and constructive comments.

Conflicts of Interest: The authors declare no conflict of interest.

References

1. Forzieri, G.; Alkama, R.; Miralles, D.G.; Cescatti, A. Satellites reveal contrasting responses of regional climate to the widespread greening of Earth. *Science* **2017**, *356*, 1180–1184. [CrossRef]
2. Pan, N.; Feng, X.; Fu, B.; Wang, S.; Ji, F.; Pan, S. Increasing global vegetation browning hidden in overall vegetation greening: Insights from time-varying trends. *Remote Sens. Environ.* **2018**, *214*, 59–72. [CrossRef]
3. Kafy, A.A.; Faisal, A.-A.; Al Rakib, A.; Fattah, M.A.; Rahaman, Z.A.; Sattar, G.S. Impact of vegetation cover loss on surface temperature and carbon emission in a fastest-growing city, Cumilla, Bangladesh. *Build. Environ.* **2022**, *208*, 108573. [CrossRef]
4. Piao, S.; Wang, X.; Ciais, P.; Zhu, B.; Wang, T.A.O.; Liu, J.I.E. Changes in satellite-derived vegetation growth trend in temperate boreal Eurasia from 1982 to 2006. *Glob. Chang. Biol.* **2011**, *17*, 3228–3239. [CrossRef]
5. De Jong, R.; Verbesselt, J.; Zeileis, A.; Schaepman, M. Shifts in Global Vegetation Activity Trends. *Remote Sens.* **2013**, *5*, 1117–1133. [CrossRef]
6. Rahaman, Z.A.; Kafy, A.A.; Saha, M.; Rahim, A.A.; Almulhim, A.I.; Rahaman, S.N.; Fattah, M.A.; Rahman, M.T.; Kalaivani, S.; Al Rakib, A. Assessing the impacts of vegetation cover loss on surface temperature, urban heat island and carbon emission in Penang city, Malaysia. *Build. Environ.* **2022**, *222*, 109335. [CrossRef]
7. Zhao, L.; Dai, A.; Dong, B. Changes in global vegetation activity and its driving factors during 1982–2013. *Agric. Forest. Meteorol.* **2018**, *249*, 198–209. [CrossRef]
8. Zhu, Z.; Piao, S.; Myneni, R.B.; Huang, M.; Zeng, Z.; Canadell, J.G.; Ciais, P.; Sitch, S.; Friedlingstein, P.; Arneth, A.; et al.. Greening of the Earth and its drivers. *Nat. Clim. Chang.* **2016**, *6*, 791–795. [CrossRef]
9. De Jong, R.; Schaepman, M.E.; Furrer, R.; De Bruin, S.; Verburg, P.H. Spatial relationship between climatologies and changes in global vegetation activity. *Glob. Chang. Biol.* **2013**, *19*, 1953–1964. [CrossRef]
10. Fensholt, R.; Horion, S.; Tagesson, T.; Ehammer, A.; Grogan, K.; Tian, F.; Huber, S.; Verbesselt, J.; Prince, S.D.; Tucker, C.J.; et al.. *Assessing Drivers of Vegetation Changes in Drylands from Time Series of Earth Observation Data, Remote Sensing Time Series;* Springer: Cham, Switzerland, 2015; pp. 183–202.
11. Wei, H.; Zhao, X.; Liang, S.; Zhou, T.; Wu, D.; Tang, B. Effects of Warming Hiatuses on Vegetation Growth in the Northern Hemisphere. *Remote Sens.* **2018**, *10*, 683. [CrossRef]
12. Peng, S.; Piao, S.; Ciais, P.; Myneni, R.B.; Chen, A.; Chevallier, F.; Dolman, A.J.; Janssens, I.A.; Peñuelas, J.; Zhang, G.; et al.. Asymmetric effects of daytime and night-time warming on Northern Hemisphere vegetation. *Nature* **2013**, *501*, 88–92. [CrossRef] [PubMed]
13. Xia, J.; Chen, J.; Piao, S.; Ciais, P.; Luo, Y.; Wan, S. Terrestrial carbon cycle affected by non-uniform climate warming. *Nat. Geosci.* **2014**, *7*, 173–180. [CrossRef]
14. Cao, R.; Shen, M.; Zhou, J.; Chen, J. Modeling vegetation green-up dates across the Tibetan Plateau by including both seasonal and daily temperature and precipitation. *Agric. Forest. Meteorol.* **2018**, *249*, 176–186. [CrossRef]
15. Shen, X.; Liu, B.; Henderson, M.; Wang, L.; Wu, Z.; Jiang, M.; Lu, X.; Wu, H. Asymmetric effects of daytime and nighttime warming on spring phenology in the temperate grasslands of China. *Agr. Forest. Meteorol.* **2018**, *259*, 240–249. [CrossRef]
16. Signarbieux, C.; Toledano, E.; Sangines de Carcer, P.; Fu, Y.H.; Schlaepfer, R.; Buttler, A.; Vitasse, Y. Asymmetric effects of cooler and warmer winters on beech phenology last beyond spring. *Glob. Chang. Biol.* **2017**, *23*, 4569–4580. [CrossRef]

17. Tan, J.; Piao, S.; Chen, A.; Zeng, Z.; Ciais, P.; Janssens, I.A.; Mao, J.; Myneni, R.B.; Peng, S.; Penuelas, J.; et al.. Seasonally different response of photosynthetic activity to daytime and night-time warming in the Northern Hemisphere. *Glob. Chang. Biol.* **2015**, *21*, 377–387. [CrossRef]
18. Wu, C.; Wang, X.; Wang, H.; Ciais, P.; Peñuelas, J.; Myneni, R.B.; Desai, A.R.; Gough, C.M.; Gonsamo, A.; Black, A.T.; et al.. Contrasting responses of autumn-leaf senescence to daytime and night-time warming. *Nat. Clim. Chang.* **2018**, *8*, 1092–1096. [CrossRef]
19. Wu, X.; Liu, H.; Li, X.; Liang, E.; Beck, P.S.; Huang, Y. Seasonal divergence in the interannual responses of Northern Hemisphere vegetation activity to variations in diurnal climate. *Sci. Rep.* **2016**, *6*, 19000. [CrossRef]
20. Zhao, J.; Du, Z.; Wu, Z.; Zhang, H.; Guo, N.; Ma, Z.; Liu, X. Seasonal variations of day- and nighttime warming and their effects on vegetation dynamics in China's temperate zone. *Acta Geogr. Sin.* **2018**, *73*, 395–404.
21. He, B.; Chen, A.; Jiang, W.; Chen, Z. The response of vegetation growth to shifts in trend of temperature in China. *J. Geogr. Sci.* **2017**, *27*, 801–816. [CrossRef]
22. Briffa, K.R.; Schweingruber, F.H.; Jones, P.D.; Osborn, T.J.; Shiyatov, S.G.; Vaganov, E.A. Reduced sensitivity of recent northern tree-growth to temperature at northern high latitudes. *Nature* **1998**, *391*, 678–682. [CrossRef]
23. Piao, S.; Nan, H.; Huntingford, C.; Ciais, P.; Friedlingstein, P.; Sitch, S.; Peng, S.; Ahlstrom, A.; Canadell, J.G.; Cong, N.; et al.. Evidence for a weakening relationship between interannual temperature variability and northern vegetation activity. *Nat. Commun.* **2014**, *5*, 5018. [CrossRef] [PubMed]
24. Andreu-Hayles, L.; Arrigo, R.D.; Kevin JAnchukaitis Beck, P.S.A. Varying boreal forest response to Arctic environmental change at the Firth River, Alaska. *Environ. Res. Lett.* **2011**, *6*, 45503. [CrossRef]
25. Fu, Y.H.; Zhao, H.; Piao, S.; Peaucelle, M.; Peng, S.; Zhou, G.; Ciais, P.; Huang, M.; Menzel, A.; Penuelas, J.; et al.. Declining global warming effects on the phenology of spring leaf unfolding. *Nature* **2015**, *526*, 104–107. [CrossRef]
26. Cong, N.; Shen, M.; Yang, W.; Yang, Z.; Zhang, G.; Piao, S. Varying responses of vegetation activity to climate changes on the Tibetan Plateau grassland. *Int. J. Biometeorol.* **2017**, *61*, 1433–1444. [CrossRef]
27. Dhakhwa, G.B.; Campbell, C.L. Potential Effects of Differential Day-Night Warming in Global Climate Change on Crop Production. *Clim. Change* **1998**, *40*, 647–667. [CrossRef]
28. Karl, T.R.; Kukla, G.; Razuvayev, V.N.; Changery, M.J.; Quayle, R.G.; Heim, R.R.; Easterling, D.R.; Fu, C.B. Global warming: Evidence for asymmetric diurnal temperature change. *Geophys. Res. Lett.* **1991**, *18*, 2253–2256. [CrossRef]
29. D'Arrigo, R.; Wilson, R.; Liepert, B.; Cherubini, P. On the 'Divergence problem' in northern forests a review of the tree-ring evidence and possible causes. *Global Planet. Change* **2008**, *60*, 289–305. [CrossRef]
30. Wu, X.; Liu, H.; Li, X.; Piao, S.; Ciais, P.; Guo, W.; Yin, Y.; Poulter, B.; Peng, C.; Viovy, N.; et al.. Higher temperature variability reduces temperature sensitivity of vegetation growth in Northern Hemisphere. *Geophys. Res. Lett.* **2017**, *44*, 6173–6181. [CrossRef]
31. Yang, Z.; Jiang, L.; Su, F.; Zhang, Q.; Xia, J.; Wan, S. Nighttime warming enhances drought resistance of plant communities in a temperate steppe. *Sci. Rep.* **2016**, *6*, 23267. [CrossRef]
32. Beck, P.S.A.; Goetz, S.J. Satellite observations of high northern latitude vegetation productivity changes between 1982 and 2008: Ecological variability and regional differences. *Environ. Res. Lett* **2011**, *6*, 45501. [CrossRef]
33. Jolly, W.M.; Cochrane, M.A.; Freeborn, P.H.; Holden, Z.A.; Brown, T.J.; Williamson, G.J.; Bowman, D.M. Climate-induced variations in global wildfire danger from 1979 to 2013. *Nat. Commun.* **2015**, *6*, 7537. [CrossRef] [PubMed]
34. Wen, Y.; Liu, X.; Pei, F.; Li, X.; Du, G. Non-uniform time-lag effects of terrestrial vegetation responses to asymmetric warming. *Agric. Forest. Meteorol.* **2018**, *252*, 130–143. [CrossRef]
35. Holben, B.N. Characteristics of maximum-value composite images from temporal AVHRR Data. *Int. J. Remote Sens.* **1986**, *7*, 1417–1434. [CrossRef]
36. Davy, R.; Esau, I.; Chernokulsky, A.; Outten, S.; Zilitinkevich, S. Diurnal asymmetry to the observed global warming. *Int. J. Climatol.* **2017**, *37*, 79–93. [CrossRef]
37. Shen, M.; Piao, S.; Chen, X.; An, S.; Fu, Y.H.; Wang, S.; Cong, N.; Janssens, I.A. Strong impacts of daily minimum temperature on the green-up date and summer greenness of the Tibetan Plateau. *Glob. Chang. Biol.* **2016**, *22*, 3057–3066. [CrossRef]
38. Du, Z.; Zhao, J.; Liu, X.; Wu, Z.; Zhang, H. Recent asymmetric warming trends of daytime versus nighttime and their linkages with vegetation greenness in temperate China. *Environ. Sci. Pollut. Res. Int.* **2019**, *26*, 35717–35727. [CrossRef]
39. Du, Z.; Zhao, J.; Pan, H.; Wu, Z.; Zhang, H. Responses of vegetation activity to the daytime and nighttime warming in Northwest China. *Environ. Monit. Assess.* **2019**, *191*, 721. [CrossRef]

plants

MDPI

Article

Comparison of Leaf Shape between a *Photinia* Hybrid and One of Its Parents

Xiao Zheng [1,2], Karl J. Niklas [3,*], David A. Ratkowsky [4], Yabing Jiao [2], Hui Ding [1] and Peijian Shi [2,*]

1 Research Center for Biodiversity Conservation and Biosafety, Nanjing Institute of Environmental Sciences, Ministry of Ecology and Environment of China, Nanjing 210042, China
2 Bamboo Research Institute, College of Biology and the Environment, Nanjing Forestry University, Nanjing 210037, China
3 School of Integrative Plant Science, Cornell University, Ithaca, NY 14853, USA
4 Tasmanian Institute of Agriculture, University of Tasmania, Private Bag 98, Hobart 7001, Australia
* Correspondence: kjn2@cornell.edu (K.J.N.); pjshi@njfu.edu.cn (P.S.); Tel.: +86-25-8542-7231 (P.S.)

Abstract: Leaf shape and size can vary between hybrids and their parents. However, this has seldom been quantitatively tested. *Photinia* × *fraseri* is an important landscaping plant in East Asia as a hybrid between evergreen shrubs *P. glabra* and *P. serratifolia*. Its leaf shape looks like that of *P. serratifolia*. To investigate leaf shape, we used a general equation for calculating the leaf area (A) of broad-leaved plants, which assumes a proportional relationship between A and product of lamina length (L) and width (W). The proportionality coefficient (which is referred to as the Montgomery parameter) serves as a quantitative indicator of leaf shape, because it reflects the proportion of leaf area A to the area of a rectangle with L and W as its side lengths. The ratio of L to W, and the ellipticalness index were also used to quantify the complexity of leaf shape for elliptical leaves. A total of >4000 leaves from *P.* × *fraseri* and *P. serratifolia* (with >2000 leaves for each taxon) collected on a monthly basis was used to examine: (i) whether there is a significant difference in leaf shape between the two taxa, and (ii) whether there is a monotonic or parabolic trend in leaf shape across leaf ages. There was a significant difference in leaf shape between the two taxa ($p < 0.05$). Although there were significant differences in leaf shape on a monthly basis, the variation in leaf shape over time was not large, i.e., leaf shape was relatively stable over time for both taxa. However, the leaf shape of the hybrid was significantly different from its parent *P. serratifolia*, which has wider and more elliptical leaves than the hybrid. This work demonstrates that variations in leaf shape resulting from hybridization can be rigorously quantified and compared among species and their hybrids. In addition, this work shows that leaf shape does not changes as a function of age either before or after the full expansion of the lamina.

Keywords: leaf ellipticalness index; leaf length; leaf width; Montgomery equation; Montgomery parameter

Citation: Zheng, X.; Niklas, K.J.; Ratkowsky, D.A.; Jiao, Y.; Ding, H.; Shi, P. Comparison of Leaf Shape between a *Photinia* Hybrid and One of Its Parents. *Plants* **2022**, *11*, 2370. https://doi.org/10.3390/plants11182370

Academic Editor: Yasutomo Hoshika

Received: 12 August 2022
Accepted: 8 September 2022
Published: 11 September 2022

1. Introduction

Leaves are the primary photosynthetic organs of the majority of land plants. Prior research has shown that leaf area and structure are correlated with photosynthetic rates [1–4]. It is important therefore to accurately calculate leaf area. Prior studies have shown that leaf area follows a general equation called the Montgomery equation (denoted henceforth as ME) for many broad-leaved species with different leaf shapes. This equation assumes a simple proportional relationship between leaf area (A) and the product of leaf length (L) and width (W) [5–13]. Leaf age and area have been demonstrated to have little influence on the validity of ME in calculating leaf area, although leaf age and size can influence to some degree the proportionality coefficient of ME among different leaf age or size groups [13,14].

Leaf shape and area are adaptively correlated and therefore can vary as a function of local climatic conditions even among conspecifics [15–18]. In previous studies, the leaf roundness index [19–22] has been widely used to measure leaf shape. However, this index is more suitable for leaves whose length and width are nearly equal. Consequently, Li et al. [23] proposed the 'leaf ellipticalness index' for elliptical and oblong leaves. For an elliptical leaf, the leaf ellipticalness index approximates 1. Additionally, in spite of the simplicity of using the leaf width/length ratio to quantify leaf shape, Shi et al. [24] have shown that the leaf width/length ratio is significantly positively correlated with the fractal dimension of the lamina perimeter based on a large sample of leaves differing in size and shape among nine Magnoliaceae species.

Hybridization across closely related plant species can also result in significant variation in leaf shape. However, the effect of hybridization on leaf size and shape has seldom been quantitatively examined, particularly as leaves mature during the growing season. In order to examine this phenomenology, *Photinia* × *fraseri*, which is a hybrid between the evergreen shrubs *P. glabra* and *P. serratifolia*, and one of its parents *P. serratifolia*, were examined to determine (i) whether there is a significant difference in leaf shape between *P.* × *fraseri* and *P. serratifolia*, and (ii) whether there is a monotonic or parabolic trend in leaf shape as a function of age. Because both the hybrid and one of its parents have elliptical or obovate leaves (Figure 1), we used the ratio of leaf width to length and the leaf ellipticalness index to quantify leaf shape in addition to the ME.

Figure 1. Representative leaves for the two *Photinia* taxa (the hybrid *P.* × *fraseri* and one of its parents *P. serratifolia*) at different times in the growing season during 2021.

2. Materials and Methods

2.1. Plant Materials

We marked newly emerging leaves on 36 trees of *P.* × *fraseri* (the hybrid) and 3 trees of *P. serratifolia* (one of the two parents) growing at the Nanjing Forestry University campus (118°48′35″ E, 32°4′67″ N) from late February to early March 2021. Representative specimens of the second parent (*P. glabra*) were not found for study on the University campus, which was under quarantine. Each month, 320–380 leaves were randomly sampled from three individual trees of *P.* × *fraseri* from mid-March to mid-August 2021 and 320–350 leaves were randomly sampled from three individual trees of *P. serratifolia* from mid-April to mid-November 2021. Figure 1 provides images of representative leaves across the different months for the two taxa. We had planned to sample the leaves of *P. serratifolia* from spring to autumn. However, the marked branches in early March were pruned by gardeners at the beginning of September 2021. Fortunately, the leaves growing from late February to early March 2021 were fully mature in August 2021, and the leaves sampled from mid-March to mid-August manifest temporal changes in leaf shape from young to mature leaves of this hybrid. When a leaf matures, its leaf shape seldom changes significantly.

2.2. Data Acquisition

The fresh leaves were scanned by a photo scanner (V550, Epson, Batam, Indonesia) to obtain .tiff images at a 600-dpi resolution, which were transferred to black-white .bmp files using the Photoshop software (CS6, version: 13.0; Adobe, San Jose, CA, USA). The Matlab (version $\geq$ 2009a; MathWorks, Natick, MA, USA) procedures developed by refs. [25,26] were used to extract the planar coordinates of each leaf boundary. The 'biogeom' package (version 1.1.1) [27] based on the statistical software R (version 4.2.0) [28] was then used to calculate leaf area, length, and width using the planar coordinates.

2.3. Methods

The Montgomery equation (ME) [5] was used to describe the relationship between leaf area (A) and the product of leaf length (L) and width (W):

$$A = m_p \cdot LW, \tag{1}$$

where m_p is the proportionality coefficient, which is the Montgomery parameter. To normalize the data, we log-transformed the data of both sides of Equation (1):

$$y = a + x, \tag{2}$$

where y = ln(A); x = ln(LW); a = ln(m_p). We used ordinary least-squares protocols to estimate the parameter a, and its 95% confidence interval (95% CI). To test whether there is a significant difference in m_p between the two taxa, we used the bootstrap percentile method [29,30] to calculate the 95% CI of the differences in the 4000 bootstrapping replicates of m_p between *P.* × *fraseri* and *P. serratifolia*. If the 95% CI of the differences includes 0, there is no significant difference in m_p between the two taxa; if it does not include 0, a significant difference is detected. We used the Pearson's product moment correlation coefficient (r) to measure the validity (i.e., statistical robustness) of the linear relationship between x and y:

$$r = \frac{\sum_{i=1}^{n}(x_i - \overline{x})(y_i - \overline{y})}{\sqrt{\sum_{i=1}^{n}(x_i - \overline{x})^2}\sqrt{\sum_{i=1}^{n}(y_i - \overline{y})^2}}, \tag{3}$$

where $\overline{x}$ and $\overline{y}$ represent the means of x- and y-values, respectively; and n is the sample size. The test of the significance of the correlation coefficient is based upon a statistic that theoretically follows a t distribution with $n - 2$ degrees of freedom if the samples of x and y are normally distributed.

The leaf ellipticalness index (EI) was used to quantify leaf shape [23]:

$$\mathrm{EI} = \frac{A}{(\pi/4)LW}. \tag{4}$$

The EI has a close relationship with m_p, i.e., EI = $m_p/(\pi/4)$. A linear regression method was used to estimate the m_p value based on the number of leaves. However, EI can also be calculated directly using A, L and W. The EI quantifies the extent of a leaf shape deviating from an ellipse regardless of the eccentricity of the ellipse. Thus, we used the W/L ratio as another leaf-shape index. ANOVA with Tukey's honestly significant difference (HSD) test [31] and a 0.05 significance level were used to test the significance of the differences in EI as well as the W/L ratio between the two taxa at the combined data level and at the intraspecific level across different dates of collection. For any two groups, we calculated the 95% confidence interval (CI) of the Tukey's test confidence interval, which equals the difference in the means between the two groups ± HSD, and we determined its significance by checking whether the 95% CI included 0. We also used the linear and parabolic equations to fit the EI (and the W/L ratio) vs. collection month data, where month was set to a numeric variable to test whether leaf shape has a monotonic or parabolic trend across different investigation dates (i.e., leaf ages).

$$\mathrm{LS} = \beta_0 + \beta_1\mathrm{Month} + \beta_2\mathrm{Month}^2, \tag{5}$$

where LS is the leaf-shape index of interest (EI or the W/L ratio); Month is the month of sampling (ranging from 3 to 8 [i.e., March to August 2021] for *P.* × *fraseri*, and from 4 to 11 [i.e., April to November 2021] for *P. serratifolia*); and β_0, β_1 and β_2 are the parameters to be fitted (referred to as regression coefficients). For the simple linear regression, β_2 was set to be zero. Each regression coefficient's statistic in Equation (5) follows a t distribution with $n - p$ degrees of freedom, where n is the sample size and p is the number of parameters [32].

The statistical software R (version 4.2.0) [28] was used to carry out calculations and to draw figures. The 'agricolae' package (version 1.3-5) was used to implement the Tukey's HSD test.

3. Results

The Montgomery equation (ME) was valid in calculating leaf area. For the pooled intraspecific data, the correlation coefficient was greater than 0.99 ($p < 0.001$). The estimated values of the Montgomery parameters of *P.* × *fraseri* and *P. serratifolia* were 0.6494 (95% CI = 0.6482, 0.6506) and 0.6981 (95% CI = 0.6971, 0.6990), respectively (Figure 2). There was a significant difference in leaf shape between the two taxa, i.e., the leaves of *P. serratifolia* were on average wider and more elliptical in shape (Figure 3). The means of the W/L ratios for the two taxa were 0.38 (*P.* × *fraseri*) and 0.40 (*P. serratifolia*); the means of the leaf elliptical indices (EIs) for the two taxa were 0.8276 (*P.* × *fraseri*) and 0.8894 (*P. serratifolia*).

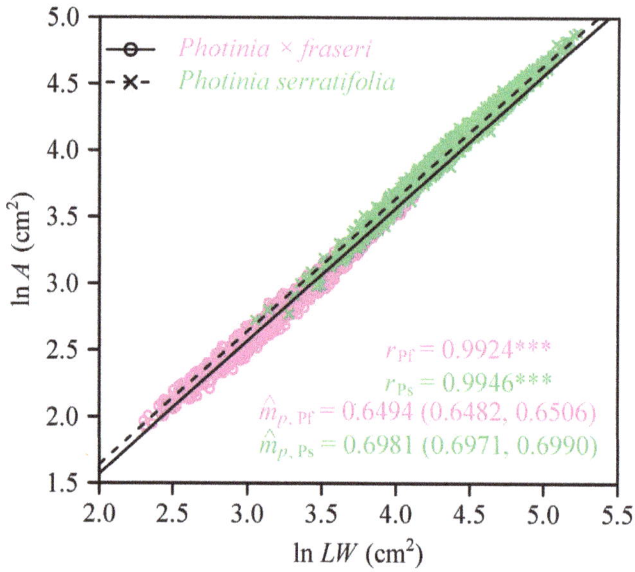

Figure 2. Log-log bivariate plot of leaf area (A) vs. the product leaf length (L) and width (W). The symbol $\hat{m}_p$ represents the estimated value of the Montgomery parameter, i.e., the estimated proportionality coefficient of the Montgomery equation. The subscripts Pf and Ps represent the hybrid *P.* × *fraseri*, and one of its parents *P. serratifolia*, respectively. *** denotes $p < 0.001$ for a correlation coefficient.

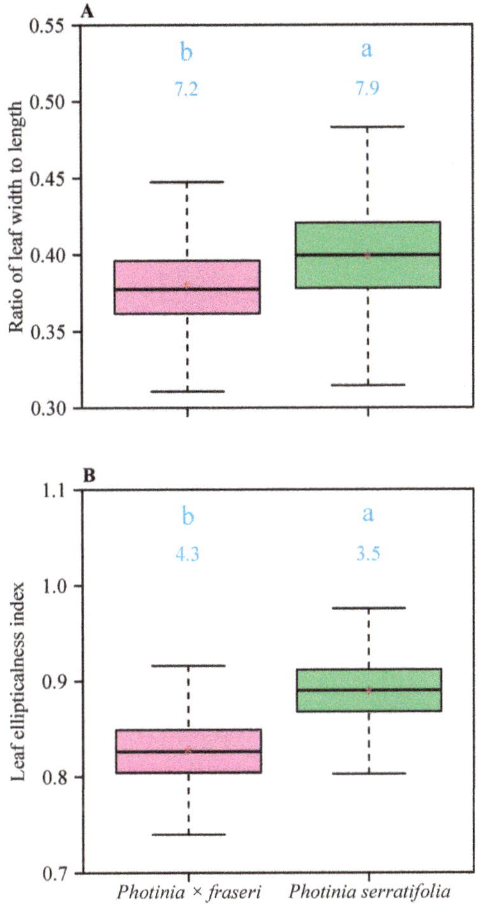

Figure 3. Comparison of leaf shape between the two *Photinia* taxa for the pooled data. (**A**) Ratio of leaf width to length; (**B**) Leaf ellipticalness index. In each panel, the letters on the whiskers show differences between the two taxa (i.e., taxa with different letters are significantly different at the $\alpha = 0.05$ significance level using Tukey's HSD); the values at the top of whiskers represent the coefficients of variation (%) for each taxon; the horizontal solid line represents the median; and the red asterisk represents the mean. The whiskers extend to the most extreme data point, which is no more than 1.5 times the interquartile range from the box. The difference in mean leaf *W/L* ratios is equal to 0.0193 with HSD = 0.0017, and the corresponding 95% Tukey's test confidence interval is 0.0176, 0.0210, which does not include 0, indicating a significant difference; the difference in mean leaf ellipticalness indices is equal to 0.0617 with HSD = 0.0019, and the corresponding 95% Tukey's test confidence interval is 0.0598, 0.0636, which also does not include 0, indicating a significant difference.

There were significant differences in leaf shape (measured as either the *W/L* ratio or the EI) among the different months of collecting (Figure 4). Three of the four slopes of the linear model of leaf shape vs. month were statistically significant (with three p values < 0.05) with the exception of the EIs of *P. serratifolia*. However, the four coefficients of determination (i.e., r^2 values) were all less than 0.06, i.e., a very weak linear relationship was observed between leaf shape and sampling month. Thus, there is no robust evidence for a monotonic increase or decrease in leaf shape across leaf ages. The linear and quadratic coefficients in the parabolic model were statistically significant (p values < 0.05) for the leaf *W/L* ratio vs. month relationship and the EI vs. month relationship for *P.* × *fraseri*. However, the two coefficients were not statistically significant (p values > 0.05) for any leaf-shape indices for *P. serratifolia*. Three of the four coefficients of determination of the parabolic regression were smaller than 0.03, with the exception of the EI vs. month data of *P.* × *fraseri* (r^2 = 0.1452). Thus, there was also no strong evidence for a parabolic trend in leaf shape across the different leaf ages (see Table 1 for details).

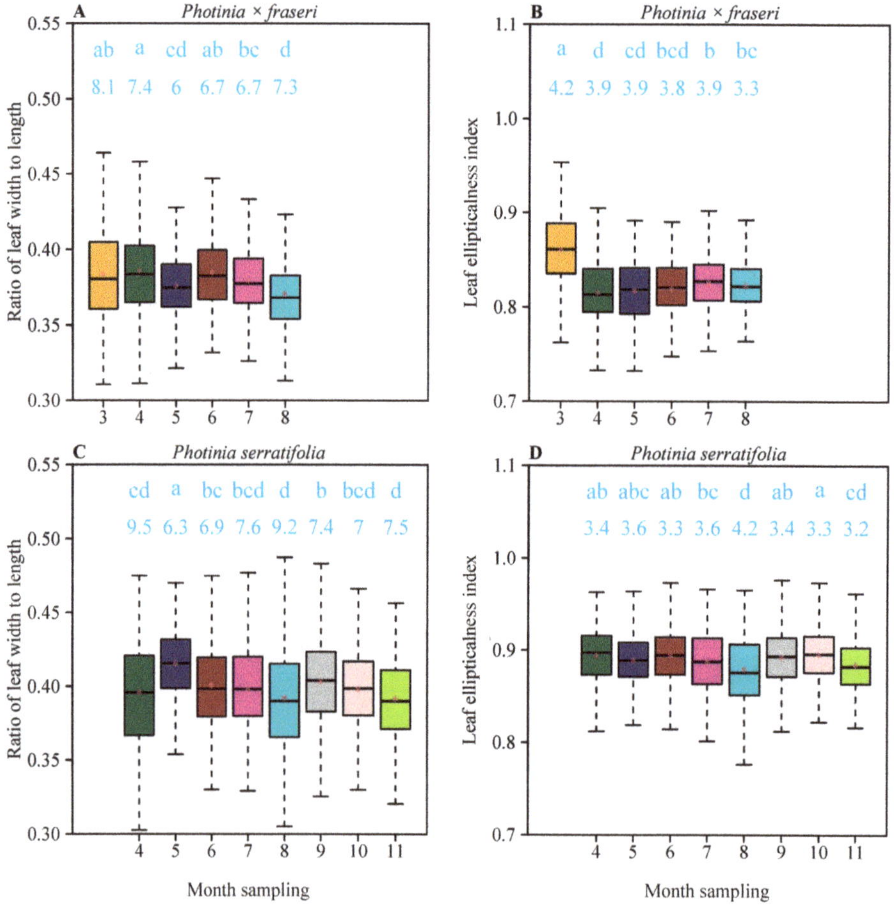

Figure 4. Comparison of leaf shapes (reflected by the ratio of leaf width to length and leaf ellipticalness index) across different sampling months for *P.* × *fraseri* (**A**,**B**) and *P. serratifolia* (**C**,**D**). In each panel, the letters on the whiskers represent the significance of the differences in leaf shape between any two sampling months among which the letters a, b, c, and d are used to represent the significance of differences in leaf shape among different sampling months; groups sharing a common letter are not significantly different at the 0.05 significance level; the values at the top of whiskers represent the coefficients of variation (%) for each sampling month; the horizontal solid line represents the median; the red asterisk represents the mean. The whiskers extend to the most extreme data point, which is no more than 1.5 times the interquartile range from the box. For the label of the x-axis, the months 3 to 11 represent March to November 2021, respectively. The differences in the means of leaf-shape indices, HSD values, and the corresponding 95% Tukey's test confidence intervals between any two groups for *P.* × *fraseri* and *P. serratifolia* are provided in Tables S3 and S4, respectively.

Table 1. Regression statistics of the leaf-shape index vs. sampling month.

Taxon/Leaf-Shape Index	Equation †	Parameter	Estimate	Significance	r^2
Photinia × *fraseri* Leaf width/length ratio	$y = a + bx$	a	0.3920	<0.001	0.0192
		b	−0.0022	<0.001	
	$y = a + bx + cx^2$	a	0.3733	<0.001	0.0233
		b	0.0054	0.0365	
		c	−0.0006	0.0031	
Photinia × *fraseri* Leaf ellipticalness index	$y = a + bx$	a	0.8549	<0.001	0.0591
		b	−0.0050	<0.001	
	$y = a + bx + cx^2$	a	0.9668	<0.001	0.1452
		b	0.0504	<0.001	
		c	0.0041	<0.001	
Photinia serratifolia Leaf width/length ratio	$y = a + bx$	a	0.4096	<0.001	0.0099
		b	−0.0014	<0.001	
	$y = a + bx + cx^2$	a	0.3974	<0.001	0.0111
		b	0.0022	0.268	
		c	−0.0002	0.071	
Photinia serratifolia Leaf ellipticalness index	$y = a + bx$	a	0.8932	<0.001	0.0014
		b	−0.0005	0.0562	
	$y = a + bx + cx^2$	a	0.9010	<0.001	0.0018
		b	−0.0028	0.166	
		c	0.0002	0.252	

† Here, y represents a specific leaf-shape index, and x represents the sampling month.

4. Discussion and Conclusions

Shi et al. [33] used the Montgomery parameter (m_p) to fit leaf A vs. LW of 101 bamboo taxa using > 10,000 bamboo leaves. In their study, the estimated value of m_p was 0.6959 (95% CI = 0.6952, 0.6966). Li et al. [23] estimated a m_p value of 0.6840 (95% CI = 0.6827, 0.9855) using > 2200 leaves from nine Magnoliaceae species, whereas Ma et al. [13] estimated a m_p value of 0.7710 (95% CI = 0.7696, 0.7714) using > 6500 leaves of an alpine oak species (*Quercus pannosa*) from different tree size groups. The present study shows that the estimated value of m_p of the two *Photinia* taxa also approximates 0.7, i.e., the leaf area approximately accounts for 70% of the area of a hypothetical rectangle with the leaf length and width as its two side lengths. This result is significantly different from the numerical value of the alpine oak provided in ref. [13], because the leaf shape of *Q. pannosa* is a special superellipse with a mean leaf ellipticalness index greater than unity [34–36], which is larger than those of the two *Photinia* taxa examined in this study. Since leaf shape is found to be closely related to climatic factors [18,37], it would be worth studying the link between the m_p and climate for some widely distributed plants in the future.

The leaf shapes of two *Photinia* taxa were quantified using different metrics, and found to significantly differ in spite of the fact that the two types of leaves appear to be superficially very similar in appearance. The leaves of *P. serratifolia* are wider and more elliptical than those of its hybrid. In addition, leaf shape significantly varied across the different leaf ages, but manifested no monotonic or parabolic tendency across the different leaf ages. The variations in leaf shape among the different months of sampling may reflect variations resulting from random samplings of individual plants. This variation, however, does not mask the fact that the leaf shape of the *Photinia* taxa is relatively stable, and that the Montgomery equation, which assumes a proportional relationship between leaf area and the product of leaf length and width, is valid at different leaf growth stages. These results highlight an approach that permits the non-destructive estimation of leaf area, i.e., direct field measurements of leaf length and width are used to estimate lamina area by multiplying their product by a proportionality coefficient (m_p).

Supplementary Materials: The following supporting information can be downloaded at: https://www.mdpi.com/article/10.3390/plants11182370/s1, Table S1: The raw leaf data of *Photinia* × *fraseri*; Table S2: The raw leaf data of *P. serratifolia*; Table S3: Tukey multiple comparisons of the means of leaf-shape indices between any sampling months for *P.* × *fraseri*; Table S4: Tukey multiple comparisons of the means of leaf-shape indices between any sampling months for *P. serratifolia*.

Author Contributions: Methodology, K.J.N., D.A.R. and P.S.; formal analysis, K.J.N., D.A.R. and P.S.; investigation, X.Z., K.J.N., D.A.R., Y.J., H.D. and P.S.; writing—original draft preparation, X.Z. and P.S.; writing—review and editing, K.J.N. and D.A.R.; funding acquisition, X.Z. and H.D. All authors have read and agreed to the published version of the manuscript.

Funding: X.Z. and H.D. were funded by the Project of Biological Resources Survey in Wuyishan National Park (grant number: HXQT2020120701).

Institutional Review Board Statement: Not applicable.

Informed Consent Statement: Not applicable.

Data Availability Statement: The data used in the present work have been listed in the online Supplementary Materials.

Acknowledgments: We thank Kexin Yu for her valuable help during the preparation of this work.

Conflicts of Interest: The authors declare no conflict of interest.

References

1. Bhagsari, A.S.; Brown, R.H. Leaf photosynthesis and its correlation with leaf area. *Crop Sci.* **1986**, *26*, 127–132. [CrossRef]
2. Reich, P.B.; Ellsworth, D.S.; Walters, M.B. Leaf structure (specific leaf area) modulates photosynthesis–nitrogen relations: Evidence from within and across species and functional groups. *Funct. Ecol.* **2002**, *12*, 948–958. [CrossRef]
3. Wu, Y.; Gong, W.; Wang, Y.; Yong, T.; Yang, F.; Liu, W.; Wu, X.; Du, J.; Shu, K.; Liu, J.; et al. Leaf area and photosynthesis of newly emerged trifoliolate leaves are regulated by mature leaves in soybean. *J. Plant Res.* **2018**, *131*, 671–680. [CrossRef]
4. Shi, P.; Miao, Q.; Niinemets, Ü.; Liu, M.; Li, Y.; Yu, K.; Niklas, K.J. Scaling relationships of leaf vein and areole traits versus leaf size for nine Magnoliaceae species differing in venation density. *Am. J. Bot.* **2022**, *109*, 899–909. [CrossRef]
5. Montgomery, E.G. *Correlation Studies in Corn, Annual Report No.24*; Nebraska Agricultural Experimental Station: Lincoln, NB, USA, 1911; pp. 108–159.
6. Kemp, C.D. Methods of estimating leaf area of grasses from linear measurements. *Ann. Bot.* **1960**, *24*, 491–499. [CrossRef]
7. Stickler, F.C.; Wearden, S.; Pauli, A.W. Leaf area determination in grain sorghum. *Agronony* **1961**, *53*, 187–188. [CrossRef]
8. Jani, T.C.; Misra, D.K. Leaf area estimation by linear measurements in *Ricinus communis*. *Nat.* **1966**, *212*, 741–742. [CrossRef]
9. Palaniswamy, K.M.; Gomez, K.A. Length-width method for estimating leaf area of rice. *Agron. J.* **1974**, *66*, 430–433. [CrossRef]
10. Shi, P.; Liu, M.; Ratkowsky, D.A.; Gielis, J.; Su, J.; Yu, X.; Wang, P.; Zhang, L.; Lin, Z.; Schrader, J. Leaf area-length allometry and its implications in leaf-shape evolution. *Trees Struct. Funct.* **2019**, *33*, 1073–1085. [CrossRef]
11. Yu, X.; Shi, P.; Schrader, J.; Niklas, K.J. Nondestructive estimation of leaf area for 15 species of vines with different leaf shapes. *Am. J. Bot.* **2020**, *107*, 1481–1490. [CrossRef]
12. Schrader, J.; Shi, P.; Royer, D.L.; Peppe, D.J.; Gallagher, R.V.; Li, Y.; Wang, R.; Wright, I.J. Leaf size estimation based on leaf length, width and shape. *Ann. Bot.* **2021**, *128*, 395–406. [CrossRef] [PubMed]
13. Ma, J.; Niklas, K.J.; Liu, L.; Fang, Z.; Li, Y.; Shi, P. Tree size influences leaf shape but does not affect the proportional relationship between leaf area and the product of length and width. *Front. Plant Sci.* **2022**, *13*, 850203. [CrossRef] [PubMed]
14. Huang, L.; Niinemets, Ü.; Ma, J.; Schrader, J.; Wang, R.; Shi, P. Plant age has a minor effect on non-destructive leaf area calculations in moso bamboo (*Phyllostachys edulis*). *Symmetry* **2021**, *13*, 369. [CrossRef]
15. Webb, L.J. Environmental relationships of the structural types of Australian rain forest vegetation. *Ecology* **1968**, *49*, 296–311. [CrossRef]
16. Lewis, M.C. The physiological significance of variation in leaf structure. *Sci. Prog.* **1972**, *60*, 25–51.
17. Givnish, T. On the adaptive significance of leaf form. In *Topics in Plant Population Biology*; Solbrig, O.T., Jain, S., Johnson, G.B., Raven, P.H., Eds.; Columbia University Press: New York, NY, USA, 1979; pp. 375–407.
18. Royer, D.L.; Wilf, P. Why do toothed leaves correlate with cold climates? Gas exchange at leaf margins provides new insights into a classic paleotemperature proxy. *Int. J. Plant Sci.* **2006**, *167*, 11–18. [CrossRef]
19. Baxes, G.A. *Digital Image Processing: Principles and Applications*; John Wiley and Sons, Inc.: New York, NY, USA, 1994.
20. Niinemets, Ü.; Cescatti, A.; Christian, R. Constraints on light interception efficiency due to shoot architecture in broad-leaved Nothofagus species. *Tree Physiol.* **2004**, *24*, 617–630. [CrossRef]
21. Niinemets, Ü.; Sparrow, A.; Cescatti, A. Light capture efficiency decreases with increasing tree age and size in the southern hemisphere gymnosperm Agathis australis. *Trees Struct. Funct.* **2005**, *19*, 177–190. [CrossRef]

22. Roth-Nebelsick, A.; Konrad, W. Fossil leaf traits as archives for the past—And lessons for the future? *Flora* **2019**, *254*, 59–70. [CrossRef]
23. Li, Y.; Quinn, B.K.; Niinemets, Ü.; Schrader, J.; Gielis, J.; Liu, M.; Shi, P. Ellipticalness index—A simple measure of the complexity of oval leaf shape. *Pak. J. Bot.* **2022**, *54*, 2233–2240. [CrossRef]
24. Shi, P.; Yu, K.; Niinemets, Ü.; Gielis, J. Can leaf shape be represented by the ratio of leaf width to length? Evidence from nine species of *Magnolia* and *Michelia* (Magnoliaceae). *Forests* **2021**, *12*, 41. [CrossRef]
25. Shi, P.; Ratkowsky, D.A.; Li, Y.; Zhang, L.; Lin, S.; Gielis, J. A general leaf-area geometric formula exists for plants—Evidence from the simplified Gielis equation. *Forests* **2018**, *9*, 714. [CrossRef]
26. Su, J.; Niklas, K.J.; Huang, W.; Yu, X.; Yang, Y.; Shi, P. Lamina shape does not correlate with lamina surface area: An analysis based on the simplified Gielis equation. *Glob. Ecol. Conserv.* **2019**, *19*, e00666. [CrossRef]
27. Shi, P.; Gielis, J.; Quinn, B.K.; Niklas, K.J.; Ratkowsky, D.A.; Schrader, J.; Ruan, H.; Wang, L.; Niinemets, Ü. 'biogeom': An R package for simulating and fitting natural shapes. *Ann. N. Y. Acad. Sci.* 2022; *in press*. [CrossRef] [PubMed]
28. R Core Team. *R: A Language and Environment for Statistical Computing*; R Foundation for Statistical Computing: Vienna, Austria, 2022. Available online: https://www.R-project.org/ (accessed on 1 June 2022).
29. Efron, B.; Tibshirani, R.J. *An Introduction to the Bootstrap*; Chapman and Hall/CRC: New York, NY, USA, 1993.
30. Sandhu, H.S.; Shi, P.; Kuang, X.; Xue, F.; Ge, F. Applications of the bootstrap to insect physiology. *Fla. Entomol.* **2011**, *94*, 1036–1041. [CrossRef]
31. Hsu, J.C. *Multiple Comparisons: Theory and Methods*; Chapman and Hall/CRC: New York, NY, USA, 1996.
32. Xue, Y.; Chen, L. *Statistical Models and R Software*; Tsinghua University Press: Beijing, China, 2007.
33. Shi, P.; Li, Y.; Niinemets, Ü.; Olson, E.; Schrader, J. Influence of leaf shape on the scaling of leaf surface area and length in bamboo plants. *Trees Struct. Funct.* **2021**, *35*, 709–715. [CrossRef]
34. Lamé, G. *Examen des Différentes Méthodes Employées Pour Résoudre les Problèmes de Géométrie*; V. Courcier: Paris, France, 1818.
35. Shi, P.; Huang, J.; Hui, C.; Grissino-Mayer, H.D.; Tardif, J.C.; Zhai, L.; Wang, F.; Li, B. Capturing spiral radial growth of conifers using the superellipse to model tree-ring geometric shape. *Front. Plant Sci.* **2015**, *6*, 856. [CrossRef]
36. Huang, W.; Li, Y.; Niklas, K.J.; Gielis, J.; Ding, Y.; Cao, L.; Shi, P. A superellipse with deformation and its application in describing the cross-sectional shapes of a square bamboo. *Symmetry* **2020**, *12*, 2073. [CrossRef]
37. Peppe, D.J.; Royer, D.L.; Gariglino, B.; Oliver, S.Y.; Newman, S.; Leight, E.; Enikolopov, G.; Fernandez-Burgos, M.; Herrera, F.; Adams, J.M.; et al. Sensitivity of leaf size and shape to climate: Global patterns and paleoclimatic applications. *New Phytol.* **2011**, *190*, 724–739. [CrossRef]

plants

Article

Growth and Development Responses of the Rhizome-Root System in *Pleioblastus pygmaeus* to Light Intensity

Weiwei Huang [1,2,3,*], Yongyan Ding [2,4], Shucong Wang [2], Chao Song [5] and Fusheng Wang [1,2]

1 Co-Innovation Center for Sustainable Forestry in Southern China, Nanjing Forestry University, 159 Longpan Road, Nanjing 210037, China
2 Bamboo Research Institute, Nanjing Forestry University, 159 Longpan Road, Nanjing 210037, China
3 Department of Geosciences and Natural Resource Management, The University of Copenhagen, Rolighedsvej 23, DK-1958 Frederiksberg, Denmark
4 College of Biology and the Environment, Nanjing Forestry University, 159 Longpan Road, Nanjing 210037, China
5 College of Field Engineering, Army Engineering University of PLA, 88 Houbiaoying Road, Nanjing 210001, China
* Correspondence: wh@njfu.edu.cn; Tel.: +86-25-8542-7231

Abstract: Light, as a primary source of energy, directly or indirectly influences virtually all morphological modifications occurring in both shoots and roots. A pot experiment was conducted to assess the growth patterns of one-year-old *Pleioblastus pygmaeus* plants' rhizome-root systems and their responses to different light intensities from 11 March to 26 December 2016. The experiment design scheme was 3.87% (L1), 11.25% (L2), 20.25% (L3), 38.76% (L4), 60.70% (L5), and 100% full sunlight (control CK). The results indicated that along the growing period from March to December, eight of the eleven studied parameters of the rhizome-root system showed significant variability and diverse growth patterns. In addition, light intensity is a key factor for determining *P. pygmaeus* plants' rhizome and root growth. Specifically, the light intensity had a significant, positive, and linear/or almost linear impact on the number of old and new rhizomes, old rhizome length, new rhizome diameter, as well as the culm root diameter. A nonlinear and positive relationship was found between light intensity and the listed three parameters, i.e., new rhizome length, new rhizome internode length, and rhizome root length. The value of the above-mentioned three parameters significantly increased when affected from 0% to 40–60% of full sunlight and then gradually increased until 100% of full sunlight. The ratio of aboveground dry weight to underground dry weight (A/U ratio) showed a single peak curve with increasing light intensity and presented the highest value under ca. 55% full sunlight. Furthermore, 40% full sunlight (equal to an average light of 2232 lux) might be the threshold for *P. pygmaeus* rhizome-root system growth. When the light intensity was below 40%, the generalized additive models (GAMs) predicted value of most studied parameters decreased to lower than zero. In conclusion, current study provides a solid basis for understanding the dynamic growth and development of *P. pygmaeus* rhizome-root system, and its responses to different light conditions, which could be used as inputs to *P. pygmaeus* plant cultivation.

Keywords: dwarf bamboo; light conditions; morphological features; growth pattern

Citation: Huang, W.; Ding, Y.; Wang, S.; Song, C.; Wang, F. Growth and Development Responses of the Rhizome-Root System in *Pleioblastus pygmaeus* to Light Intensity. *Plants* **2022**, *11*, 2204. https://doi.org/10.3390/plants11172204

Academic Editor: Daniel K. Gladish

Received: 21 June 2022
Accepted: 19 August 2022
Published: 25 August 2022

1. Introduction

Bamboo has approximately 1300 species under 150 genera worldwide and is distributed across a wide range of tropical and subtropical areas, from alluvial plains to high mountains [1,2]. China has more than 500 bamboo species in 35 genera, accounting for 46% of the world's bamboo species [3]. *Pleioblastus pygmaeus* (Miq.) Nakai is a dwarf bamboo which prefers to grow in moderate moisture and well-drained soil, and was introduced from Japan to China in the early 20th century. *P. pygmaeus* is an excellent ground cover ornamental plant, which has slender and erect green stalks, purplish nodes, and palm-shaped and emerald green leaves. In addition, for its well-developed and intricate rhizome-root system, it shows a strong water and fertilizer retention. Nowadays, the application demand for *P. pygmaeus* in landscaping is continually increasing in China, so its economic value is also getting an increase.

Light, as a primary source of energy, is one of the most important environmental factors for plant growth and survival. In the presence of light, the photosynthesis of green plants combines carbon dioxide and water to form carbohydrates and oxygen [4,5]. In order to survive and grow in a range of light conditions, plants dynamically adjust their architecture to optimize growth and performance in response to different light intensity [6]. A variety of photomorphogenic responses have been investigated extensively in the aboveground parts of plants [7–12]. Under deep shade, plants reduce their leaf dry matters and stem diameters as well as photosynthetic rate [7–9,13,14]. However, under low light intensity, tolerant species try to increase stem height and leaf area in order to increase the ability of capturing light and the net assimilation rate [15–18]. Under natural growth conditions, aboveground parts are directly exposed to light whereas root systems develop underground, shielded from direct illumination [19]. Notably, accumulating evidence demonstrates that underground roots are able to directly or indirectly perceive light signals to trigger photomorphogenic responses and experience dramatic changes in morphology and development under changing light conditions [6,9,14,19–21]. Concerning *P. pygmaeus*, as an excellent ground cover ornamental plant for soil consolidation and slope protection, it is very important to understand the responses of its rhizome-root morphological and developmental characteristics to changing light conditions [22].

Based on the rhizome growth patterns, bamboos were divided into three main types: the amphipodial, monopodial, and sympodial [23,24]. *P. pygmaeus* has an amphipodial rhizome, which processes a mixed culm morphology both aggregated and scattered. The detailed features are: (i) some of the underground rhizomes are sympodial with short rhizomes and internodes, and cannot spread in soil for a long distance; top buds unearth to generate new culms and stalks grow in dense clusters. (ii) The other part of the underground system is monopodial with stems spreading underground called "whipper root"; lateral buds unearth to form new culms and the stalks are scattered. Rhizomes together with roots generated from rhizome nodes form a rhizome-root system which have a decisive role in bamboo growth [25,26]. The rhizome-root system functions in food storage, fluid transport, and vegetative reproduction [27]. Young bamboos start to grow the new culms in height first, and then expand branches and leaves. Therefore, there is little leaf photosynthesis by new bamboo shoot systems during its culm height growth. Rhizome-root systems widely spread and connect the young culms with other mature bamboo culms to transport carbohydrates and nutrients for new culm height growth [27–30]. A strong translocation function of the *Phyllostachys pubescens* Mazel ex H. de Lehaie rhizome was that more than 20% of the compensative water used during summer was transferred from older culms through connected rhizomes [31]. As well as the carbohydrate storage and fluid transport function, a reticular and extensive underground rhizome-root system of bamboo can effectively bind and hold topsoil against soil erosion [2,32–34].

Very limited published evidence exists in the area of bamboo rhizome-root system responses to different light intensities. Our previous study found that low light intensity decreased underground dry weight, number of all rhizomes, mean length, and diameter of rhizomes of *P. pygmaeus* [22]. However, limited attention has been given to fully understanding the growth status of bamboo rhizomes and roots under changing light conditions. Thus, a pot experiment was conducted to assess the temporal dynamic growth of *P. pygmaeus* old and new rhizomes, culm roots, and rhizome roots during growing season from April to December, and to examine how the growth of its rhizome-root system is affected by different light intensity.

2. Results

2.1. Light Intensity Variation

Across all the survey times, the average light intensity was decreased corresponding to the decreasing aperture of the cover plates, showing CK > L5 > L4 > L3 > L2 > L1 (Figure 1A). For the diurnal variation, the light intensity increased from 9:00 o'clock, then reached the maximum value between 11:30 and 12:30 o'clock, and thereafter decreased until the last records at 16:00 o'clock (Figure 1C). The maximum average light illumination of control (CK) during a day was 8372.73 lux, which was 25.10, 7.56, 4.41, 1.99, and 1.41 times higher than that of treatment L1, L2, L3, L4, and L5, respectively. In addition, the minimum mean light intensity of the control (CK) was 2135.87 lux, which was 21.97, 9.46, 5.28, 2.96, and 1.97 times higher than treatment L1, L2, L3, L4, and L5, respectively.

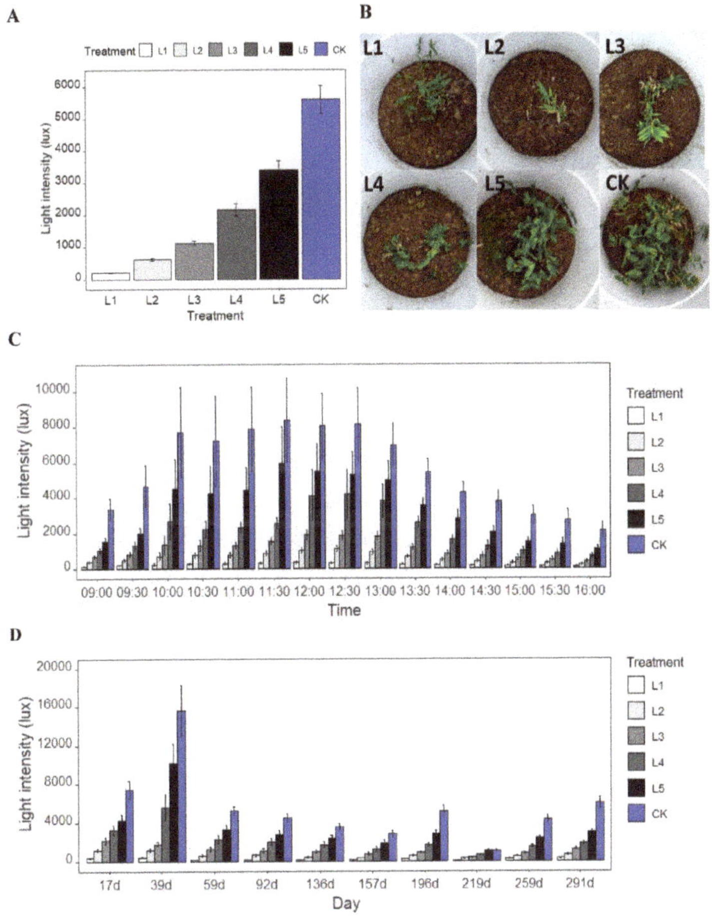

Figure 1. The plots of (**A**) mean light intensity under six light levels; (**B**) typical phenotype of *P. pygmaeus* on November 30, 2016; (**C**) diurnal variation of light intensity under six light levels; (**D**) light intensity along the investigation time under six light levels. The average light intensity of different treatments is L1 = 3.87%, L2 = 11.25%, L3 = 20.25%, L4 = 38.76%, L5 = 60.70%, and CK = 100% of full sunlight.

During the treatment from April to December in 2016, the light intensity of treatment L1, L4, L5, and CK firstly increased from 17 d to 39 d, reaching the maximum value, and then decreased and reached the minimum value at 219 d, and thereafter increased again (Figure 1D). As the investigation time continued, the light intensity of treatment L2 and L3 decreased from 17 d, reaching the minimum value at 219 d, and thereafter increased until the last records at 291 d. At 39 d, the light illumination of control (CK) was 15,619.33 lux, which was 39.03, 13.50, 8.84, 2.79, and 1.54 times higher than that of treatment L1, L2, L3, L4, and L5, respectively. At 219 d, the light illumination of the control (CK) was 1046.73 lux, which was 11.80, 3.51, 2.72, 1.71, and 0.99 times larger than that of treatment L1, L2, L3, L4, and L5, respectively.

Overall, the average illumination intensity under treatment L1, L2, L3, L4, and L5 was 3.87%, 11.25%, 20.25%, 38.76%, and 60.70% of the control (CK), respectively.

2.2. Rhizome and Root Growth Variation during Growing Period

From the results of the generalized additive models (GAMs), it is obvious that the investigated eleven parameters showed diverse growth patterns along the growing period from April to December 2016 (Table 1; Figure 2). Non-significant dynamic growth changes were found in old rhizome length and old and new rhizome diameter during growing season (Figure 2b,c,f, Figure S1 and Figure S2 in the online supplementary data).

Table 1. Results from the generalized additive models (GAMs) in explaining the influence of light intensity and investigation time on the rhizomes' and roots' growth of *P. pygmaeus* plants. *** $p < 0.001$; ** $p < 0.01$; * $p < 0.005$.

Parameters	Independent Variable	Degrees of Freedom	*F* Value	Pr (>|t|)	R^2_{adj}
Number of old rhizomes	*S* (light intensity)	1.000	13.45	<0.001 ***	0.134
	S (invesitigation time)	7.504	5.85	<0.001 ***	
Old rhizome length	*S* (light intensity)	1.000	8.36	<0.05 *	0.038
	S (invesitigation time)	4.233	1.68	0.134	
Old rhizome diameter	*S* (light intensity)	1.000	0.75	0.388	0.004
	S (invesitigation time)	2.140	1.28	0.394	
Number of new rhizomes	*S* (light intensity)	1.928	74.57	<0.001 ***	0.418
	S (invesitigation time)	2.712	12.08	<0.001 ***	
New rhizome length	*S* (light intensity)	3.043	23.11	<0.001 ***	0.345
	S (invesitigation time)	8.107	5.70	<0.001 ***	
New rhizome diameter	*S* (light intensity)	1.329	7.74	<0.05 *	0.071
	S (invesitigation time)	6.171	1.38	0.230	
New rhizome internode length	*S* (light intensity)	1.825	23.90	<0.001 ***	·0.292
	S (invesitigation time)	6.302	6.79	<0.001 ***	
Culm root length	*S* (light intensity)	1.000	0.25	0.617	0.034
	S (invesitigation time)	3.068	4.34	<0.05 *	
Culm root diameter	*S* (light intensity)	1.000	5.00	<0.05 *	0.029
	S (invesitigation time)	1.331	5.12	<0.01 **	
Rhizome root length	*S* (light intensity)	1.913	24.91	<0.001 ***	0.203
	S (invesitigation time)	5.761	7.17	<0.001 ***	
Ratio of aboveground dry weight to underground dry weight (A/U ratio)	*S* (light intensity)	1.965	14.15	<0.001 ***	0.118
	S (invesitigation time)	5.072	3.85	<0.001 ***	

As the investigated time increased from one to twelve, the culm root diameter significantly reduced (degrees of freedom = 1.331, $p < 0.01$, Table 1, Figure 2i). However, the number of new rhizomes showed an opposite relationship with investigation time. The number of new rhizomes significantly increased from April to December (degrees of freedom = 2.712, $p < 0.001$, Figure 2d). In addition, as the light treatment continued, the gap between the control (CK) and treatments of the new rhizome number significantly widened, especially after the 9th investigation time (Figure S2 in the online supplementary data).

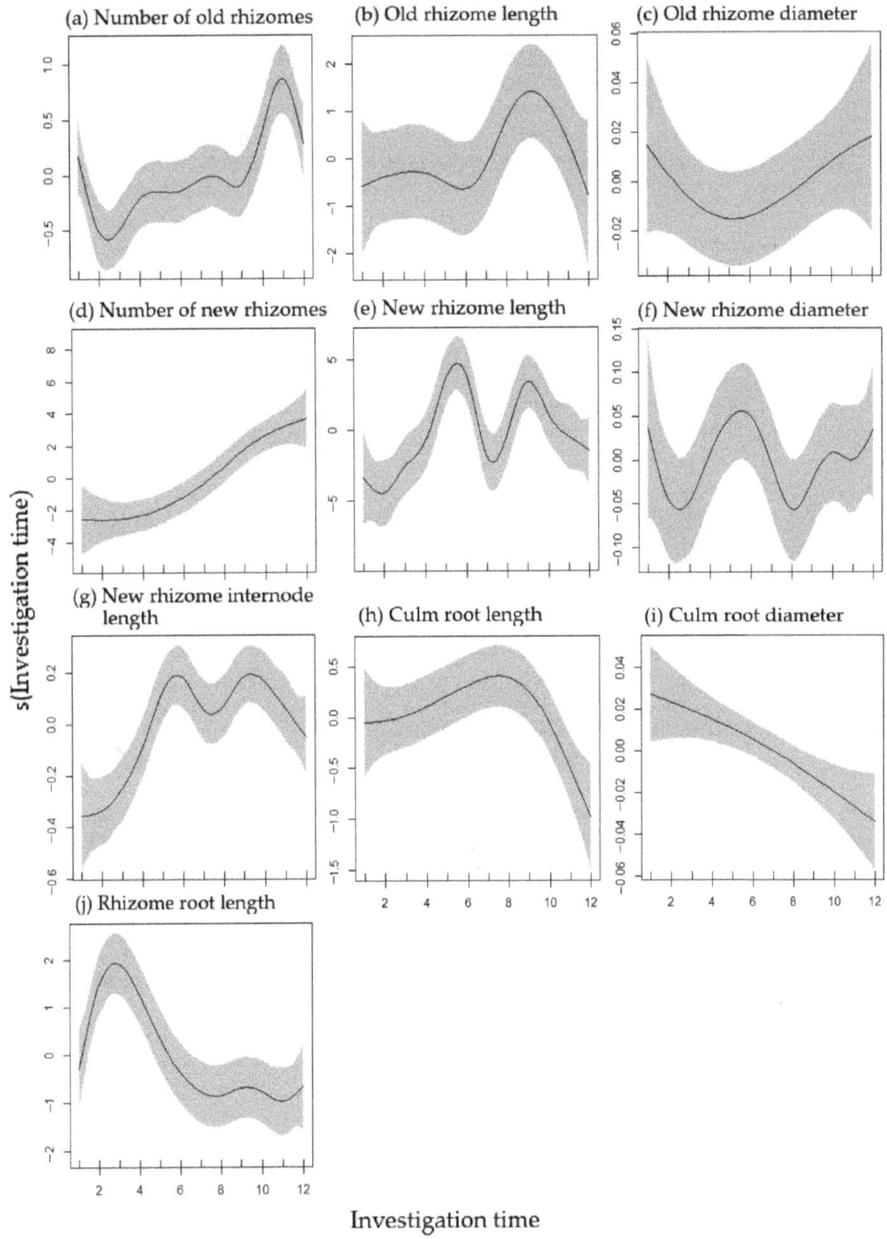

Figure 2. The plots of the GAMs smooth function for indicating the effects of investigation time on the number of old rhizomes (**a**), old rhizome length (**b**), old rhizome diameter (**c**), number of new rhizomes (**d**), new rhizome length (**e**), new rhizome diameter (**f**), new rhizome internode length (**g**), culm root length (**h**), culm root diameter (**i**) and rhizome root length (**j**). The grey ribbon shadow indicates the 95% confidence intervals of the fitted smoothers. The investigation times 1–12 in 2016 are 1 = 15 April, 2 = 9 May, 3 = 3 June, 4 = 27 June, 5 = 21 July, 6 = 12 August, 7 = 5 September, 8 = 23 September, 9 = 17 October, 10 = 10 November, 11 = 2 December, 12 = 26 December.

The new rhizome length and new rhizome internode length indicated a bimodal trend along the investigation time (Figure 2e,g). The new rhizome length and new rhizome internode length reached a peak at around the 6th and 9th survey times, and decreased to a trough at the 7th survey time. In addition, the value of the new rhizome length showed a lowest value at the 2nd survey time and a highest value at around the 6th survey time, whereas the value of the new rhizome internode length showed a lowest value at the 1st survey time and a highest value at around the 6th and 9th survey times.

The culm root length significantly correlated with investigation time, which increased from the first survey time and reached a peak at around the 8th survey time, then sharply declined ($p < 0.05$, Figure 2h, Figure S3 in the online supplementary data). The maximum value of the culm root length along the investigation time predicted from GAMs is the same as the actual value.

A significant change was found in the old rhizome number and rhizome root length along the survey time ($p < 0.001$). The GAMs predicted that the number of old rhizomes would decrease first, reach the minimum value at the 2nd sampling, then increase and reach a peak at the 11th sampling time. Thereafter, the number of old rhizomes would decrease again (Figure 2a). However, the rhizome root length tended to increase first and reached a peak at the 3rd survey time, and then rapidly decreased until the last survey time (Figure 2j).

2.3. *Effects of Light Intensity on Rhizome-Root System Growth*

The results of the GAMs indicated that light intensity had a significant effect on the growth of the rhizome-root system of *P. pygmaeus* plants (Table 1, Figure 3). The relationships of old rhizome number vs. light intensity ($p < 0.001$), old rhizome length vs. light intensity ($p < 0.05$), and culm root diameter vs. light intensity ($p < 0.05$) were linear and positive (degree of freedom = 1.000, Figure 3a,b,i). In addition, the new rhizome diameter showed a significant, positive, and almost linear relationship with increasing light intensity ($p < 0.05$, degrees of freedom = 1.329, Figure 3f). All the above-mentioned four parameters were significantly increased as the light intensity increased from 0 to 100% full sunlight.

The relationship of new rhizome number vs. light intensity was almost linear and positive ($p < 0.001$, degrees of freedom = 1.928). The new rhizome number significantly decreased as the light intensity was reduced from 100% full sunlight to around 40%, and then gradually decreased between 40% and 0% full sunlight (Figure 3d).

Non-linear, positive, and significant relationships were observed of new rhizome length vs. light intensity, new rhizome internode length vs. light intensity, and rhizome root length vs. light intensity ($p < 0.001$, Table 1 and Figure 3e,g,j). The new rhizome length rapidly increased from 0% to ca. 40% full sunlight, and the new rhizome internode length and rhizome root length dramatically increased from 0% to ca. 60% full sunlight, and all the three parameters then gradually increased until 100% full sunlight.

Non-significant relationships were observed of old rhizome diameter vs. light intensity and culm root length vs. light intensity ($p > 0.05$, Figure 3c,h).

2.4. *A/U Ratio-Light Intensity and Investigation Time Analysis*

The ratio of aboveground dry weight to underground dry weight (A/U ratio) was significantly affected by light intensity and investigation time (Figure 4). As the light intensity decreased to 38.76 and 20.25% full sunlight (L4 and L3), the A/U ratio was significantly increased, and then dramatically decreased under treatment L2 and L1 (11.25 and 3.87% full sunlight, Figure 4B). From the results of the GAMs, the A/U ratio presented a single peak curve along the light intensity and a prolonged investigation time (Figure 4D,E) with the highest value under around 55% of full sunlight and at around the 7th sampling time.

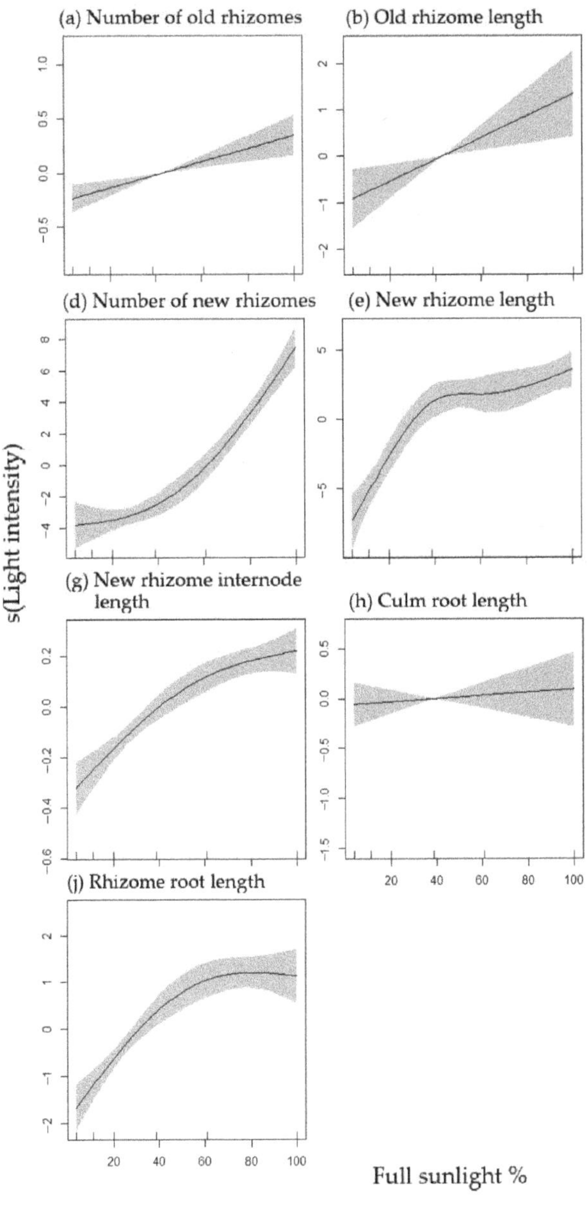

Figure 3. The plots of the GAMs' smooth function for indicating the effects of light intensity on the number of old rhizomes (**a**), old rhizome length (**b**), old rhizome diameter (**c**), number of new rhizomes (**d**), new rhizome length (**e**), new rhizome diameter (**f**), new rhizome internode length (**g**), culm root length (**h**), culm root diameter (**i**) and rhizome root length (**j**). The grey ribbon shadow indicates the 95% confidence intervals of the fitted smoothers.

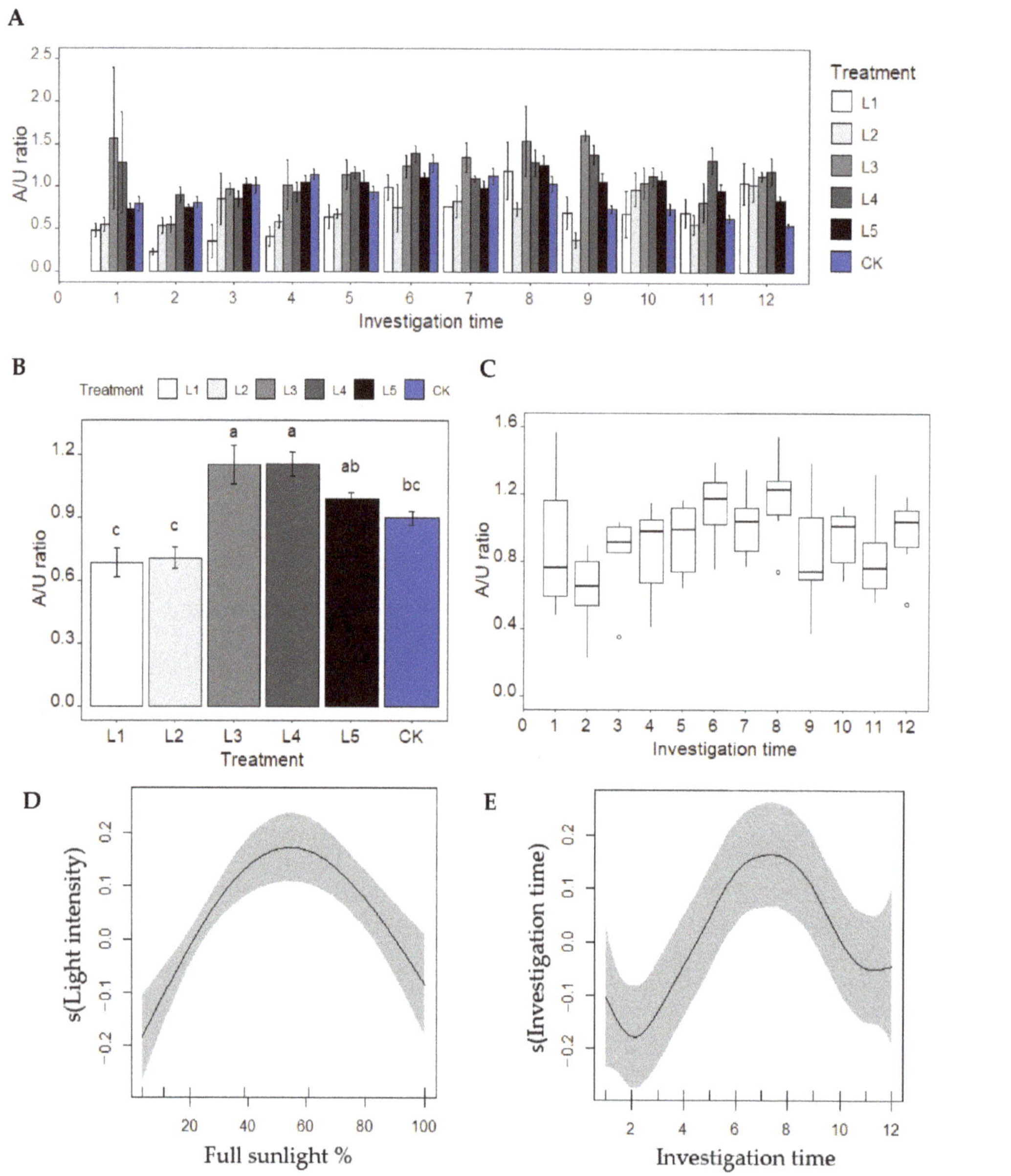

Figure 4. Effects of light intensity and investigation time on the ratio of aboveground to underground dry weight (A/U ratio). (**A**) The A/U ratio under different light intensity along the growing period from April to December; (**B**) average A/U ratio under different light treatment. A Different letter indicates significant differences between treatments at $p < 0.05$ level; (**C**) box plots of A/U ratio under different investigation time; (**D**) the plots of the GAMs' smooth function for illustrating the relationship of A/U ratio vs. light intensity; (**E**) the plots of the GAMs' smooth function for illustrating the relationship of A/U ratio vs. investigation time.

3. Discussion

3.1. Growth Characteristics of Rhizome-Root System

Bamboo rhizome-root systems are the basis of a bamboo forest's growth and development [2]. Along the duration of the experiment from April to December, there were non-significant dynamic growth changes in the old rhizome length and old and new rhizome diameter of *P. pygmaeus*. However, the results of the GAMs predicted that the number of old and new rhizomes significantly increased from April to December. Under full sunlight (CK), the number of new rhizomes at the 9th survey time was 3.1 times higher than that at the 8th survey time, which means that a lot of new rhizomes emerged between September and October (Figure S2 in the online supplementary data). This trend was similar with leaf dry biomass, which at the 9th survey time was 3.8 times larger than that at the 8th survey time [22]. We speculate that the growth of leaves, as the key component of plant's photosynthetic apparatus [35–37], significantly affects new rhizome production but influences old rhizomes' growth less. In other words, carbon used for new rhizome production might be mainly supported by leaf photosynthesis rather than stored non-structural carbonhydrates (NSCs), which is contrary to the proposal that stored NSCs are principally used for promoting new shoots growth [26,38]. The elongation of a new rhizome and its internodes showed a bimodal curve along the growing period, which reached the maximum value around late July and October. After the first peak (late July), the length of new rhizomes and new rhizome internodes significantly decreased until early of September. We speculate that during summer, *P. pygmaeus* adapts its traits to produce shorter rhizomes, which could shorten the nutrient and water transport distance in order to adapt to increasing temperatures and drought. After October, rhizome length decreases since *P. pygmaeus* enters the underground growth stage of bamboo shoots, which is similar to a previous study on *P. pubescens* [39]. However, the rhizome and culm root length increased first and reached a maximum value in June and September, respectively. Based on the results of our previous study, leaf- and aboveground-dry-biomass of *P. pygmaeus* plants were not fully developed when the length of new rhizomes, culm roots, and rhizome roots reached a peak [22]. Thus, we speculate that photosynthetic capacity is insufficient, and the new rhizome and root length growth might require a lot of NSCs transported from old rhizomes. This result is consistent with previous studies on *P. pubescens* that NSCs of leaves, branches, trunks, and rhizomes of attached mature bamboos were transformed and utilized into young bamboo growth [26,38]. During the growing season under full sunlight, the underground biomass (i.e., rhizome-root system) of *P. pygmaeus* plants was 1.11–1.95 times higher than the aboveground biomass. However, opposite results were found in other bamboo species, e.g., *Phyllostachys heterocycla* (Carr.) Mitford cv. Pubescens, *Phyllostachys praecox* C. D. Chu et C. S. Chao 'Prevernalis' and *Fargesia denudata* Yi, which indicated higher aboveground biomass than in the underground [40–42]. In addition, Umemura and Takenaka reported that the *P. pubescens* aboveground/underground biomass ratio is different at differing sites [43]. Specifically, moso bamboo growing on a mountainside with lower water content showed a significantly higher ratio of aboveground to underground biomass than that growing at mountain bases and near water flow. In addition, the culm roots significantly reduced their diameter from April to December, which was similar with culm that decreased its diameter from April until September [22]. However, it is still unknown why culm and culm root decrease its diameter during the growing period, which needs further anatomical and plant physiological study for this phenomenon to be revealed.

3.2. Effects of Light Intensity on Rhizome-Root System Growth

Light directly or indirectly influences virtually all morphological modifications occurring in both shoots and roots [6]. Previous studies confirmed that roots experienced significant changes in morphology and development in response to light, such as root length elongation, root cell elongation, lateral root formation, root nodule formation, nitrate uptake, and carbon assimilation [19,20,44–46]. Most of the mentioned studies used

Arabidopsis as subject. However, the underlying morphological mechanisms that control how light influences bamboo rhizome-root growth remain poorly understood. Here, we report the effects of light illumination on *P. pygmaeus* rhizome-root system growth. We found that light intensity had a significant, positive, and linear/or almost linear impact on the number of old and new rhizomes, old rhizome length, new rhizome diameter, as well as the culm root diameter of *P. pygmaeus* plants. A nonlinear and positive relationship was found between light intensity and the listed three parameters, i.e., new rhizome length, new rhizome internode length, and rhizome root length. The value of the mentioned three parameters significantly increased from 0% to 40–60% full sunlight and then gradually increased until 100% of full sunlight. We found that when the light intensity went lower than ca. 40% of full sunlight, the GAMs predicted that the value of most parameters was lower than zero, e.g., the number of old and new rhizomes, old rhizome length, etc. Therefore, 40% of full sunlight might be the threshold for *P. pygmaeus* rhizome-root system growth. When light was lower than 40%, which is equal to an average light of 2232 lux, the underground growth of *P. pygmaeus* was inhibited.

Light, as one of the most important environmental factors, is not only involved in the shoot growth but also influences root growth and development. The aboveground- and underground-dry-biomass of *P. pygmaeus* were drastically reduced as the light intensity decreased [22]. However, the rate of decline was different. Under minor light control (60.76% of full sunlight), the ratio of aboveground to underground dry weight (A/U ratio) was not significantly different compared to being under full sunlight. Median light limitation (20.25–38.76% of full sunlight) reduced aboveground growth less than underground biomass, while extreme light limitation (<11.25%) reduced aboveground growth more than underground growth. This is consistent with our previous study on *P. pygmaeus* that the leaf/underground dry weight ratio showed a unimodal curve with the decreasing of light intensity [22]. Similar results were found in root/shoot ratio responses of *Eugenia uniflora* L., *Lactuca sativa* L. Var. youmaicai, *Lactuca sativa* L., *Fragaria* × *ananassa* Duch. cv. Benihoppe to changing light intensity [11,47–49]. Light can regulate COP1 (constitutively photomorphogenic 1) to stabilize some proteins (e.g., HY5, Elongated hypocotyl 5). Under deep shade, COP1-mediated light signaling can enhance the primary root elongation through modulating auxin transporter and activating PIN1 expression, increasing PIN1 and PIN2 localization on root-cell plasma membrane [50]. At the same time, HY5 moves from shoots to roots, regulating the expression of the auxin transporters PIN3 and LAX3 to coordinate shoot and root growth [51]. We speculate that under a median limitation (20.25–38.76% of full sunlight), *P. pygmaeus* might try to adjust its plasticity through increasing leaf area and stem height to optimize growth and performance in response to decreasing light intensity like other species [11,17,22,48]. The extreme light limitation (<11.25% full sunlight) might exceed the low light threshold for *P. pygmaeus* growth and result in A/U ratio dramatically reduction. Especially after October, leaf and new rhizome growth were significantly suppressed by low light, and the gap between control (CK) and treatments was significantly widened. A previous study on dwarf bamboos *Sinarundinaria nitida* (Mitford) Nakai found no difference in the ratio of leaf length to width under different light conditions, indicating that leaves in shade prefer to alter their size rather than change the shape [52]. In shade, woody dwarf bamboo can reduce its leaf vein density, which is the pathway for water flow and strongly relates to leaf hydraulic conductance [52–54]. Therefore, reduced leaf water supplied by lower vein density, lower stomatal density, and stomatal conductance could decrease the photosynthetic rate of *S. nitida* [52]. We speculate that the carbon assimilation products of *P. pygmaeus* might be reduced by decreasing light intensity, which consequently results in the reduction of rhizome-root biomass.

4. Materials and Methods

4.1. Plant Materials, Soil, and Container

In March 2016, one-year-old *P. pygmaeus* plants from seeds were obtained from Baima Resource Nursery, Nanjing Forestry University, Nanjing, Jiangsu Province, China

(119°9′15″ E, 31°36′49″ N). Soil was prepared by mixing nutrient soil and loess (1:2 volume ratio). The plastic containers with 51 cm outer-diameter, 44 cm inner-diameter, and 35 cm tall were selected.

4.2. Experimental Design and Light Intensity Recording

The pot experiment was performed in a greenhouse at Baima Resource Nursery. The top shed of the greenhouse was provided with two layers of inner and outer shading nets. A water curtain was set on one side of the greenhouse ventilation and an exhaust fan was set on the other side. In early May, we opened the outer shading net; in late May, the outer shading net and exhaust fan were opened; in mid-June, both of the inner and outer shading nets were opened; in early July, two layers of shading nets and water curtains were opened; in mid-October, the inner shading net was closed; and in mid-November, the outer shading net was also closed.

The two-color plates were used to cover the pot to control light intensity, and the control plots (CK) were not covered by plates. The two-color plates were 60 cm × 60 cm in size and 0.5 cm in thickness. The tope side was wood color and opposite side was black. The light intensity was controlled by punching circular holes with different diameters in the center of the cover plates. The diameters of the punching circular holes were 5 cm, 10 cm, 15 cm, 20 cm, 25 cm, and assigned as treatment L1, L2, L3, L4, and L5 (Figure 5). The illuminance percentages of the pore diameter area to the inner diameter area of the pot were 1.29%, 5.16%, 11.62%, 20.66%, 32.28%, and 100% for treatment L1, L2, L3, L4, L5, and CK, respectively. Two bushes of one-year-old *P. pygmaeus* plants were transplanted into the center of each plastic pot in early March 2016. In total, 216 plants were transplanted with 36 replicates for each treatment.

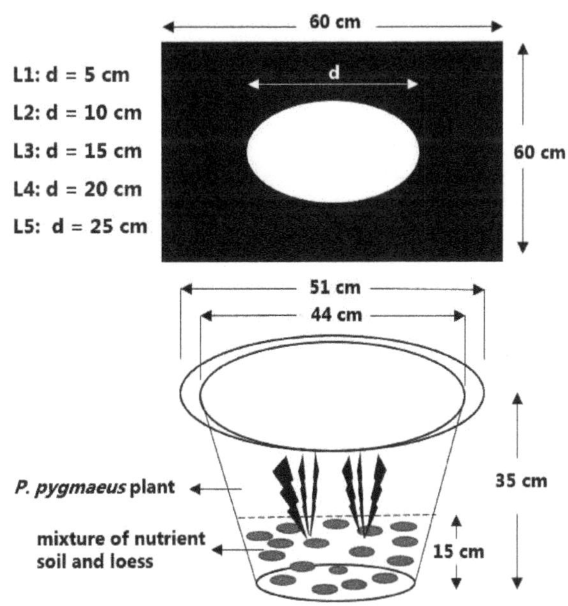

Figure 5. Experiment design and set-up. This drawing was obtained from Huang et al. [22].

One week after confirming the plants' survival, two-color plates were used to cover the containers on 11 March 2016. The sampling and measurement were started on 15 April 2016, and then the next 11 samplings were conducted around every 25 days until December 2016. At each survey time, three pots of each treatment were randomly selected for direct measurement. The specific dates of the 12 investigation times in 2016 are 15 April, 9 May,

3 June, 27 June, 21 July, 12 August, 5 September, 23 September, 17 October, 10 November, 2 December, and 26 December. The specific sampling dates were equal to 36 d, 60 d, 85 d, 109 d, 133 d, 155 d, 179 d, 197 d, 221 d, 245 d, 267 d, and 291 d after the first day of light treatment, respectively.

During light treatment, a sunny and cloudless day was regularly selected each month to record the light intensity every half hour between 9:00 and 16:00 o'clock by a TES-1332A digital illuminance meter (TES Electrical Electronic Corp., Taipei). Three pots of each treatment were randomly selected to measure the light intensity. During measurement, the probe was set between the aperture of the cover plate and the canopy of the plant. When the value of the illuminance meter remained stable for more than three seconds, the value was recorded.

4.3. Growth Measurement

From April 2016 to December 2016, at each investigation time, various parameters to access the rhizome-root growth status were measured including: (i) old rhizome: number of old rhizomes, old rhizome length and old rhizome diameter, (ii) new rhizome: number of new rhizomes, new rhizome length, new rhizome diameter, and new rhizome internode length, (iii) culm root: length and diameter, (iv) rhizome root length.

After morphological measurement, the samples were dried at 80 °C for at least 72 h to a constant dry weight by a ventilated oven (Type: XMTD-8222, Jinghong Experimental Equipment Co., Ltd., Shanghai, China). After drying, the aboveground and underground parts were separated and weighed by an electronic balance (ME204/02, Mettler Toledo Company, Greifensee, Switzerland).

4.4. Statistical Analysis

Generalized additive models (GAMs) are a non-parametric extension of generalized linear models. The GAMs were used to access the correlation coefficient test of the joint effects of light intensity and investigation time on surveyed parameters:

$$g(E(Y_i)) = \beta_0 + s_1(x_i) + s_2(x_i) + e_i \quad (1)$$

where g is a link function, $E(Y_i)$ is the estimate for the responsible variable Y_i, s_1 is the smooth function of x_i over different light treatments, s_2 is the smooth function of x_i along investigation time, x_i (i = 1, 2, 3, . . . , 12) are the explanatory variables, and they are number of new rhizomes, new rhizome length, new rhizome diameter, etc. β_0 is constant term and e_i is error term. All calculations were conducted within the R environment using the "mgcv" package (version 3.6.3) [55].

5. Conclusions

In summary, this study provides a solid basis for understanding *P. pygmaeus* rhizome-root system growth patterns, and its growth responses to different light intensity. The results demonstrate that along the growing period from March to December, eight of the eleven studied rhizome-root parameters showed significant variability and diverse growth patterns. In addition, decreasing light intensity significantly reduced *P. pygmaeus* rhizome-root system growth. As the increasing of light intensity from 0% to 100% full sunlight, the number of old and new rhizomes, old rhizome length, new rhizome diameter and culm root diameter significantly and linearly/or almost linearly increased. When light availability was lower than 40% full sunlight, the new rhizome length significantly decreased; while when light intensity was lower than 60% full sunlight, the new rhizome internode length and rhizome root length were drastically reduced. Based on the prediction value of GAMs, 40% full sunlight (equal to average light of 2232 lux) might be the threshold for *P. pygmaeus* rhizome-root system growth. When light was lower than 40% full sunlight, the rhizome-root system growth of *P. pygmaeus* was inhibited. The A/U ratio showed a unimodal curve with decreasing light intensity. Under median limitation (20.25–38.76% of full sunlight), *P. pygmaeus* could try to adjust its plasticity through increasing stem height to optimize growth and performance. However, extreme light limitation (<11.25% full sunlight) might exceed its low light threshold and significantly reduce A/U ratio. Detailed analyses on how bamboo culms and rhizomes-roots coordinate their responses to light through light-signaling components and pathways need further investigation.

Supplementary Materials: The following supporting information can be downloaded at: https://www.mdpi.com/article/10.3390/plants11172204/s1, Figure S1: The plots of (A) number of old rhizomes, (B) old rhizome length and (C) old rhizome diameter under six different light intensity along the investigation time; Figure S2: The plots of (A) number of new rhizomes, (B) new rhizome length, (C) new rhizome diameter, and (D) new rhizome internode length under six different light intensity along the investigation time; Figure S3: The plots of (A) culm root length, (B) culm root diameter, and (C) rhizome root length under six different light intensity along the investigation time.

Author Contributions: Conceptualization, W.H.; experiment implementation, S.W.; validation, W.H.; formal analysis, W.H.; writing—original draft preparation, W.H.; writing—review and editing, W.H., Y.D., S.W. and C.S.; visualization, W.H.; project administration, F.W.; funding acquisition, W.H. All authors have read and agreed to the published version of the manuscript.

Funding: The study was supported by the Natural Science Foundation of Jiangsu Province, China (grant No. BK20220413), the Nanjing Forestry University Landmark Achievements Cultivation Project, China (grant number: no number), and the Research Start-up Fund of Metasequoia Teachers of Nanjing Forestry University, China (grant number: no number).

Data Availability Statement: Not applicable.

Acknowledgments: We are deeply thankful to Peijian Shi for his invaluable suggestion and guideline on the design of this experiment.

Conflicts of Interest: The authors declare no conflict of interest. The funders had no role in the design of the study; in the collection, analyses, or interpretation of data; in the writing of the manuscript; or in the decision to publish the results.

References

1. FAO. *Global Forest Resources Assessment 2010. Food and Agriculture Organization of the United Nations, Rome, 340p*. Available online: http://www.fao.org/forestry/fra/fra2010/en/ (accessed on 20 September 2020).
2. Liese, W.; Köhl, M. *Bamboo: The Plant and Its Uses*; Springer: Berlin/Heidelberg, Germany, 2015; pp. 44–48.
3. Li, P.; Zhou, G.; Du, H.; Lu, D.; Mo, L.; Xu, X.; Shi, Y.; Zhou, Y. Current and potential carbon stocks in Moso bamboo forests in China. *J. Environ. Manag.* **2015**, *156*, 89–96. [CrossRef] [PubMed]
4. Kunstler, G.; Coomes, D.A.; Canham, C.D. Size-dependence of growth and mortality influence the shade tolerance of trees in a lowland temperate rain forest. *J. Ecol.* **2009**, *97*, 685–695. [CrossRef]
5. Sterck, F.J.; Duursma, R.A.; Pearcy, R.W.; Valladares, F.; Cieslak, M.; Weemstra, M. Plasticity influencing the light compensation point offsets the specialization for light niches across shrub species in a tropical forest understorey. *J. Ecol.* **2013**, *101*, 971–980. [CrossRef]
6. Lee, H.J.; Park, Y.J.; Ha, J.H.; Baldwin, I.T.; Park, C.M. Multiple routes of light signaling during root photomorphogenesis. *Trends Plant Sci.* **2017**, *22*, 803–812. [CrossRef]
7. Schumann, T.; Paul, S.; Melzer, M.; Dörmann, P.; Jahns, P. Plant growth under natural light conditions provides highly flexible short-term acclimation properties toward high light stress. *Front. Plant Sci.* **2017**, *8*, 681. [CrossRef]
8. Baligar, V.C.; Elson, M.K.; He, Z.; Li, Y.; Paiva, A.D.Q.; Almeida, A.A.F.; Ahnert, D. Light intensity effects on the growth, physiological and nutritional parameters of tropical perennial legume cover crops. *Agronomy* **2020**, *10*, 1515. [CrossRef]
9. Raffo, A.; Mozzanini, E.; Nicoli, S.F.; Lupotto, E.; Cervelli, C. Effect of light intensity and water availability on plant growth, essential oil production and composition in *Rosmarinus officinalis* L. *Eur. Food Res. Technol.* **2020**, *246*, 167–177. [CrossRef]
10. Dufour, L.; Guérin, V. Low light intensity promotes growth, photosynthesis and yield of *Anthurium andreanum* Lind. in tropical conditions. *Adv. Hort. Sci.* **2003**, *17*, 9–14.
11. Johkan, M.; Shoji, K.; Goto, F.; Hahida, S.; Yoshihara, T. Effect of green light wavelength and intensity on photomorphogenesis and photosynthesis in *Lactuca sativa*. *Environ. Exp. Bot.* **2012**, *75*, 128–133. [CrossRef]
12. Vialet-Chabrand, S.; Matthews, J.S.A.; Simkin, A.J.; Raines, C.A.; Lawson, T. Importance of fluctuations in light on plant photosynthetic acclimation. *Plant Physiol.* **2017**, *173*, 2163–2179. [CrossRef]
13. Greer, D.H.; Weedon, M.M. Interactions between light and growing season temperatures on, growth and development and gas exchange of Semillon (*Vitis vinifera* L.) vines grown in an irrigated vineyard. *Plant Physiol. Bioch.* **2012**, *54*, 59–69. [CrossRef] [PubMed]

14. Cheng, L.; Tang, X.; Vance, C.P.; White, P.J.; Zhang, F.; Shen, J. Interactions between light intensity and phosphorus nutrition affect the phosphate-mining capacity of white lupin (*Lupinus albus* L.). *J. Exp. Bot.* **2014**, *65*, 2995–3003. [CrossRef] [PubMed]
15. Gottschalk, K.W. Effects of shading on growth and development of northern red oak, black oak, black cherry, and red maple seedlings. I. height, diameter, and root/shoot ratio. In *Fifth Central Hardwood Forest Conference*; Dawson, J.O., Majerus, K.A., Eds.; Society of American Foresters: Washington, DC, USA, 1985; pp. 189–195.
16. Falster, D.S.; Westoby, M. Plant height and evolutionary games. *Trends Ecol. Evol.* **2003**, *18*, 337–343. [CrossRef]
17. Hou, J.; Li, W.; Zheng, Q.; Wang, W.; Xiao, B.; Xing, D. Effects of low light intensity on growth and accumulation of secondary metabolites in roots of *Glycyrrhiza uralensis* Fisch. *Biochem. Syst. Ecol.* **2010**, *38*, 160–168. [CrossRef]
18. Weraduwage, S.M.; Chen, J.; Anozie, F.C.; Morales, A.; Weise, S.E.; Sharkey, T.D. The relationship between leaf area growth and biomass accumulation in *Arabidopsis thaliana*. *Front. Plant Sci.* **2015**, *6*, 167. [CrossRef]
19. Miotto, Y.E.; da Costa, C.T.; Offringa, R.; Kleine-Vehn, J.; dos Santos Maraschin, F. Effects of light intensity on root development in a D-root growth system. *Front. Plant Sci.* **2021**, *12*, 778382. [CrossRef]
20. Dyachok, J.; Zhu, L.; Liao, F.; He, J.; Huq, E.; Blancaflor, E.B. SCAR mediates light-induced root elongation in *Arabidopsis* through photoreceptors and proteasomes. *Plant Cell* **2011**, *23*, 3610–3626. [CrossRef]
21. Suzuki, A.; Suriyagoda, L.; Shigeyama, T.; Tominaga, A.; Sasaki, M.; Hiratsuka, Y.; Yoshinaga, A.; Arima, S.; Agarie, S.; Sakai, T.; et al. *Lotus japonicus* nodulation is photomorphogenetically controlled by sensing the red/far red (R/FR) ratio through jasmonic acid (JA) signaling. *Proc. Natl. Acad. Sci. USA* **2011**, *108*, 16837–16842. [CrossRef]
22. Huang, W.; Olson, E.; Wang, S.; Shi, P. The growth and mortality of *Pleioblastus pygmaeus* under different light availability. *Glob. Ecol. Conserv.* **2020**, *24*, e01262. [CrossRef]
23. Rivieres, A.; Rivieres, C. Les bambous. *Bull. Soc. Natl. Acclim.* **1897**, *5*, 221–253.
24. Jiang, Z. *Bamboo and Rattan in the World*; China Forestry Publishing House: Beijing, China, 2007.
25. Embaye, K.; Weih, M.; Ledin, S.; Christersson, L. Biomass and nutrient distribution in a highland bamboo forest in southwest Ethiopia: Implications for management. *Forest Ecol. Manag.* **2005**, *204*, 159–169. [CrossRef]
26. Song, X.; Peng, C.; Zhou, G.; Gu, H.; Li, Q.; Zhang, C. Dynamic allocation and transfer of non-structural carbohydrates, a possible mechanism for the explosive growth of Moso bamboo (*Phyllostachys heterocycla*). *SCI REP-UK* **2016**, *6*, 25908. [CrossRef] [PubMed]
27. Li, R.; Werger, M.; During, H.; Zhong, Z. Carbon and nutrient dynamics in relation to growth rhythm in the giant bamboo *Phyllostachys pubescens*. *Plant Soil* **1998**, *201*, 113–123. [CrossRef]
28. Li, R.; Werger, M.J.A.; de Kroon, H.; During, H.J.; Zhong, Z.C. Interactions between shoot age structure, nutrient availability and physiological integration in the giant bamboo *Phyllostachys pubescens*. *Plant Boil.* **2000**, *2*, 437–446. [CrossRef]
29. Zhou, G.; Jiang, P.; Xu, Q. *Carbon Fixing and Transition in the Ecosystem of Bamboo Stands*, 2nd ed.; Science Press: Beijing, China, 2010.
30. Mohamed, J.; Hamid, H.A.; Nuruddin, A.A.; Majid, N.M.N.A. Chemical attributes of *Gigantochloa scortechinii* bamboo rhizome in relation with hydraulic conductance. *BioResources* **2019**, *14*, 8155–8173.
31. Zhao, X.; Zhao, P.; Zhang, Z.; Zhu, L.; Niu, J.; Ni, G.; Hu, Y.; Ouyang, L. Sap flow-based transpiration in *Phyllostachys pubescens*: Applicability of the TDP methodology, age effect and rhizome role. *Trees-Struct. Funct.* **2017**, *31*, 765–779.
32. Zhou, B.; Fu, M.; Xie, J.; Yang, X.; Li, Z. Ecological functions of bamboo forest: Research and application. *J. For. Res.* **2005**, *16*, 143–147.
33. Lin, D.; Liu, W.; Lin, S. Estimating the effect of shear strength increment due to root on the stability of Makino bamboo forest slopeland. *J. GeoEng.* **2011**, *6*, 73–88.
34. Li, C.; Cai, Y.; Xiao, L.; Gao, X.; Shi, Y.; Zhou, Y.; Du, H.; Zhou, G. Rhizome extension characteristics, structure and carbon storage relationships with culms in a 10-year moso bamboo reforestation period. *Forest Ecol. Manag.* **2021**, *498*, 119556. [CrossRef]
35. Green, D.S.; Kruger, E.L. Light-mediated constraints on leaf function correlate with leaf structure among deciduous and evergreen tree species. *Tree Physiol.* **2001**, *21*, 1341–1346. [CrossRef]
36. Li, Y.; Zhang, Y.; Zhang, X.; Korpelainen, H.; Berninger, F. Effects of elevated CO_2 and temperature on photosynthesis and leaf traits of an understory dwarf bamboo in subalpine forest zone, China. *Physiol. Plantarum* **2013**, *148*, 261–272. [CrossRef] [PubMed]
37. Huang, W.; Ratkowsky, D.A.; Hui, C.; Wang, P.; Su, J.; Shi, P. Leaf fresh weight versus dry weight: Which is better for describing the scaling relationship between leaf biomass and leaf area for broad-leaved plants? *Forests* **2019**, *10*, 256. [CrossRef]
38. Wang, S.; Chen, T.; Liu, E.; Liu, C. Accessing the nursing behaviour of moso bamboo (*Phyllostachys edilus*) on carbohydrates dynamics and photosystems. *SCI REP-UK* **2020**, *10*, 1015. [CrossRef]
39. Chen, C.; Dong, W.; Zheng, J. Growth and rhythm of rhizome system for natural forest of *Phyllostachys pubescens* clonal population in Haiziping. *For. Sci. Technol. Dev.* **2012**, *26*, 26–29. (In Chinese)
40. Chen, H.; Hong, W.; Lan, B.; Zheng, Y.; He, D. Study on biomass and productivity of *Phyllostachys heterocycala* cv. *pubescens forest in the north of Fujian*. *Sci. Silvae Sin.* **1998**, *34*, 60–64. (In Chinese)
41. Hong, C.; Fang, J.; Jin, A.; Cai, J.; Guo, H.; Ren, J.; Shao, Q.; Zheng, B. Comparative growth, biomass production and fuel properties among different perennial plants, bamboo and Miscanthus. *Bot. Rev.* **2011**, *77*, 197–207. [CrossRef]
42. Liu, C.; Wang, Y.; Pan, K.; Zhu, T.; Li, W.; Zhang, L. Carbon and nitrogen metabolism in leaves and roots of dwarf bamboo (*Fargesia denudata* Yi) subjected to drought for two consecutive years during sprouting period. *J. Plant Growth Regul.* **2014**, *33*, 243–255. [CrossRef]

43. Umemura, M.; Takenaka, C. Biological cycle of silicon in moso bamboo (*Phyllostachys pubescens*) forests in central Japan. *Ecol. Res.* **2014**, *29*, 501–510. [CrossRef]
44. Correll, M.J.; Kiss, J.Z. The roles of phytochromes in elongation and gravitropism of roots. *Plant Cell Physiol.* **2005**, *46*, 317–323. [CrossRef]
45. Chen, X.; Yao, Q.; Gao, X.; Jiang, C.; Harberd, N.P.; Fu, X. Shoot-to-root mobile transcription factor HY5 coordinates plant carbon and nitrogen. *Curr. Biol.* **2016**, *26*, 640–646. [CrossRef]
46. Silva-Navas, J.; Moreno-Risueno, M.A.; Manzano, C.; Téllez-Robledo, B.; Navarro-Neila, S.; Carrasco, V.; Pollmann, S.; Gallego, F.J.; del Pozo, J.C. Flavonols mediate root phototropism and growth through regulation of proliferation-to-differentiation transition. *Plant Cell* **2016**, *28*, 1372–1387. [CrossRef] [PubMed]
47. Mielke, M.S.; Schaffer, B. Photosynthetic and growth responses of *Eugenia uniflora* L. seedlings to soil flooding and light intensity. *Environ. Exp. Bot.* **2010**, *68*, 113–121. [CrossRef]
48. Fu, Y.; Li, H.; Yu, J.; Liu, H.; Cao, Z.; Manukovsky, N.S.; Liu, H. Interaction effects of light intensity and nitrogen concentration on growth, photosynthetic characteristics and quality of lettuce (*Lactuca sativa* L. Var. youmaicai). *Sci. Hortic.* **2017**, *214*, 51–57. [CrossRef]
49. Zheng, J.; Ji, F.; He, D.; Niu, G. Effect of light intensity on rooting and growth of hydroponic strawberry runner plants in a LED plant factory. *Agronomy* **2019**, *9*, 875. [CrossRef]
50. Sassi, M.; Lu, Y.; Zhang, Y.; Wang, J.; Dhonukshe, P.; Blilou, I.; Dai, M.; Li, J.; Gong, X.; Jaillais, Y.; et al. COP1 mediates the coordination of root and shoot growth by light through modulation of PIN1- and PIN2-dependent auxin transport in Arabidopsis. *Development* **2012**, *139*, 3402–3412. [CrossRef]
51. Yang, Y.; Liu, H. Coordinated shoot and root responses to light signaling in Arabidopsis. *Plant Commun.* **2020**, *1*, 100026. [CrossRef]
52. Yang, S.; Sun, M.; Zhang, Y.; Cochard, H.; Cao, K. Strong leaf morphological, anatomical, and physiological responses of a subtropical woody bamboo (*Sinarundinaria nitida*) to contrasting light environments. *Plant Ecol.* **2014**, *215*, 97–109. [CrossRef]
53. Brodribb, T.J.; Field, T.S.; Jordan, G.J. Leaf maximum photosynthetic rate and venation are links by hydraulics. *Plant Physiol.* **2007**, *144*, 1890–1898. [CrossRef]
54. Sack, L.; Scoffoni, C. Leaf venation: Structure, function, development, evolution, ecology and applications in the past, present, and future. *New Phytol.* **2013**, *198*, 983–1000. [CrossRef]
55. Wood, S.N. *Generalized Additive Models: An Introduction with R*, 2nd ed.; Chapman and Hall/CRC: New York, NY, USA, 2017.

plants

MDPI

Article

Climatic Factors Determine the Distribution Patterns of Leaf Nutrient Traits at Large Scales

Xianxian Wang [1], Jiangfeng Wang [1], Liuyang Zhang [2], Chengyu Lv [3], Longlong Liu [4], Huixin Zhao [1,*] and Jie Gao [1,5,*]

[1] College of Life Sciences, Xinjiang Normal University, Urumqi 830054, China
[2] Hubei Forestry Investigation and Planning Institute, No. 4, Zhuodaoquan South Road, Hongshan District, Wuhan 430070, China
[3] Taiyuan Ecology and Environment Monitoring & Science Research Center, Taiyuan 030002, China
[4] East China Survey and Planning Institute, National Forestry and Grassland Administration, Hangzhou 310019, China
[5] Institute of Ecology and Key Laboratory of Earth Surface Processes of Ministry of Education, College of Urban and Environmental Sciences, Peking University, Beijing 100871, China
* Correspondence: zhaohuixin101@sina.com (H.Z.); jiegao72@gmail.com (J.G.)

Abstract: Leaf nutrient content and its stoichiometric relationships (N/P ratio) are essential for photosynthesis and plant growth and development. Previous studies on leaf nutrient-related functional traits have mainly focused on the species level and regional scale, but fewer studies have investigated the distribution patterns of the leaf N and P contents (LN, LP) and N/P ratios (N/P) in communities and their controlling factors at a large scale; therefore, we used LN, LP, and N/P data at 69 sites from 818 forests in China. The results showed significant differences ($p < 0.05$) in the LN, LP, and N/P at different life forms (tree, shrub, and herb). Neither LN, LP, nor N/P ratios showed significant patterns of latitudinal variation. With the increase in temperature and rainfall, the LN, LP, and leaf nutrient contents increased significantly ($p < 0.001$). Across life forms, LN at different life forms varied significantly and was positively correlated with soil P content ($p < 0.001$). The explanatory degree of climatic factors in shaping the spatial variation patterns of LN and N/P was higher than that of the soil nutrient factors, and the spatial variation patterns of the leaf nutrient traits of different life forms were shaped by the synergistic effects of climatic factors and soil nutrient factors.

Keywords: leaf nutrients; functional traits; life forms; climate change; soil nutrients

Citation: Wang, X.; Wang, J.; Zhang, L.; Lv, C.; Liu, L.; Zhao, H.; Gao, J. Climatic Factors Determine the Distribution Patterns of Leaf Nutrient Traits at Large Scales. *Plants* **2022**, *11*, 2171. https://doi.org/10.3390/plants11162171

Academic Editors: Martina Pollastrini and Oleg Chertov

Received: 8 August 2022
Accepted: 18 August 2022
Published: 21 August 2022

1. Introduction

The functional traits of plants reflect their survival strategies in response to environmental changes [1,2]. Nitrogen (N) and phosphorus (P) are important components of the basic structure of plant cells and their levels. The stoichiometric relationships between N and P drive photosynthesis, plant growth, reproduction, and other life processes [3]. Leaf nitrogen (LN), phosphorus (LP), and leaf N/P ratios (N/P) are key plant traits that influence the productivity of forest communities and regulate carbon cycling [4]. Numerous studies have found that climatic factors, soil nutrient factors, and community genealogical structure influence the plant functional traits by affecting plant metabolism [5]. Exploring the distribution patterns of functional traits related to leaf nutrients on a large scale has important ecological significance for quantifying the impact of environmental changes on plant functional traits [6]. However, the key drivers that determine the distribution patterns of these key functional traits remain elusive.

Climate factors influence the functional traits of plant leaves by controlling plant metabolism, growth, and development processes [7]. A global data-based study found that climate change significantly affected the leaf nutrient content [8]. With increasing temperature, the LN decreases significantly and LP increases significantly [9]. Higher LN enhances photosynthesis under low-temperature stress, and when temperatures are higher, plants coordinate metabolic processes by reducing the LN and increasing LP [10]. Other studies have found that the N/P of plant leaves increased significantly with increasing temperature [11]. Numerous previous studies have shown that temperature is a key factor in determining changes in leaf nutrient traits, especially the mean annual temperature (MAT) [4,12,13].

Plant leaf nutrients are also influenced by precipitation factors [14]. Precipitation factors can alter soil nutrient availability, and when rainfall is low and soils are subject to drought stress, plants respond to water stress by changing functional traits [15]. Temperature and precipitation factors jointly determine the nutrient changes in plant leaves [16]. In specific forest communities, the key climatic factors that determine the leaf nutrients in different species remain elusive due to differences in biological adaptations and plant nutrient uptake strategies. Therefore, it is particularly important to quantify the effects of climatic factors in the leaves of different plant forms at a macroscopic scale.

Not only do climatic factors have an important impact on the nutrient profiles of plant leaves, but soil nutrient factors also play a key role in shaping the spatiotemporal pattern of leaf nutrient traits [17]. Climate affects the plant nutrient profiles by influencing soil nutrient redistribution [18]. Plant vessels absorb inorganic salts and water from the soil and transport them to the leaves, so the soil is the main source of nitrogen and phosphorus for plant leaves [19]. The nutrient profiles of leaves are limited by the nutrient availability of the soil [20]. Some studies have found that the N/P of Chinese plant leaves is significantly lower than the global average due to the low effective phosphorus content of Chinese soils [21]. It has been shown that the N and P content of plant leaves increases with increasing soil N content and soil pH [22], and that soil nutrient factors directly affect the nutrient profiles of leaves.

Forest communities are divided into tree, shrub, and herb levels according to species composition, structure, and production, with each plant having its life form, and vertical differentiation provides a good indication of the community's adaptation to environmental conditions [23]. The nutrient profiles of leaves vary significantly among different life forms due to differences in their survival strategies [24]. Comparisons of the leaf nutrient traits of different plant forms are mostly based on the species scale or local scale, and few studies have been conducted on a larger spatial scale based on the community scale. Community trait variation is more effective than species trait variation in predicting the plant response to environmental change [25]. However, due to interspecific competition and intraspecific struggle in a given forest community, there is a great theoretical risk in predicting the functional traits of individual plants [26]. Therefore, the effects of environmental factors on community leaf nutrient traits at a macro scale can reduce the impact of certain local deterministic processes (e.g., competition, mutualism).

Based on field survey data from 89 sites in China, an attempt was made to identify the key drivers influencing the leaf nutrient traits at different life forms in forest communities at the macro scale. To explain this, the following hypotheses were made: (1) there are significant differences in leaf nutrient traits at different life forms at the macro scale; and (2) climatic factors are the dominant environmental factors affecting leaf nutrient traits at different life forms of the community, while soil nutrient factors also play a coordinating role that cannot be ignored.

2. Results

2.1. Variability in LN, LP, and N/P of Different Life Forms and Distribution Patterns in China

Leaf nutrient traits showed significant geographical variation at different life forms, with overall higher LN and LP in northeastern China and relatively higher LP content in northwestern China (Figure 1). However, neither LN, LP, nor N/P (Figure 2) showed a significant latitudinal pattern. The LN, LP, and N/P of the herb levels showed significant differences ($p < 0.05$). The leaf nutrient traits of the arbor layer and the herb layer were significantly different ($p < 0.001$), and the herb layer had relatively higher LN and lower LP.

Leaf nutrient traits were generally significantly different between the shrub and herb levels ($p < 0.05$), but LN was not significant in the herb and shrub levels ($p > 0.05$).

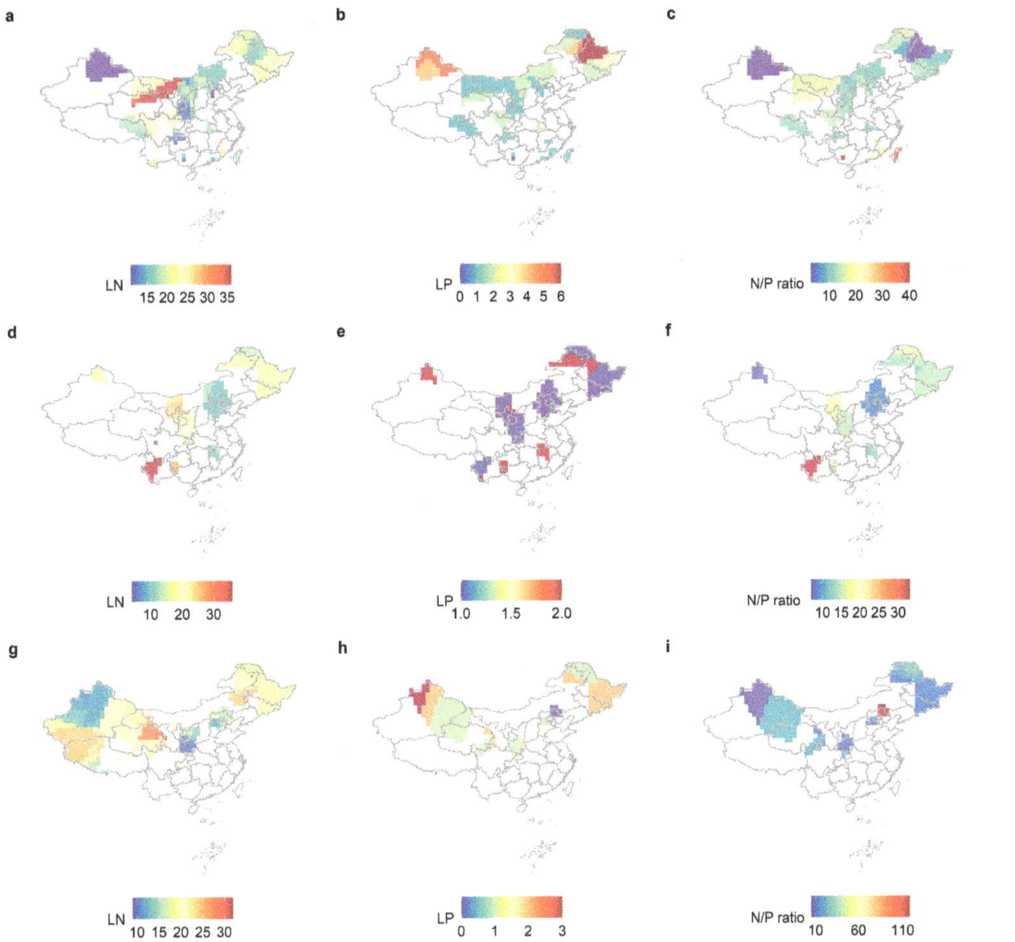

Figure 1. The distribution patterns of LN, LP, and N/P at different life forms in China with a spatial resolution of 1 × 1 km were studied by kernel density estimation. (**a**) LN of the tree levels; (**b**) LP of the tree levels; (**c**) N/P of the tree levels; (**d**) LN of the shrub levels; (**e**) LP of the shrub levels; (**f**) leaf N/P of the shrub levels; (**g**) LN of the herb levels; (**h**) herb levels LP; (**i**) herb levels N/P.

2.2. Correlations between Climatic Factors and LN, LP, and N/P at Different Life Forms

The LN and LP of the tree, shrub, or herb levels increased significantly with increasing MAT and mean annual precipitation (MAP) (Figures 3a,d and 4a,d) and decreased significantly with increasing annual sunlight duration (ASD) (Figures 3e,f and 4e). Whereas LP in both the tree and shrub levels increased significantly with increasing mean coldest monthly temperature (MCMT) and mean warmest monthly temperature (MWMT), LN in the herb level decreased significantly. The ASD and mean annual evaporation (MAE) had better predictive power for LN in the tree level compared to other climatic factors ($R^2 = 0.20$, $p < 0.001$; $R^2 = 0.16$, $p < 0.001$) and better predictive power for LP in the tree layer ($R^2 = 0.46$, $p < 0.001$; $R^2 = 0.50$, $p < 0.001$; $R^2 = 0.37$, $p < 0.001$) were MWMT, MAP, and ASD.

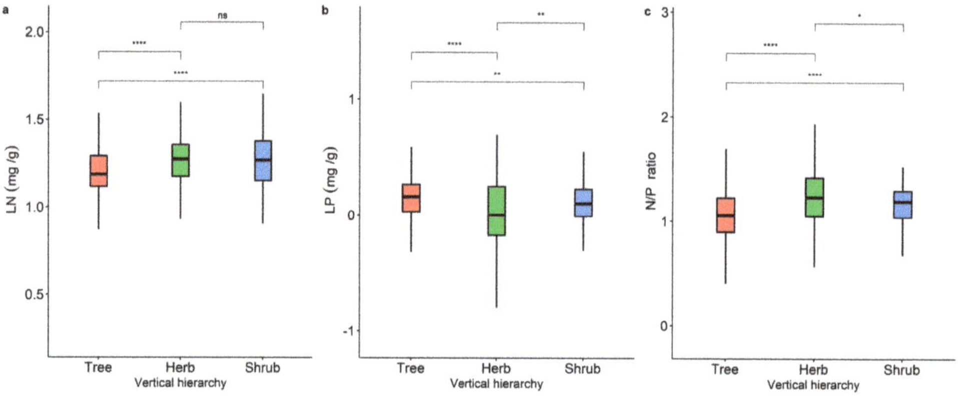

Figure 2. A comparison of the differences in the LN (**a**), LP (**b**), and N/P (**c**) at different life forms. (**a**) Variability of LN among the trees, shrubs, and herbs; (**b**) variability of LP among the trees, shrubs, and herbs; (**c**) variability of N/P among the trees, shrubs, and herbs. Tree represents the tree levels, Shrub represents the shrub levels, and Herb represents the herb levels. Levels are grouped where ns represents non−significant ($p > 0.05$) at the 0.05 level, * represents $p < 0.05$, ** represents $p < 0.01$, **** represents $p < 0.0001$.

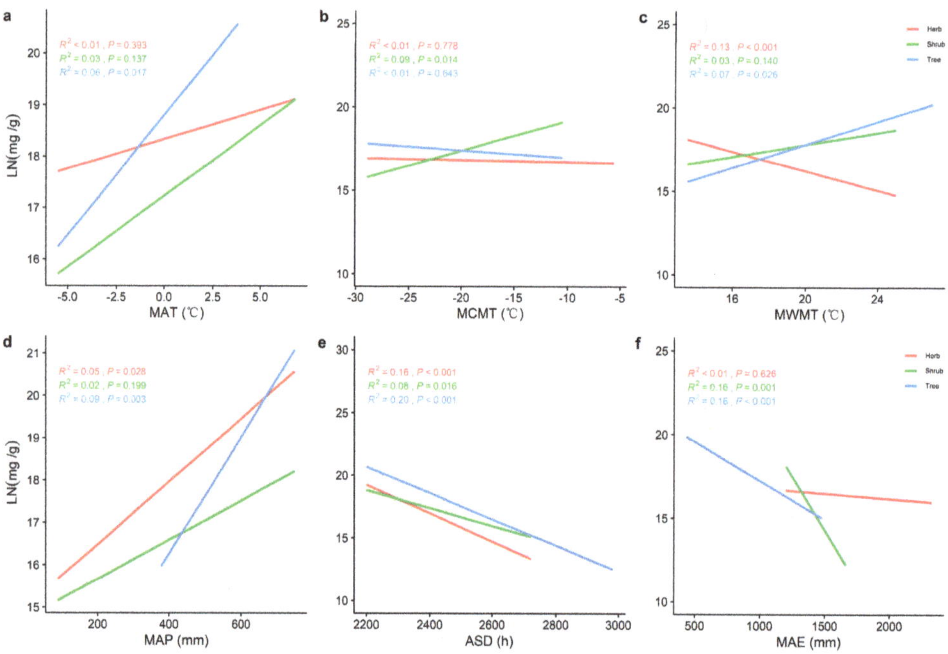

Figure 3. The general linear correlation analysis of climate factors with LN at different life forms. (**a**) General linear relationship between the MAT and LN of plants in the trees, shrubs, and herbs. (**b**) General linear relationship between the MCMT and LN of plants in the trees, shrubs, and herbs. (**c**) General linear relationship between the MWMT and LN of plants in the trees, shrubs, and herbs (**d**) General linear relationship between the MAP and LN of plants in the trees, shrubs, and herbs (**e**) General linear relationship between the ASD and LN of plants in the trees, shrubs, and herbs (**f**).

General linear relationship between the MAE and LN of plants in the trees, shrubs, and herbs. MAT represents the mean annual temperature, MCMT represents the mean coldest monthly temperature, MWMT represents the mean warmest monthly temperature, MAP represents the mean annual precipitation, ASD represents the annual sunlight duration, and MAE represents the mean annual evaporation. The red line represents the tree level, the green line is the shrub level, and the blue line is the herb level. R^2 represents how well the model fits the variables studied and the *p*-value represents the significance level.

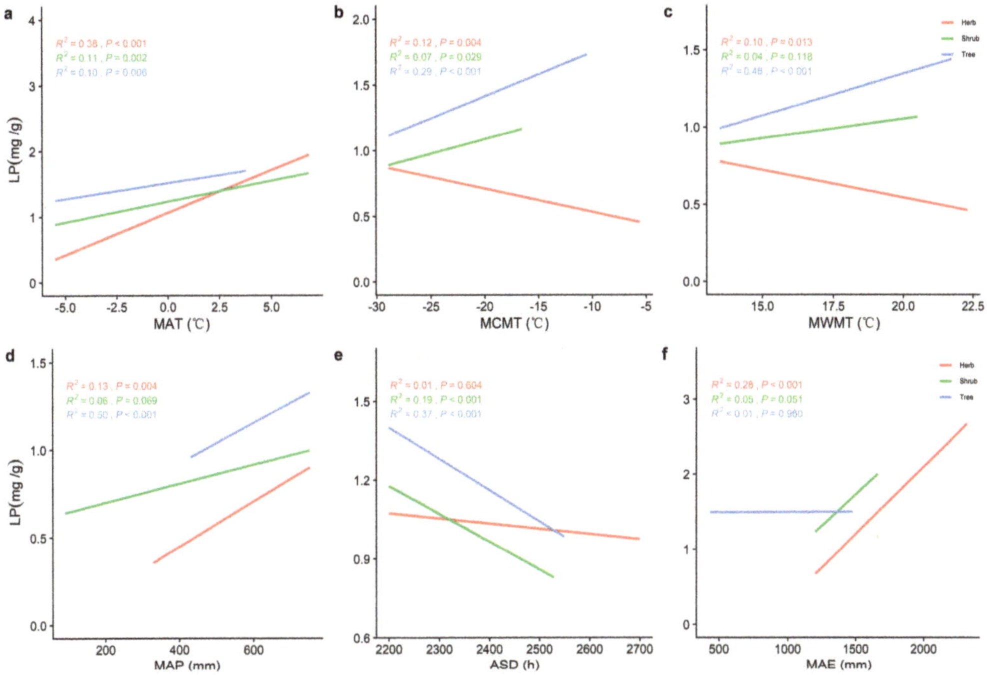

Figure 4. The general linear correlation analysis of climate factors with LP at different life forms. (**a**) General linear relationship between the MAT and LP of plants in the trees, shrubs, and herbs. (**b**) General linear relationship between the MCMT and LP of plants in the trees, shrubs, and herbs. (**c**) General linear relationship between the MWMT and LP of plants in the trees, shrubs, and herbs. (**d**) General linear relationship between the MAP and LP of plants in the trees, shrubs, and herbs. (**e**) General linear relationship between the ASD and LP of plants in the trees, shrubs, and herbs. (**f**) General linear relationship between the MAE and LP of plants in the trees, shrubs, and herbs. MAT represents the mean annual temperature, MCMT represents the mean coldest monthly temperature, MWMT represents the mean warmest monthly temperature, MAP represents the mean annual precipitation, ASD represents the annual sunlight duration, and MAE represents the mean annual evaporation. The red line represents the tree level, the green line is the shrub level, and the blue line is the herb level. R^2 represents how well the model fits the variables studied and the *p*-value represents the significance level.

The N/P of the herb level increased with the increase in MCMT and MWMT (Figure 5b,c), and decreased with the increase in MAT, MAP, ASD, and MAE (Figure 5a,d–f). MAT, MCMT, and MWMT (Figure 5a–c) were significantly correlated with the N/P of the shrub layer ($p < 0.001$). MAT had the best fitting effect on the N/P of the herbal layer (Figure 5b; $R^2 = 0.59$, $p < 0.001$).

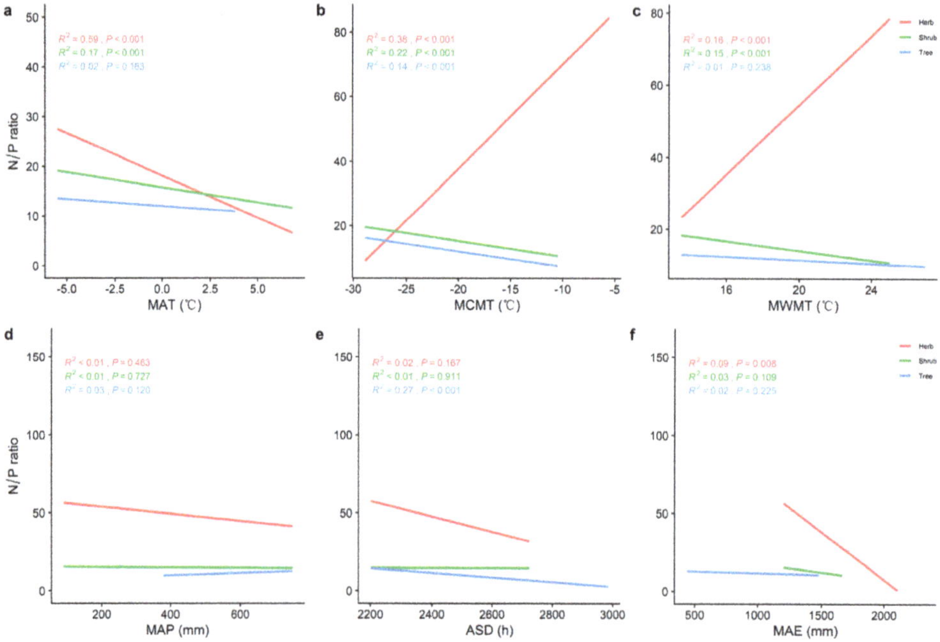

Figure 5. The general linear correlation analysis of climate factors with N/P at different life forms. (**a**) General linear relationship between the MAT and N/P of plants in the trees, shrubs, and herbs. (**b**) General linear relationship between the MCMT and N/P of plants in the trees, shrubs, and herbs. (**c**) General linear relationship between the MWMT and N/P of plants in the trees, shrubs, and herbs. (**d**) General linear relationship between the MAP and N/P of plants in the trees, shrubs, and herbs. (**e**) General linear relationship between the ASD and N/P of plants in the trees, shrubs, and herbs. (**f**): General linear relationship between the MAE and N/P of plants in the trees, shrubs, and herbs. MAT represents the mean annual temperature, MCMT represents the mean coldest monthly temperature, MWMT represents the mean warmest monthly temperature, MAP represents the mean annual precipitation, ASD represents the annual sunlight duration, and MAE represents the mean annual evaporation. The red line represents the tree level, the green line is the shrub level, and the blue line is the herb level. R^2 represents how well the model fits the variables studied and the *p*-value represents the significance level.

2.3. *Effect of Soil Factors on the Relationship between LN, LP, and Leaf N/P at Different Life Forms*

The LN of the different life forms was significantly and positively correlated with the soil P (Figure 6b), while the trends of LP and N/P varied with the soil nutrient factors. The LN of the three life forms increased significantly with increasing soil N and P (Figure 6). Soil pH (Figure 7c) all showed a significant positive correlation ($p < 0.001$) with LP for the different life forms, whereas LN decreased significantly with increasing soil pH (Figure 6). The LP of the shrub levels increased significantly with the increasing soil N and P. The trend between the herb and shrub levels was reversed. The N/P in the herb level was significantly positively correlated with soil N ($R^2 = 0.30$, $p < 0.001$) (Figure 8a). Soil pH (Figure 7c) was the best predictor of LP in the herb and tree levels ($R^2 = 0.25$, $p < 0.001$; $R^2 = 0.29$, $p < 0.001$) (Figure 7c), and the best predictor of N/P in the tree level (Figure 8c) ($R^2 = 0.29$, $p < 0.001$) (Figure 8c). In contrast, the best prediction of N/P for the shrub level was for soil P ($R^2 = 0.22$, $p < 0.001$; Figure 8b).

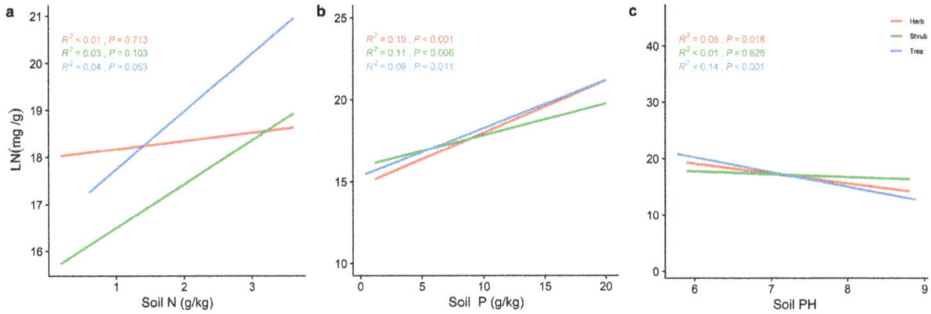

Figure 6. The general linear analysis of soil factors with different life forms of LN. (**a**) General linear relationship between the soil N and LN of plants in the trees, shrubs, and herbs. (**b**) General linear relationship between the soil P and LN of plants in the trees, shrubs, and herbs. (**c**) General linear relationship between the soil pH and LN of plants in the trees, shrubs, and herbs. The red line represents the tree level, the green line is the shrub level, and the blue line is the herb level. R^2 represents how well the model fits the variables studied and *p*-value represents the level of significance.

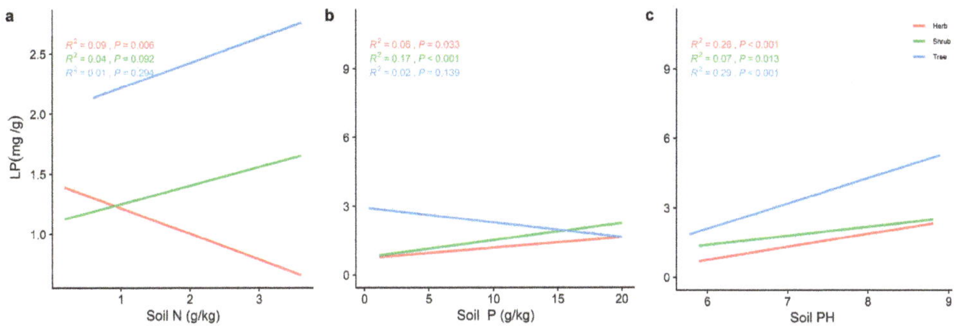

Figure 7. The general linear analysis of soil factors with different life forms of LP. (**a**) General linear relationship between the soil N and LP of plants in the trees, shrubs, and herbs (**b**) General linear relationship between the soil P and LP of plants in the trees, shrubs, and herbs. (**c**) General linear relationship between the soil pH and LP of plants in the trees, shrubs, and herbs. The red line represents the tree level, the green line is the shrub level, and the blue line is the herb level. R^2 represents how well the model fits the variables studied and the *p*-value represents the level of significance.

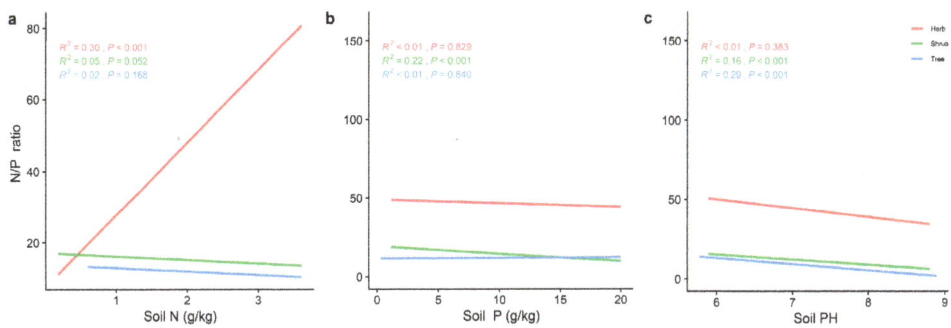

Figure 8. The general linear analysis of soil factors with different life forms of N/P. (**a**) General linear

relationship between the soil N and N/P of plants in the trees, shrubs, and herbs (**b**) General linear relationship between the soil P and N/P of plants in the trees, shrubs, and herbs. (**c**) General linear relationship between the soil pH and N/P of plants in the trees, shrubs, and herbs. The red line represents the tree level, the green line is the shrub level, and the blue line is the herb level. R^2 represents how well the model fits the variables studied and the p-value represents the level of significance.

2.4. Climatic and Soil Factors Dominate Changes in the Functional Traits of Different Communities

We analyzed the effects of environmental factors on the leaf LN, LP, and N/P at different life forms based on a generalized additive model with non−metric multidimensional scaling (NMDS) ranking. Overall, the leaf N/P showed strong environmental plasticity, and environmental factors generally explained the spatial variation in leaf N/P more than LN and LP at different life forms (Figures 9–11). Climate factors played a greater role than the soil nutrient factors in shaping the spatial variation in the leaf LN (Figure 9d−f; de = 42.2%, 41%, 22.8%) and LP (Figure 10d−f; de = 17.2%, 17.1%, 50.6%).

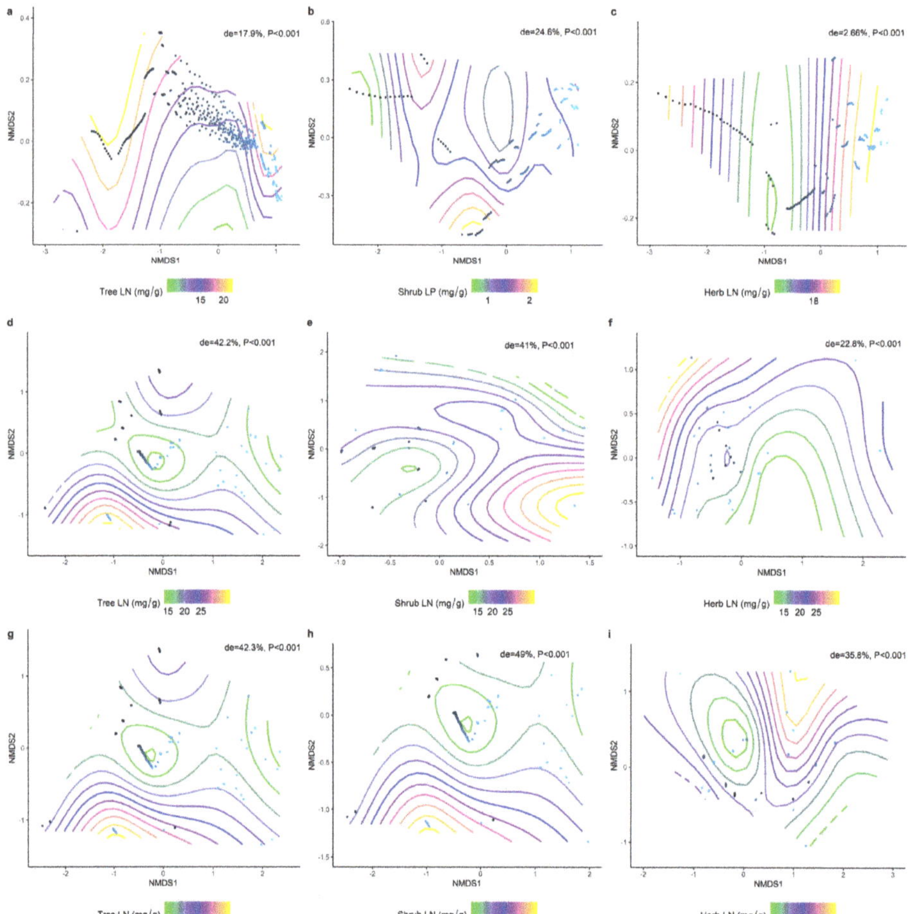

Figure 9. The NMDS ranking of climatic and soil factors with different life forms of LN. Value of de

represents the deviation explained by the corresponding model. (**a**) NMDS ranking of soil factors with tree levels LN; (**b**) NMDS ranking of soil factors with shrub levels LN; (**c**) NMDS ranking of soil factors with herb levels LN; (**d**) NMDS ranking of climatic factors with tree levels LN; (**e**) NMDS ranking of climatic factors with shrub levels LN; (**f**) NMDS ranking of climate factors and herb levels LN; (**g**) NMDS ranking of the sum of soil factors and climate factors and tree levels LN; (**h**) NMDS ranking of the sum of soil factors and climate factors and shrub levels LN; (**i**) NMDS ranking of the sum of soil factors and climate factors and herb levels LN. Trait stacking indicates that abiotic factors, indicated by points on the NMD, are associated with higher or lower trait values, consistent with a colored trait gradient. Note that if the relationship between the LN and abiotic factors is linear, the gradient splines will be parallel. Nonlinear relationships between LN and abiotic factors are represented by curve splines.

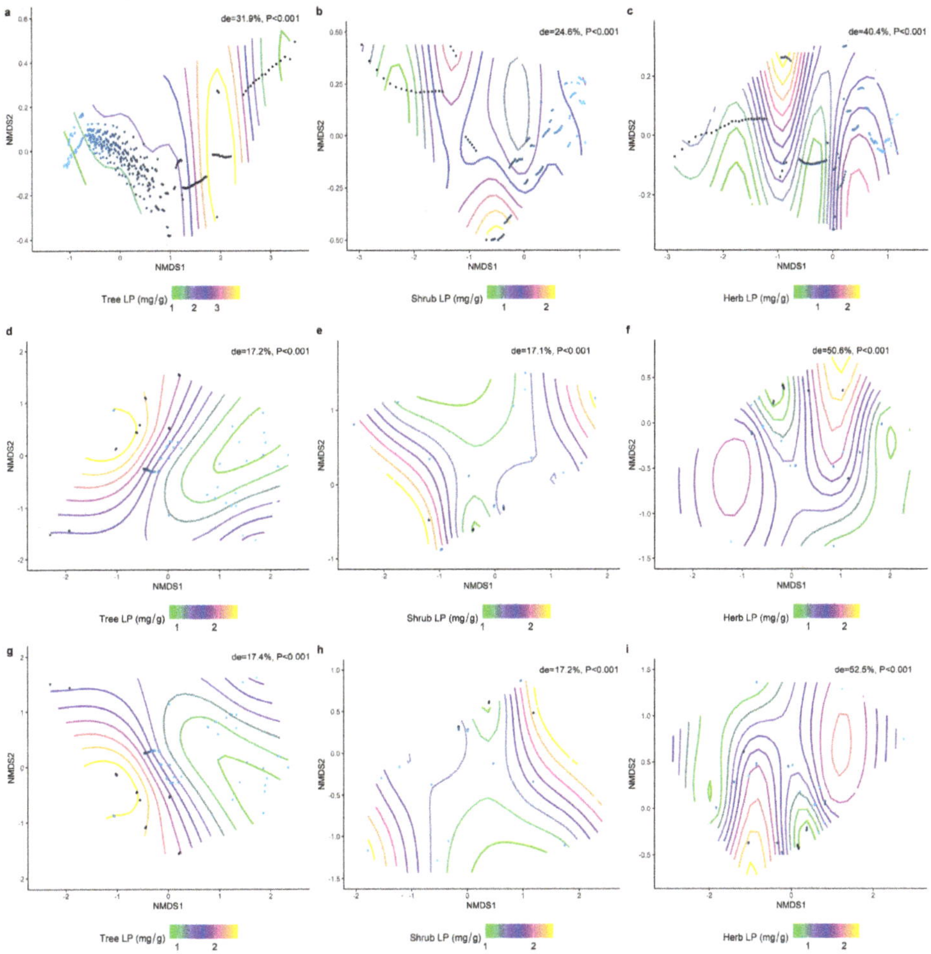

Figure 10. The NMDS ranking of climatic and soil factors with different life forms of LP. Value of de represents the deviation explained by the corresponding model. (**a**) NMDS ranking of soil factors with tree levels LP; (**b**) NMDS ranking of soil factors with shrub levels LP; (**c**) NMDS ranking of soil factors with herb levels LP; (**d**) NMDS ranking of climatic factors with tree levels LP; (**e**) NMDS ranking of climatic factors with shrub levels LP; (**f**) NMDS ranking of climate factors and herb levels

LP; (**g**) NMDS ranking of the sum of soil factors and climate factors and tree levels LP; (**h**): NMDS ranking of the sum of soil factors and climate factors and shrub levels LP; (**i**): NMDS ranking of the sum of soil factors and climate factors and herb levels LP. Trait stacking indicates that abiotic factors, indicated by points on the NMD, are associated with higher or lower trait values, consistent with a colored trait gradient. Note that if the relationship between LP and abiotic factors is linear, the gradient splines will be parallel. Nonlinear relationships between LP and abiotic factors are represented by curve splines.

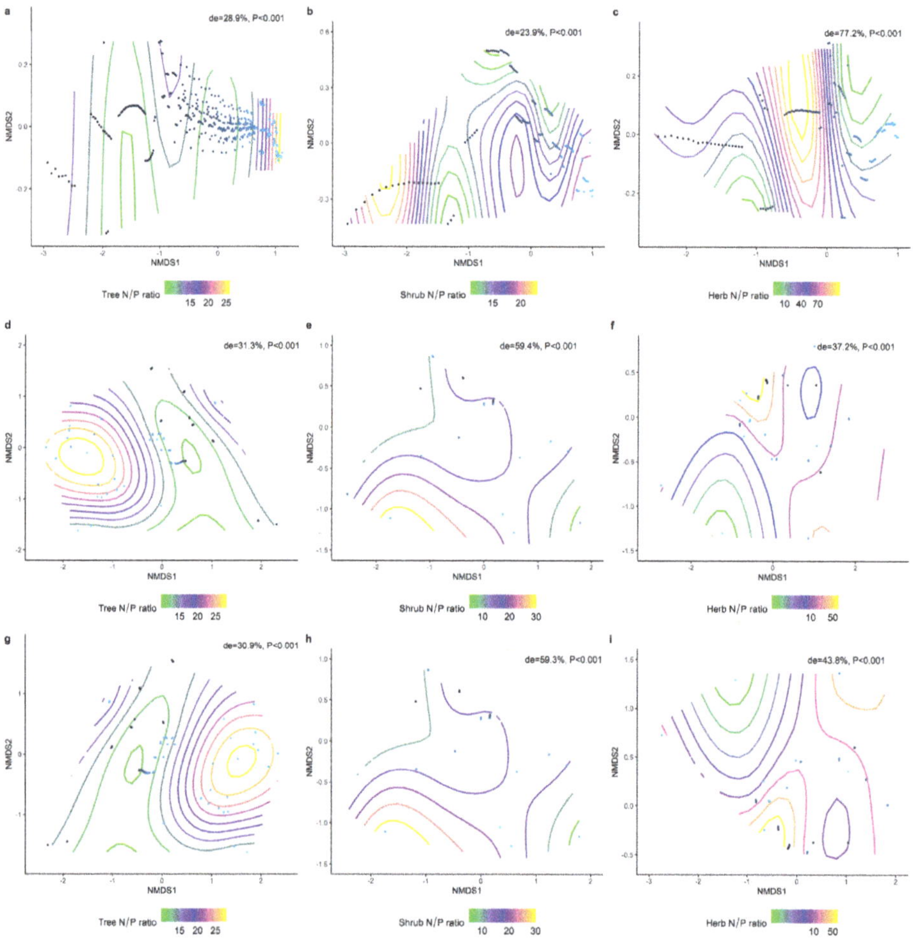

Figure 11. The NMDS ranking of climatic and soil factors with different life forms of N/P. Value of de represents the deviation explained by the corresponding model. (**a**) NMDS ranking of soil factors with tree levels N/P; (**b**) NMDS ranking of soil factors with shrub levels N/P; (**c**) NMDS ranking of soil factors with herb levels N/P; (**d**) NMDS ranking of climatic factors with tree levels N/P; (**e**) NMDS ranking of climatic factors with shrub levels N/P; (**f**) NMDS ranking of climate factors and herb levels N/P; (**g**) NMDS ranking of the sum of soil factors and climate factors and tree levels N/P; (**h**) NMDS ranking of the sum of soil factors and climate factors and shrub levels N/P; (**i**) NMDS ranking of the sum of soil factors and climate factors and herb levels N/P. Trait stacking indicates that abiotic factors, indicated by points on the NMD, are associated with higher or lower trait values,

consistent with a colored trait gradient. Note that if the relationship between N/P and abiotic factors is linear, the gradient splines will be parallel. Nonlinear relationships between N/P and abiotic factors are represented by curve splines.

3. Discussion

The leaf nutrient traits (LN, LP, and N/P) of different plant forms showed significant geographical differences [27], and some studies have found that the LN of shrub levels is significantly higher than that of the tree and herb levels in northwestern China, while the LP of shrub levels is significantly lower than that of herb levels [28], and that the LP and N/P of shrubs are higher than those of herbs in the desertification zone of southwest Hunan [29]. Within a given community, differences in the ecological niches of species and differences in resource use patterns can lead to differences in the leaf nutrient traits between species [17]. Leaf nutrient traits are widely used to quantify plant adaptations to the environment, and even to predict ecosystem function [6,30]. Thus, the differences in LN and LP of different life forms of plants result from differences in plant ecological niches. Compared to herb plants, woody plants have lower LP, which is aligned with their low growth rate strategy [31]. Quantifying the geographical distribution patterns of leaf nutrient traits is important for our scientific assessment of forest development dynamics [24].

Temperature can have a significant effect on the LN, LP, and N/P [32], and the temperature–biogeochemical hypothesis also suggests that LN and LP increase monotonically with temperature on a global scale [33]. It has been found that LN and LP are strongly correlated with effective soil N and P content [32], and that climatic factors can act directly on soil nutrients [25]. Soil enzymes are beneficial to maintain soil fertility, and increasing the soil temperature and moisture can promote the activity of soil enzymes and significantly improve the decomposition efficiency of soil organic matter [34]. Low temperatures not only have an inhibitory effect on organic matter decomposition and mineralization, but also limit microbial activity and affect the decomposition of apoplastic matter, thus reducing the availability of soil N and P [33]. Conversely, as the temperature increases, soil microbial respiration is enhanced, the efficiency of organic matter mineralization and decomposition increases, and the effective soil N, and P content rises [35]. Therefore, MAT is a key limiting factor for changes in leaf nutrient traits [12,13]. It has also been found that ASD significantly affects LN and LP [36]. Plant N and P elements are involved in the light reaction process and have an impact on photosynthesis, which is enhanced with increasing light duration and LN and LP are heavily utilized [37]. The nutrient content of plant leaves gradually decreased with increasing light time. Our results demonstrate the important influence of temperature on the nutrients of different life forms of leaves.

Precipitation is also one of the main factors of nutrient traits in plant leaves [14]. Previous studies have found that in water–limited ecosystems, plant N and P uptake of elements is mainly limited by water [38]. On one hand, most soil water comes from rainfall, and soil water content can directly affect the activity of soil enzymes, thereby affecting the release of soil nutrients [34]. On the other hand, rainfall promotes plant N uptake by promoting soil element availability, and plant N uptake is positively correlated with soil N availability [39]. In addition, rainfall had a significant effect on the soil microbial respiration and accelerated the leaching and transformation of P, which contributed to the transformation of soil P and thus increased the effective soil P content. With increased precipitation, the effectiveness of the soil nutrient factors increased, and plants were able to take up more N and P, significantly increasing the LN and LP [40]. Thus, precipitation is an important effect on the spatial and temporal distribution patterns of plant nutrient traits.

Soil N and P are basic nutrients for plant growth and are closely related to plant leaf nutrient, and most of the LN and LP of plants originate from soil [9]. The availability of soil N and P elements directly determines the growth and development process of plants, and available soil nutrients are positively correlated with leaf nutrients [18]. It was found that fertilizing the soil with N and P was beneficial to maintaining and improving the activity of soil enzymes and releasing more micronutrients and nutrients [41]. The higher the effective soil N and P content, the higher the plant root N and P uptake efficiency, and the higher the LN and N/P [22]. Soil microbial activity is positively correlated with soil pH, especially the activity of soil microbial enzymes related to the breakdown of soil nutrients. As the pH increases, soil microbial metabolic activity decreases, inhibiting the soil N mineralization and plant uptake of soil N, and LN is subsequently reduced [42]. However, it was also found that soil microbial biomass increased with the increasing soil pH, effectively driving organic P mineralization and the dissolution of fixed P [43], and LP in the arboreal and herbaceous layers increased. Thus, soil nutrient factors significantly influenced the changes in the leaf nutrient traits.

At the macro scale, climate factors have a stronger ability to shape leaf traits than soil factors [44]. Studies have shown that plant leaves are more sensitive to changes in climate and that plants can respond to changing climate by changing their traits [1]. Soil nutrient elements are closely related to leaf nutrient elements and provide many nutrients for plant growth and development, but the availability of soil nutrients is limited by temperature and precipitation [33]. Thus, climatic factors play a stronger role in shaping leaf nutrient distribution patterns than soil factors [45], but the influence of soil nutrient factors on leaf traits is not negligible.

The research area of this thesis included most of areas, which together with the relatively large amount of data ensures the plausibility of our experimental results. In addition, the results of this study are presented using a variety of data analysis methods. Overall, our results quantify the relative contribution of different environmental factors in shaping the functional traits of leaf nutrients and are important for a better understanding of the impact of global climate change on plant physiology [10]. In future studies, more biological factors such as community biodiversity and stand density need to be further explored for their effects on the distribution patterns of community leaf nutrients.

4. Materials and Methods

4.1. Study Area and Sample Data

The relatively wide distribution of forests in China facilitates studies on large scales. Data from 818 forest at 89 sites plots surveyed between 2005 and 2020 were used to investigate the spatial distribution and driving factors of LN, LP, and N/P at different life forms (tree levers, shrub levels, herb levels) of forest communities (Figure S1). The study sites ranged from 19.1° to 53.5° N and 79.7° to 129.3° E.

4.2. Functional Data

Community nutrient profiles can better reflect the adaptability of local vegetation to different environments than individual nutrient profiles [46]. Data for the selected nutrient profiles in this study include LN, LP, and N/P.

In this experiment, LN was determined using the national standard method, the Kjeldahl method [47], LP was determined using the vanadium−molybdenum yellow absorbance method [48], and N/P was equal to LN/LP.

Theoretical risk in predicting functional traits in individual plants is due to intraspecific and interspecific struggles [49]. Community−weighted mean traits (CWM_i) represents the forest mean trait values.

$$CWM_i = \frac{\sum_i^n D_i \times Trait}{\sum_i^n D_i}$$

where CWM_i represents the community−weighted functional trait identity value and D_i represents the abundance of dominant tree species. $Trait_i$ represents the selected functional trait [50].

4.3. Environmental Data

A study found that changes in the functional traits of building blocks were related to climate change [51]. The mean annual temperature (MAT), mean coldest monthly temperature (MCMT), mean warmest monthly temperature (MWMT), and mean annual precipitation (MAP) were extracted from the WorldClim global climate layer at a spatial resolution of 1 km. The mean annual evaporation (MAE) was taken from the Climate Data Center of the China Meteorological Administration (http://data.cma.cn/site/index.html) (accessed on 1 April 2021), sunlight. The annual sunlight duration (ASD), a key factor in photosynthesis [52], is from the Climate Data Center of the China Meteorological Administration (http://data.cma.cn/site/index.html) (accessed on 1 April 2021). Soil pH, soil N (http://www.csdn.store) (accessed on 1 April 2021), and soil P (https://www.osgeo.cn/data/wc137) (accessed on 1 April 2021) in the top 30 cm of soil were extracted from a 250 m resolution grid.

4.4. Data Analysis

Data analysis of the LN, LP, and N/P data from our study conformed to a normal distribution (Figure S2). A significant difference test at the 0.05 significance level was used to test for significant differences in LN, LP, and N/P between the tree, shrub, and herb levels (Figure 2). Significant differences were analyzed using the R package agricolae (version 4.1.0, R Core Team, 2020).

The extent to which environmental factors explain LN, LP, and N/P was investigated using a linear regression model in the R package agricolae (version 4.1.0, R Core Team, 2020), where R^2 represents how well the model fits the variables studied.

The generalized additive model (GAM) is used to test the effects of various environmental factors on the functional traits of leaves, with data for the LN, LP, and N/P consisting of parametric and non−parametric components to reduce the model risk associated with linear models [49]. First, the GAM method was used to select the key influencing factors based on a significant difference test at the 0.05 level of significance. Then, a GAM model was constructed to measure the relationship between the environmental factors and LN, LP, and N/P and used non−metric multidimensional scaling analysis (NMDS) to reflect the results for GAM [53].

$$g[E(Y|X)] = \sum_{i} \beta iXi + \sum_{j} fi(Xi) + \varepsilon$$

where g(•) represents the connection function, the form depends on the specific form, which can be interpreted as the Y−variable distribution. ϵ is the random error term, which can be interpreted as the variable connection with the normal distribution function name identity, and the connection function has the form g(u) = u, u = E(Y | X), $E(\epsilon | X) = 0$. X_i is the explanatory variable that strictly follows the parametric form, β_i is the corresponding parameter, and $f_j(X_j)$ is the corresponding explanatory variable that follows the nonparametric form of the smoothing function. In our study, the spline smoothing function S(•) was selected for fitting, thin−plate spline smoothing was selected for function fitting between different nodes, and each smoothing function S(•) was estimated using penalized least squares [50].

5. Conclusions

This study used data at 89 sites from 818 forest plots across China over 15 years from 2005 to 2020 to verify the relative roles of climatic and soil factors in shaping LN, LP, and N/P at different life forms. Our results confirm that LN, LP, and N/P differ significantly between life forms and that climatic and soil factors in the community habitat jointly influence the distribution patterns of the N/P nutrient profiles, with MAT being the dominant factor influencing LN, LP, and N/P at different life forms. Climatic factors also indirectly influence the leaf nutrient traits by affecting soil nutrients, but climatic factors are more influential than soil factors in shaping the distribution patterns of LN, LP, and N/P.

This study has important implications for our understanding of the distribution patterns of the plant N and P nutrient profiles at different life forms in the context of climate change and ecosystem modeling.

Supplementary Materials: The following supporting information can be downloaded at: https://www.mdpi.com/article/10.3390/plants11162171/s1, Figure S1: Sample site map of the distribution of the sample sites used for this experimental study in China. The sample sites ranged from 19.1° N to 53.5° N and 79.7° E to 129.3° E.; Figure S2: The normal distribution charts for the LN, LP, and N/P data.

Author Contributions: X.W. and J.W. contributed equally to this work. Conceptualization, J.G.; Experiment implementation, J.G.; Validation, J.G.; Formal analysis, X.W. and J.W.; Writing—original draft preparation, X.W.; Writing—review and editing, X.W., J.W., C.L., L.L. and L.Z.; Visualization, J.G.; Project administration, J.G. and H.Z.; Funding acquisition, J.G. and H.Z. All authors have read and agreed to the published version of the manuscript.

Funding: The study was supported by the Xinjiang Normal University Landmark Achievements Cultivation Project, China (grant number: no number), the Scientific Research Program of Colleges and Universities in Xinjiang (No. XJEDU2021I023), the Natural Science Foundation of China (No. 32160074), and the Open Project of Key Laboratory in Xinjiang (No. 2020D4010).

Institutional Review Board Statement: Not applicable.

Informed Consent Statement: Not applicable.

Data Availability Statement: Not applicable.

Conflicts of Interest: The authors declare no conflict of interest. The funders had no role in the design of the study; in the collection, analyses, or interpretation of data; in the writing of the manuscript; or in the decision to publish the results.

References

1. Bello, F.D.; Lavorel, S.; Díaz, S.; Berg, M.P.; Cipriotti, P.; Feld, C.K.; Hering, D. Towards an assessment of multiple ecosystem processes and services via functional traits. *Biodivers. Conserv.* **2010**, *19*, 2873–2893. [CrossRef]
2. Ma, J.; Niklas, K.J.; Liu, L.; Fang, Z.D.; Li, Y.R.; Shi, P.J. TTree size influences leaf shape but does not affect the proportional relationship between leaf area and the product of length and width. *Front. Plant Sci.* **2022**, *13*, 850203. [CrossRef]
3. Huang, C.; Xu, Y.; Zang, R. Variation patterns of functional trait moments along geographical gradients and their environmental determinants in the subtropical evergreen broadleaved forests. *Front. Plant Sci.* **2021**, *12*, 1414. [CrossRef]
4. Tian, D.; Yan, Z.; Ma, S.; Ding, Y.; Luo, Y.; Chen, Y.; Du, E.; Han, W.; Kovacs, E.D.; Shen, H.; et al. Family-level leaf nitrogen and phosphorus stoichiometry of global terrestrial plants. *Sci. China Life Sci.* **2019**, *62*, 1047–1057. [CrossRef]
5. Hofhansl, F.; Chacón-Madrigal, E.; Brännström, Å.; Dieckmann, U.; Franklin, O. Mechanisms driving plant functional trait variation in a tropical forest. *Ecol. Evol.* **2021**, *11*, 3856–3870. [CrossRef]
6. Wang, Z.; Zheng, R.; Yang, L.; Tan, T.; Li, H.; Liu, M. Elevation gradient distribution of indices of tree population in a montane forest: The role of leaf traits and the environment. *For. Ecosyst.* **2022**, *9*, 100012. [CrossRef]
7. Taylor, P.G.; Cleveland, C.C.; Wieder, W.R.; Sullivan, B.W.; Doughty, C.E.; Dobrowski, S.Z.; Townsend, A.R. Temperature and rainfall interact to control carbon cycling in tropical forests. *Ecol. Lett.* **2017**, *20*, 779–788. [CrossRef]
8. Zhang, S.; Zhang, Y.; Xiong, K.; Yu, Y.; Min, X. Changes of leaf functional traits in karst rocky desertification ecological environment and the driving factors. *Glob. Ecol. Conserv.* **2020**, *24*, e01381. [CrossRef]
9. Guo, Y.; Yan, Z.; Zhang, Y.W.; Zhou, G.; Xie, Z.; Tang, Z. Environmental constraints on the inter-genus variation in the scaling relationship between leaf nitrogen and phosphorus concentrations. *J. Plant Ecol.* **2021**, *14*, 616–627. [CrossRef]
10. Biswas, D.; Ma, B.; Morrison, M.J. Changes in leaf nitrogen and phosphorus, photosynthesis, respiration, growth and resource use efficiency of a rapeseed cultivar as affected by drought and high temperature. *Can. J. Plant Sci.* **2019**, *99*, 413–419. [CrossRef]
11. Borer, E.T.; Bracken, M.E.; Seabloom, E.W.; Smith, J.E.; Cebrian, J.; Cleland, E.E.; Elser, J.J.; Fagan, W.F.; Gruner, D.S.; Harpole, W.S. Global biogeography of autotroph chemistry: Is insolation a driving force? *Oikos* **2013**, *122*, 1121–1130. [CrossRef]
12. Miatto, R.C.; Wright, I.J.; Batalha, M.A. Relationships between soil nutrient status and nutrient-related leaf traits in Brazilian cerrado and seasonal forest communities. *Plant Soil.* **2016**, *404*, 13–33. [CrossRef]
13. Wright, I.J.; Dong, N.; Maire, V.; Prentice, I.C.; Westoby, M.; Díaz, S. Global climatic drivers of leaf size. *Nature* **2017**, *357*, 917–921. [CrossRef] [PubMed]
14. Cheng, H.; Gong, Y.; Zuo, X. Precipitation variability affects aboveground biomass directly and indirectly via plant functional traits in the desert steppe of inner Mongolia, Northern China. *Front. Plant Sci.* **2021**, *12*, 674527. [CrossRef] [PubMed]

15. Picotte, J.J.; Rhode, J.M.; Cruzan, M.B. Leaf morphological responses to variation in water availability for plants in the Piriqueta caroliniana complex. *Plant Ecol.* **2009**, *200*, 267–275. [CrossRef]
16. Fang, Z.; Han, X.Y.; Xie, M.Y.; Jiao, F. Spatial Distribution Patterns and Driving Factors of Plant Biomass and Leaf N, P Stoichiometry on the Loess Plateau of China. *Plants.* **2021**, *10*, 2420. [CrossRef] [PubMed]
17. Faucon, M.P.; Houben, D.; Lambers, H. Plant Functional Traits: Soil and Ecosystem Services. *Trends Plant Sci.* **2017**, *20*, 385–394. [CrossRef]
18. Zhu, J. Abiotic Stress Signaling and Responses in Plants. *Cell.* **2016**, *167*, 313–324. [CrossRef]
19. Gong, H.; Li, Y.; Yu, T.; Zhang, S.; Sun, D. Soil and climate effects on leaf nitrogen and phosphorus stoichiometry along elevational gradients. *Glob. Ecol. Conserv.* **2020**, *23*, e01138. [CrossRef]
20. Zhu, B.; Izuta, T.; Watanabe, M. Nitrogen use efficiency of Quercus serrata seedlings under different soil nitrogen and phosphorus supplies. *J. Agric. Meteorol.* **2020**, *76*, 11–18.
21. Zhang, Y.W.; Guo, Y.; Tang, Z.; Feng, Y.; Fang, J. Patterns of nitrogen and phosphorus pools in terrestrial ecosystems in China. *Earth Syst. Sci. Data* **2021**, *13*, 5337–5351. [CrossRef]
22. Gao, J.; Song, Z.; Liu, Y. Response mechanisms of leaf nutrients of endangered plant (Acer catalpifolium) to environmental factors varied at different growth stages. *Glob. Ecol. Conserv.* **2019**, *17*, e00521. [CrossRef]
23. Latham, P.A.; Zuuring, H.R.; Coble, D.W. A method for quantifying vertical forest structure. *For. Ecol. Manag.* **1998**, *104*, 157–170. [CrossRef]
24. Cheng, X.; Ping, T.; Li, Z.; Wang, T.; Han, H.; Epstein, H.E. Effects of environmental factors on plant functional traits across different plant life forms in a temperate forest ecosystem. *New For.* **2022**, *53*, 125–142. [CrossRef]
25. Poorter, H.; Niinemets, Ü.; Poorter, L.; Wright, I.J.; Villar, R. Causes and consequences of variation in leaf mass per area (LMA): A meta-analysis. *New Phytol.* **2009**, *182*, 565–588. [CrossRef] [PubMed]
26. Taylor, P.; Asner, G.; Dahlin, K.; Anderson, C.; Knapp, D.; Martin, R.; Mascaro, J.; Chazdon, R.; Cole, R.; Wanek, W. Landscape-Scale Controls on Aboveground Forest Carbon Stocks on the Osa Peninsula, Costa Rica. *PLoS ONE* **2015**, *10*, e0126748. [CrossRef] [PubMed]
27. Tian, D.; Yan, Z.; Niklas, K.J.; Han, W.; Kattge, J.; Reich, P.B.; Luo, Y.; Chen, Y.; Tang, Z.; Hu, H. Global leaf nitrogen and phosphorus stoichiometry and their scaling exponent. *Natl. Sci. Rev.* **2018**, *5*, 728–739. [CrossRef]
28. He, M.S.; Luo, Y.; Peng, Q.W.; Yang, S.Q.; Li, K.H.; Han, W.X. Leaf C: N: P stoichiometry of 67 plant species and its relations with climate factors across the deserts in Xinjiang, China. *J. Appl. Ecol.* **2019**, *30*, 2171–2180.
29. Jing, Y.R.; Deng, X.W.; Hui, W.; Li, Y.Q.; Xiang, W.H. Leaf nitrogen and phosphorus stoichiometry of shrubland plants in the rocky desertification area of Southwestern Hunan, China. *J. Appl. Ecol.* **2017**, *28*, 415–422.
30. Guo, X.C.; Shi, P.J.; Niinemets, Ü.; Wang, R.; Liu, M.; Li, Y.R.; Dong, L. "Diminishing returns" for leaves of five age-groups of Phyllostachys edulis culms. *Am. J. Bot.* **2021**, *108*, 1662–1672. [CrossRef]
31. Jiang, P.P.; Yang, C.; Chen, Y.M.; Aamp, N. C, N, P stoichiometric characteristics of tree, shrub, herb leaves and litter in forest community of Shaanxi Province, China. *J. Appl. Ecol.* **2016**, *27*, 365–372.
32. Wei, D.P.; Zhang, J.; Zhang, D.J.; Li, C.B.; Zhao, Y.B.; Zhang, J. Leaf carbon, nitrogen, and phosphorus resorption and the stoichiometry in Pinus massoniana plantations with various canopy densities. *Chin J Appl Environ Biol.* **2017**, *23*, 560–569.
33. Tan, Q.Q.; Jia, Y.F.; Wang, G.A. Decoupling of soil nitrogen and phosphorus dynamics along a temperature gradient on the Qinghai-Tibetan Plateau. *Geoderma.* **2021**, *396*, 115084. [CrossRef]
34. Bungau, S.; Behl, T.; Aleya, L.; Bourgeade, P.; Aloui-Sossé, B.; Purza, A.L. Expatiating the impact of anthropogenic aspects and climatic factors on long-term soil monitoring and management. *Environ. Sci. Pollut. Res.* **2021**, *28*, 30528–30550. [CrossRef]
35. Mondal, S.; Ghosal, S.; Barua, R. Impact of elevated soil and air temperature on plants growth, yield and physiological interaction: A critical review. *Zhongguo Nong Ye Ke Xue* **2016**, *14*, 293–305.
36. Gong, H.D.; Wang, H.; Jiao, D.Y.; Cai, Z.Q. Phenotypic plasticity of seedlings of five tropical tree species in response to different light and nutrient availability. *Trop. Ecol.* **2016**, *57*, 727–737.
37. Wang, C.X.; He, G.X.; Song, Z.B.; Fan, B.; Shi, L.T. Effects of soil nitrogen and phosphorus levels on leaf nitrogen and phosphorus contents and photosynthesis of tamarindus indica l. In yuanjiang and yuanmou dry-hot valley. *Chin. J. Ecol.* **2019**, *38*, 710–718.
38. Jiang, D.; Geng, Q.; Li, Q.; Luo, Y.; Vogel, J.; Shi, Z.; Ruan, H.; Xu, X. Nitrogen and Phosphorus Resorption in Planted Forests Worldwide. *Forests* **2019**, *10*, 201. [CrossRef]
39. Song, W.; Chen, S.; Zhou, Y.; Lin, G. Rainfall amount and timing jointly regulate the responses of soil nitrogen transformation processes to rainfall increase in an arid desert ecosystem. *Geoderma* **2020**, *364*, 114197. [CrossRef]
40. Gao, X.L.; Li, X.G.; Zhao, L. Shrubs magnify soil phosphorus depletion in Tibetan meadows: Conclusions from C:N:P stoichiometry and deep soil profiles. *Sci. Total Environ.* **2021**, *785*, 147320. [CrossRef]
41. Samuel, A.D.; Bungau, S.; Tit, D.M.; Melinte, C.E.; Purza, L.; Badea, G.E. Effects of long term application of organic and mineral fertilizers on soil enzymes. *Rev. Chim.* **2018**, *69*, 2608–2612. [CrossRef]
42. Liu, Y.; Evans, S.E.; Friesen, M.L.; Tiemann, L.K. Root exudates shift how N mineralization and N fixation contribute to the plant-available N supply in low fertility soils. *Soil Biol. Biochem.* **2022**, *165*, 108541. [CrossRef]
43. Shen, J.; Yuan, L.; Zhang, J.; Li, H.; Bai, Z.; Chen, X.; Zhang, W.; Zhang, F. Phosphorus Dynamics: From Soil to Plant. *Plant Physiol.* **2011**, *156*, 997–1005. [CrossRef] [PubMed]

44. Hg, A.; Qc, A.; Jie, G.B. Latitudinal, soil and climate effects on key leaf traits in northeastern China. *Glob. Ecol. Conserv.* **2020**, *22*, e00904.
45. He, M.; Dijkstra, F.A.; Zhang, K.; Li, X.; Tan, H.; Gao, Y.; Li, G. Leaf nitrogen and phosphorus of temperate desert plants in response to climate and soil nutrient availability. *Sci. Rep.* **2014**, *4*, 6932. [CrossRef] [PubMed]
46. Muscarella, R.; Uriarte, M. Do community-weighted mean functional traits reflect optimal strategies? *Proc. R. Soc. B Biol. Sci.* **2016**, *283*, 20152434. [CrossRef]
47. Potarzycki, J. Effect of magnesium or zinc supplementation at the background of nitrogen rate on nitrogen management by maize canopy cultivated in monoculture. *Plant Soil Environ.* **2011**, *57*, 19–25. [CrossRef]
48. Walash, M.I.; Metwally, E.S.; Eid, M.I.; El-Shaheny, R.N. Spectrophotometric determination of risedronate and etidronate in pharmaceutical formulations via the molybdovanadate method. *Anal. Lett.* **2009**, *42*, 1571–1587. [CrossRef]
49. Wang, L.Q.; Ali, A. Climate regulates the functional traits–aboveground biomass relationships at a community-level in forests: A global meta-analysis. *Sci. Total Environ.* **2021**, *761*, 143238. [CrossRef]
50. Wang, J.F.; Wang, X.X.; Ji, Y.H.; Gao, J. Climate factors determine the utilization strategy of forest plant resources at large scales. *Front. Plant Sci.* **2022**, *13*, 990441. [CrossRef]
51. Kramer, P.J. Carbon dioxide concentration, photosynthesis, and dry matter production. *Bioscience* **1981**, *31*, 29–33. [CrossRef]
52. Zou, B.; Chen, J.; Zhai, L.; Fang, X.; Zheng, Z. Satellite based mapping of ground $PM_{2.5}$ concentration using generalized additive modeling. *Remote Sens.* **2017**, *9*, 1. [CrossRef]
53. Sweeney, C.J.; Vries, F.T.; Dongen, B.E. Root traits explain rhizosphere fungal community composition among temperate grassland plant species. *New Phytol.* **2021**, *229*, 1492–1507. [CrossRef] [PubMed]

plants

MDPI

Article

Quantifying the Variation in the Geometries of the Outer Rims of Corolla Tubes of *Vinca major* L.

Lin Wang [1], Qinyue Miao [1], Ülo Niinemets [2,3,*], Johan Gielis [4] and Peijian Shi [1,*]

[1] College of Science & College of Biology and the Environment, Nanjing Forestry University, Nanjing 210037, China; lwang@njfu.edu.cn (L.W.); qymiao@njfu.edu.cn (Q.M.)
[2] Institute of Agricultural and Environmental Sciences, Estonian University of Life Sciences, 51006 Tartu, Estonia
[3] Estonian Academy of Sciences, 10130 Tallinn, Estonia
[4] Department of Biosciences Engineering, University of Antwerp, B-2020 Antwerp, Belgium; johan.gielis@uantwerpen.be
* Correspondence: ylo.niinemets@emu.ee (Ü.N.); pjshi@njfu.edu.cn (P.S.); Tel.: +86-25-8542-7231 (P.S.)

Abstract: Many geometries of plant organs can be described by the Gielis equation, a polar coordinate equation extended from the superellipse equation, $r = a\left[\left|\cos\left(\frac{m}{4}\varphi\right)\right|^{n_2} + \left|\frac{1}{k}\sin\left(\frac{m}{4}\varphi\right)\right|^{n_3}\right]^{-1/n_1}$. Here, r is the polar radius corresponding to the polar angle φ; m is a positive integer that determines the number of angles of the Gielis curve when $\varphi \in [0 \text{ to } 2\pi)$; and the rest of the symbols are parameters to be estimated. The pentagonal radial symmetry of calyxes and corolla tubes in top view is a common feature in the flowers of many eudicots. However, prior studies have not tested whether the Gielis equation can depict the shapes of corolla tubes. We sampled randomly 366 flowers of *Vinca major* L., among which 360 had five petals and pentagonal corolla tubes, and six had four petals and quadrangular corolla tubes. We extracted the planar coordinates of the outer rims of corolla tubes (in top view) (ORCTs), and then fitted the data with two simplified versions of the Gielis equation with $k = 1$ and $m = 5$: $r = a\left[\left|\cos\left(\frac{5}{4}\varphi\right)\right|^{n_2} + \left|\sin\left(\frac{5}{4}\varphi\right)\right|^{n_3}\right]^{-1/n_1}$ (Model 1), and $r = a\left[\left|\cos\left(\frac{5}{4}\varphi\right)\right|^{n_2} + \left|\sin\left(\frac{5}{4}\varphi\right)\right|^{n_2}\right]^{-1/n_1}$ (Model 2). The adjusted root mean square error ($RMSE_{adj}$) was used to evaluate the goodness of fit of each model. In addition, to test whether ORCTs are radially symmetrical, we correlated the estimates of n_2 and n_3 in Model 1 on a log-log scale. The results validated the two simplified Gielis equations. The $RMSE_{adj}$ values for all corolla tubes were smaller than 0.05 for both models. The numerical values of n_2 and n_3 were demonstrated to be statistically equal based on the regression analysis, which suggested that the ORCTs of *V. major* are radially symmetrical. It suggests that Model 1 can be replaced by the simpler Model 2 for fitting the ORCT in this species. This work indicates that the pentagonal or quadrangular corolla tubes (in top view) can both be modeled by the Gielis equation and demonstrates that the pentagonal or quadrangular corolla tubes of plants tend to form radial symmetrical geometries during their development and growth.

Keywords: flower geometry; Gielis equation; model complexity; natural geometries; planar coordinates; polygonal structure; radial symmetry

Citation: Wang, L.; Miao, Q.; Niinemets, Ü.; Gielis, J.; Shi, P. Quantifying the Variation in the Geometries of the Outer Rims of Corolla Tubes of *Vinca major* L. *Plants* **2022**, *11*, 1987. https://doi.org/10.3390/plants11151987

Academic Editor: Yasutomo Hoshika

Received: 4 July 2022
Accepted: 26 July 2022
Published: 30 July 2022

1. Introduction

In geometry, the circle and ellipse have been demonstrated to be expressed as two special cases of the superellipse [1]. As a direct extension of the superellipse equation in the polar coordinate system, Gielis [2,3] created a highly versatile equation to reflect natural geometries, especially symmetrical geometries. We refer to it as the Gielis equation hereinafter. In the past decade, many studies have been carried out to examine the validity of the Gielis equation in fitting actual biological geometries. A simplified version of this equation was used by Shi et al. [4,5] and Lin et al. [6] to fit the boundary data of cross-sections of tree rings for five species of conifers and bamboo leaves for 46 bamboo species. These studies verified the potential of the Gielis equation to describe the shapes of tree rings and bamboo leaves and demonstrated that the Gielis equation is a valid scientific method for quantitative characterization of the size and shape of widely differing planar biological objects.

Shi et al. [7] proposed a twin version of the Gielis equation by introducing a link function and found that the twin Gielis equation was superior in depicting the shapes of some sea stars. Tian et al. [8] used the Gielis equation to fit the seed projections (in side view) of two *Gingko biloba* cultivars, and used it to quantify the morphological differences between the two cultivars. Li et al. [9] compared the original Gielis equation with its twin version in describing the planar projections of *Koelreuteria paniculata* fruits (in top view) and demonstrated that the two versions of the Gielis equation both can model the shapes of the vertical fruit projections well. Nevertheless, the twin Gielis equation predicted an axial symmetry of *K. paniculata* fruits, while the original Gielis equation tends to overfit the data. In addition, a recent study shows that a simplified Gielis equation can describe all existing egg shapes of birds and has a better goodness of fit than other egg shape models [10]. Flowers can also be modelled in the same way [2,3,11–13], but the capacity of the Gielis equation to simulate flower shape has not been quantitatively studied.

Here, we focused on the geometries of the outer rims of corolla tubes of *Vinca* flowers that are representative to the five-petal flowers with a fused pentagonal base (corolla top). The genus *Vinca* is native to western Mediterranean Europe, Asia Minor and Northern Africa, but it has been introduced as ornamental to all continents and has naturalized in many sites. It is a small evergreen ground cover plant and in addition to sexual reproduction, spreads via stolons. Taxonomically, *Vinca* belongs to the Apocynaceae family, one of the five families within the order Gentianales; together with the orders Solanales and Lamiales they form the Lamiids [14]. Lamiids are characterized by late sympetaly, the fusion of stamen filaments with the corolla tube and opposite leaves [15]. The Apocynaceae have a conserved architecture of highly synorganized flowers, and within this family *Vinca* L. is the type genus of the tribe Vinceae, in particular, of the subtribe Vincinae. The corolla is infundibuliform, and the lobe aestivation is sinistrorse [15].

The petals of *Vinca* are fused at their bottom, forming a corolla tube. When the flower opens, the distal, unfused parts of the petals fold back in a plane, whereas the upper part of the corolla tube formed by the fused parts of the petals becomes clearly delineated. The upper ridge or rim of the corolla tube has a clear pentagonal symmetry (although quadrangular or hexagonal symmetry may occur). In contrast to the purple color of the free petals and the base of the tube, the upper rim of the floral tube is white. The purple petals exhibit a high ultraviolet reflectance, whereas the corolla tube, in particular, the upper rim of the flower, strongly absorbs ultraviolet (Ultraviolet Flowers: Vinca minor. Available online: http://www.naturfotograf.com/UV_VINC_MIN.htm (accessed on 1 July 2022)).

Most flowers in this species have five petals, but flowers with four or six petals occur rarely (Figure 1). Five petals correspond to a corolla tube of pentagonal symmetry and a pentagonal rim, and four petals correspond to a quadrangular rim (Figure 1). In the remainder of the paper, we use the term the outer rim of the corolla tube (in top view) (ORCT) to denote this polygonal structure and characterize the geometry of the outer rim of the corolla tube in top view (i.e., represented by a vertical projection). The two types of ORCTs (i.e., pentagonal and quadrangular rims) both seem to exhibit a radial symmetry, whereas the distal ends of the petals are rotated counterclockwise, relative to the corolla tube. The main reason is the sinistrorse aestivation of the flowers. The *Vinca* flowers display contorted aestivation of the corolla, and each petal is asymmetric. However, the entire corolla exhibits a rotational symmetry (Figure 1).

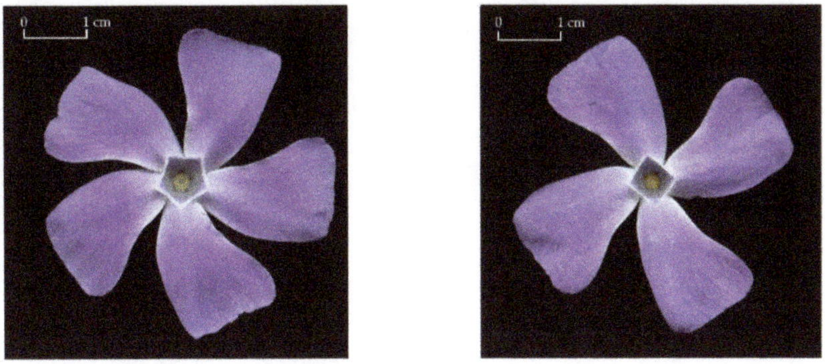

(a) **(b)**

Figure 1. Representative flowers of *Vinca major* L. with five (**a**) and with four petals (**b**).

Pentagonal symmetry is a general condition in Eudicots [16], but a clear pentagonal shape, as in the corolla tube of *Vinca*, is uncommon. In trumpet, campanulate or salverform corolla tubes, the transition from fused proximal parts of the petals to the free parts is gradually curved. In *Vinca*, on the other hand, the plane formed by the free ends of the petals is almost perpendicular to the corolla tube. Although the ORCT is never completely flat, a projection using a photography or an image scanner can be obtained, making a 2D quantitative study possible. In this study, 360 flowers with five petals and six flowers with four petals from *V. major* were sampled to examine whether the ORCTs follow the Gielis equation, and to evaluate whether a deviation from pure rotational symmetry can be found.

2. Materials and Methods

2.1. Flower Sampling and Image Processing

Vinca major flowers were randomly sampled at the Nanjing Forestry University campus (118°48′35″ E, 32°04′67″ N), Nanjing, China from 7–23 April 2022 when the peak blooming occurred. To keep the flowers fresh, each sampled flower was placed in a 10 mL beaker with 1–2 mL water until its image was scanned. The flowers were scanned by an Epson photo scanner (V550, Epson, Batam, Indonesia) at a resolution of 2400 dpi, then their images were converted into black-white images after being cropped and saved as bmp (Bitmap) format using Adobe Photoshop CS2 (version 9.0; Adobe, San Jose, CA, USA; http://www.adobe.com/products/photoshop.html, accessed on 1 July 2022).

2.2. Data Acquisition

To extract the planar coordinates of the outer rim of the corolla tube (in top view) (ORCT), we used MATLAB (version $\geq$ 2009a; MathWorks, Natick, MA, USA) with a program developed by refs. [4,17,18]. We fitted the coordinate data of each ORCT using the 'biogeom' package (version 1.0.5) [19] based on R (version 4.2.0) [20].

2.3. Models

Gielis [2] proposed a polar coordinate equation to describe natural shapes:

$$r(\varphi) = \left(\left| \frac{1}{A} \cos\left(\frac{m}{4} \varphi \right) \right|^{n_2} + \left| \frac{1}{B} \sin\left(\frac{m}{4} \varphi \right) \right|^{n_3} \right)^{-\frac{1}{n_1}} \quad (1)$$

where r and φ are the polar radius and polar angle, respectively; A, B, n_1, n_2 and n_3 are parameters to be fitted; and m is a positive integer that determines the number of angles of the Gielis curve within $[0, 2\pi)$. Given that the ORCT can be pentagonal or quadrangular, m was either five or four in this study.

Equation (1) can be rewritten as [7,8]:

$$r(\varphi) = a\left(\left|\cos\left(\frac{m}{4}\varphi\right)\right|^{n_2} + \left|\frac{1}{k}\sin\left(\frac{m}{4}\varphi\right)\right|^{n_3}\right)^{-\frac{1}{n_1}} \quad (2)$$

where $a = A^{n_2/n_1}$ and $k = B/A^{n_2/n_3}$. In this work, k was set to 1 because the ORCT of *V. major* exhibits radial symmetry. Most flowers have five petals, so m was set to 5, resulting in:

$$r(\varphi) = a\left(\left|\cos\frac{5}{4}\varphi\right|^{n_2} + \left|\sin\frac{5}{4}\varphi\right|^{n_3}\right)^{-\frac{1}{n_1}} \quad (3)$$

To test whether a radial symmetrical version of the Gielis equation is applicable to the ORCT, we set $n_2 = n_3$ in Equation (3) so that we have:

$$r(\varphi) = a\left(\left|\cos\frac{5}{4}\varphi\right|^{n_2} + \left|\sin\frac{5}{4}\varphi\right|^{n_2}\right)^{-\frac{1}{n_1}} \quad (4)$$

We refer to Equations (3) and (4) as Models 1 and 2 for convenience hereinafter.

2.4. Model Fitting and Data Analysis

We used the Nelder–Mead optimization method [21] to minimize the residual sum of square (RSS) between the observed and predicted polar radii, and obtained the estimated values of the parameters in Equations (3) and (4):

$$\text{RSS} = \sum_{i=1}^{N} (r_i - \hat{r}_i)^2 \quad (5)$$

where r_i represents the observed distance from the polar point to the i-th point on a scanned ORCT; $\hat{r}_i$ represents the predicted distance from the polar point to the i-th point on the predicted ORCT based on Equation (3) or Equation (4); and N represents the number of data points on a scanned ORCT.

Additionally, the root-mean-square error (RMSE) was calculated to reflect the goodness of fit:

$$\text{RMSE} = \sqrt{\text{RSS}/N} \quad (6)$$

However, given the influence of the ORCT size (area) on absolute values of RMSE, we used the adjusted RMSE (RMSE_{adj}) [7,9,22]. In that case, we can directly compare the differences in the model goodness of fits among different ORCT sizes.

$$\text{RMSE}_{\text{adj}} = \sqrt{\frac{\text{RSS}/N}{S/\pi}} \quad (7)$$

where S represents the area of an ORCT. The RMSE_{adj} represents the ratio of the mean absolute deviation (between the observed and predicted radii from the polar point to the ORCT) to the radius of a hypothetical circle whose area equals to that of the ORCT projection, which can standardize the prediction error regardless of the ORCT size.

There are still four parameters in Model 1, and the question is whether a less complex model can produce an approximate goodness of fit with fewer parameters. On the one hand, the difference between n_2 and n_3 in Model 1 determines the extent of symmetry. If the difference is equal to zero (i.e., Model 2), it can produce a perfectly axial symmetrical curve for the ORCT; the larger the difference, the worse the extent of the rim's symmetry. On the other hand, the ORCTs of interest appear visually to be axial symmetrical pentagons, so it is necessary to test whether Model 1 (with four model parameters) can be simplified to Model 2 (with three model parameters). Therefore, we performed a linear regression on the estimated values of n_3 and n_2 for all of the samples. To stabilize the variance of the estimated values of the two parameters and to normalize the data, the log-transformation was used [23,24]:

$$y = \alpha + \beta\, x \quad (8)$$

where $y = \ln \hat{n}_3$ and $x = \ln \hat{n}_2$, where the circumflex represents the estimated value. The parameters α and β were estimated using reduced major axis regression protocols [25,26]. The bootstrap percentile method [27,28] was used to calculate 95% confidence intervals (CIs) of the intercept and the slope of the regression line. If the CI of the intercept includes zero and the CI of the slope includes one, it can indicate that n_3 is not significantly different from n_2, which means that Model 2 is superior to Model 1 due to reduced complexity. It also suggests that ORCTs tend to be perfectly axially symmetrical. If n_3 is not statistically significant from n_2, it is unnecessary to compare the RMSE_{adj} values of the two models. It is apparent that an additional parameter can increase the goodness of fit, but it might result from the overfitting to the data with a certain measurement error. In other words, the flexibility in curve fitting of Model 1 may cause an incorrect parameter estimation due to the measurement errors. As a rule of thumb, a ≤ 0.05 RMSE_{adj} can reflect the validity of a model in curve fitting.

To compare the goodness of fit between Models 1 and 2, a paired sample *t*-test was used to compare the average values of the RMSE_{adj} values of Models 1 and 2 at 0.05 significance level. The log-transformed values of adjusted RMSEs were used to normalize the data of adjusted RMSEs.

In addition, we used nonlinear least-squares to fit the ORCTs by minimizing the sum of squares (RSS) between the observed and predicted polar radii. The distribution of the residuals in nonlinear least-squares is usually hypothesized to be normal, but the hypothesis in nonlinear least-squares seldom holds true. In that case, we calculated the mean and corresponding 95% CI of residuals to examine whether the mean equals 0 and the corresponding 95% CI includes 0. We also calculated the skewness (S_k) of residuals to see whether the skewness seriously deviates from zero. For a normal distribution, the skewness equals zero. A positive skewness represents a right skewed distribution curve, and a negative skewness represents a right-skewed distribution curve [17]:

$$S_k = E\left[\left(\frac{z-\mu}{\sigma}\right)^3\right] \tag{9}$$

where z represents the residual between the i-th observed and predicted radii; μ represents the mean of residuals; and σ represents the standard error of residuals. We used the Shapiro–Wilk test [29] to test the normality of residuals only as a reference.

3. Results

Both Models 1 and 2 provided generally a good representation of the ORCT (Figure 2 for sample fits). The RMSE_{adj} values of pentagonal ORCTs predicted by Models 1 and 2 ranged from 0.0116 to 0.0392 and from 0.0118 to 0.0481, respectively (Figure 3a). This verified the validities of the two models in describing the shapes of ORCTs of *V. major*. The RMSE_{adj} values calculated by Model 1 were smaller than those calculated by Model 2 for 346 out of 360 (ca. 96%) ORCT samples, and for the remaining 14 samples, the RMSE_{adj} values between the two models were the same. The mean $\ln(\text{RMSE}_{\text{adj}})$ of Model 1, M_1, was significantly smaller than that of Model 2, M_1 ($t = -5.5666$, df = 718, $p < 0.001$). This indicates that Model 1 with four parameters provided a better fit than Model 2 with three parameters, although the difference was significant in statistics but fairly small. The percentage error, i.e., $|(M_2 - M_1)/M_1| \times 100\%$, between the two mean $\ln(\text{RMSE}_{\text{adj}})$ values, was less than 2.7% (Figure 3b). From the tradeoff between the model complexity and goodness of fit, Model 1 is not as good as Model 2, and the latter has a more concise model structure.

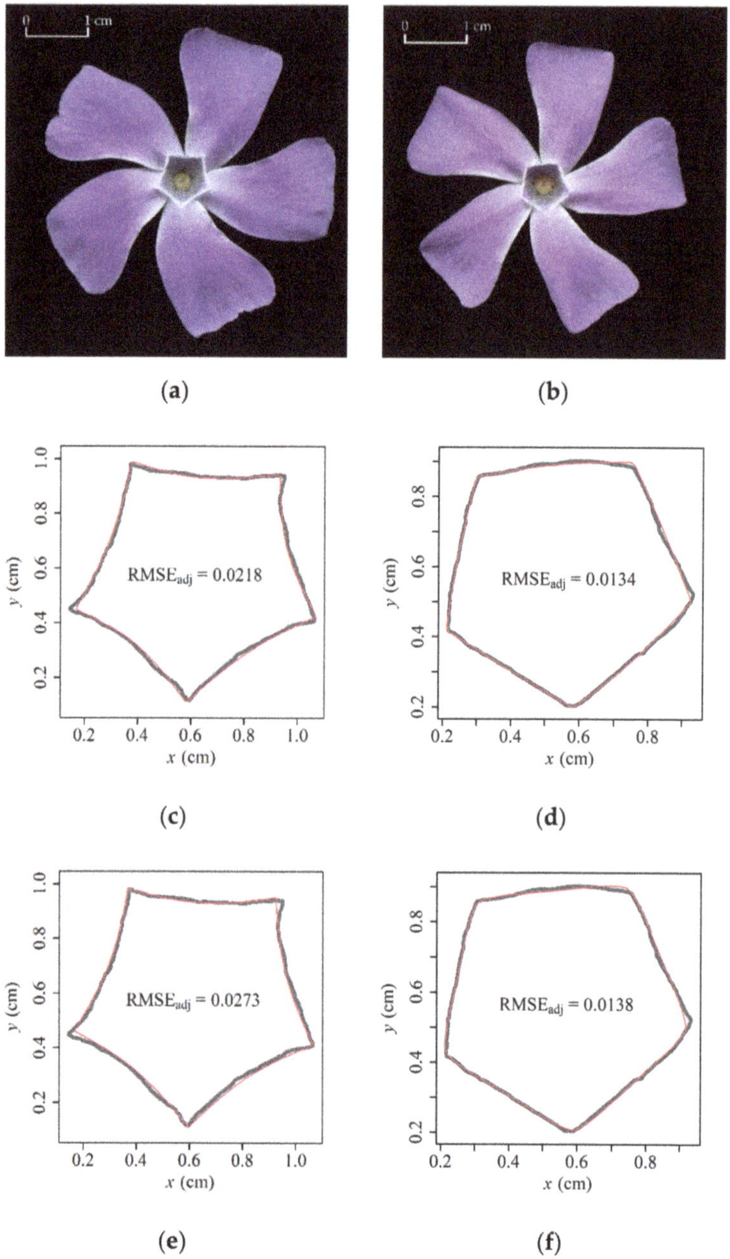

Figure 2. Representative images of *V. major* flowers with five petals (top view; **a**,**b**), and the measured (gray) and predicted (red) outer rims of corolla tubes (ORCTs) using Model 1 (**c**,**d**) and Model 2 (**e**,**f**). The boundary coordinate data of ORCTs were obtained from scanned images, and were fitted by the R package 'biogeom'. $RMSE_{adj}$ is the adjusted root-mean-square error (Equation (7)).

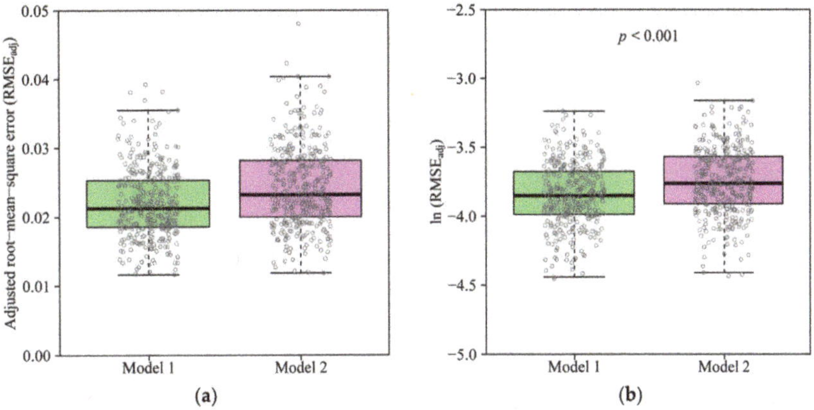

Figure 3. Comparison of adjusted root-mean-square errors (**a**, $RMSE_{adj}$, Equation (7)) and their ln-transformations between Models 1 and 2 (**b**). The thick horizontal lines represent median values in the boxes; a box's body length represents the difference between the 3/4 quantile and the 1/4 quantile; whiskers represent 1.5 times the box's body length or maximum (or minimum) values; and the small gray open circles represent the distribution of data points. The ln-transformed values in (**b**) were compared by a paired sample *t*-test.

The results of the linear regression of the data of $\ln(\hat{n}_3)$ vs. $\ln(\hat{n}_2)$ and the distribution diagrams of 2000 bootstrap replicates of the regression intercept (α) and slope (β) indicated that the 95% CI of the intercept included zero and the 95% CI of the slope included 1.0 (Figures 4 and 5). Thus, there was no significant difference between the estimated values of n_3 and that of n_2, suggesting that the three-parameter Model 2 is sufficient to depict the geometry of the ORCT of *V. major*, although the goodness of fit of Model 1 is slightly greater than that of Model 2 (Figure 3). The estimated values of parameters in Models 1 and 2 have been listed in Tables S1 and S2 in the online Supplementary Materials.

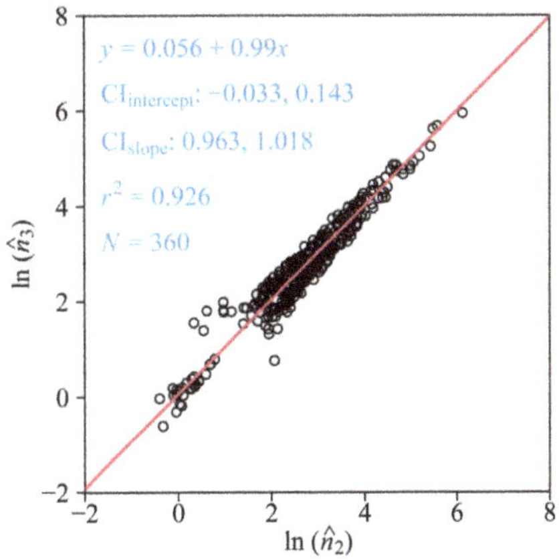

Figure 4. Fitted results to the data of the estimated values of the model parameter n_3 vs. the model parameter n_2 (Equation (3)). The data were fitted according to reduced major axis protocols on a log-log scale. y represents the ln-transformation of the estimated value of n_3; x represents the ln-transformation of the estimated value of n_2; the straight line is the regression line; $CI_{intercept}$ represents the 95% confidence interval of the intercept; CI_{slope} represents the 95% confidence interval of the slope; r^2 is the coefficient of determination that reflects the goodness of fit; and N is the sample size, i.e., the number of flowers used.

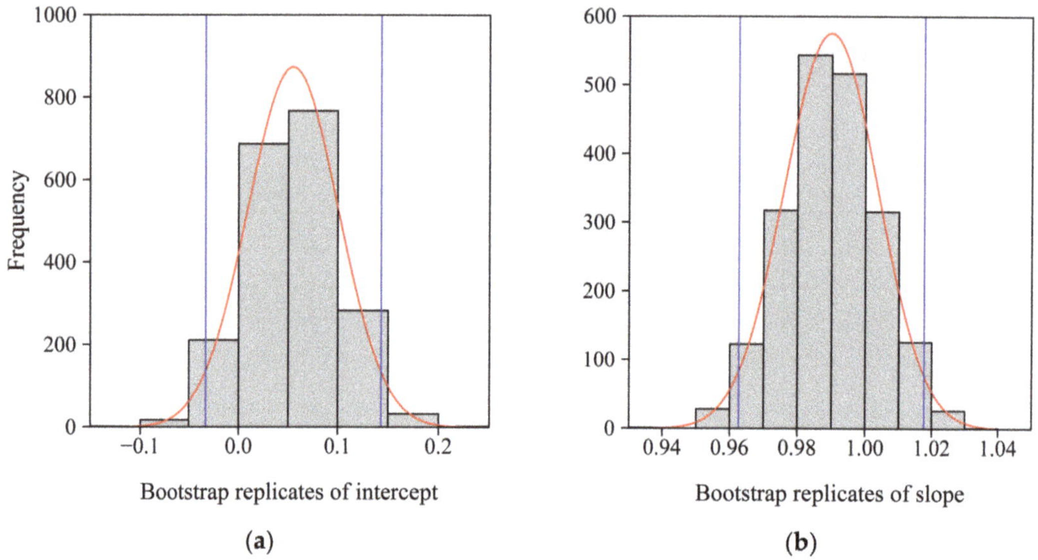

Figure 5. Frequency histograms of bootstrap replicates of the intercept (**a**) and slope (**b**) of the regression between the model parameters n_3 and n_2 (Equation (3); Figure 4). The red curves are normal density curves; and the blue vertical straight lines represent 0.025 quantile (left) and 0.975 quantile (right) that are the lower and upper bounds of the 95% confidence interval in each panel.

Although all p values of the normality test were smaller than 0.05, indicating that the distributions of residuals between the observed and predicted radii were not normal, the skewness only ranged from −1.1 to 1.1, which suggested that the distributions of residuals were not seriously skewed. The means of residuals for all samples ranged from −0.0002 to 0.0006 approximate to zero, and all of the 95% CIs of residuals included 0 (see Table S2). In other words, the distribution of residuals is approximate to be normal, and the above results at least demonstrated the validity of nonlinear least-squares in fitting the ORCTs.

4. Discussion

4.1. Analysis of the Prediction Errors

Multiple artificial measurement errors occurred when we scanned and extracted the boundaries of the 360 outer rims of corolla tubes (ORCTs). On the way from the site of plant growth to the laboratory, artificial physical pressures might have caused a certain deformation of the flowers. In addition, during scanning, the physical pressures on different contact points of a ORCT with the scanning plane could also cause a certain morphological deformation of the ORCT. Thus, there are some parts on an ORCT especially close to the vertices of the pentagon that are not well resolved in scanned images. This generates some uncertainties in extracting the ORCT. Nevertheless, given that such deformations were smaller than 0.05 mm, we conclude that the impact of such deformations on model fitting can be neglected.

In addition, the interspecific variation from a perfect symmetrical pattern in the growth process of *V. major* probably increased the deviation of the actual shape of ORCT from the predicted shape. The intraspecific variation reflects both genotypic differences as well as plastic modifications to differences in the growth microenvironment, including differences in the availability in light, water and nutrients [30–34]. Although the plants used were irrigated by sprinklers, the individuals closer to the water source formed a denser cover, indicating that water availability introduced some changes in the plant phenotype in this study. A stronger intraspecific competition for space at greater water availability caused some physical compression within adjacent individuals during flower growth; this resulted in the deformation of some flowers to a certain degree and subsequently led ORCTs to be less symmetrical. All of the factors outlined likely had some impact on the prediction errors using the two simplified Gielis models to describe the shapes of ORCTs. Nevertheless, the $RMSE_{adj}$ values of the two models were all smaller than 0.05. This verified the validity of the Gielis equation in fitting the real natural geometries of corolla tubes of *V. major*.

4.2. Comparison of the Two Models

The parameters n_2 and n_3 determine the extent of symmetry for the geometry generated by Model 1. The regression analysis indicated that there was no significant difference between n_2 and n_3, indicating that the analysis can be simplified by replacing n_3 with n_2. Therefore, we conclude that Model 2 with fewer parameters can replace Model 1 for describing the ORCT of *V. major*. Although most of the adjusted RMSE values calculated by Model 1 were slightly smaller than those calculated by Model 2 among 360 ORCTs, all of the $RMSE_{adj}$ values for both models were less than 0.05. This means that for any sample, the average absolute deviation between the observed and predicted radii from the polar point to an ORCT was less than 5% of the radius of a circle whose hypothetical area equals that of the ORCT.

The use of Model 2 has several practical advantages. In particular, as fewer parameters need to be fitted, the efficiency of parameter estimation is higher. This implies that the probability for the model parameters not to converge is lower for Model 2; also, the average computer running time to complete the optimization for each dataset using Model 2 is less than that using Model 1. In addition, Model 1 with an additional parameter n_3 tends to lead to an overfitting of the planar coordinate data of ORCTs. Accordingly, this blurs the biological meaning of the parameter values. Model 2 with parameters $n_3 = n_2$ diminishes the possibility of generating an asymmetrical curve and therefore agrees better with previous studies (e.g., refs. [7,9]).

4.3. Variation in the Number of Polygon Sides of the ORCT of *V. major*

In addition to pentagonal ORCTs, there was a very small proportion of quadrangular ORCTs among the flowers of the sampled *V. major* population (Figure 6a). In addition, flowers with hexagonal ORCTs can be observed in this species (personal observations), but were not found in the current study. The plants with pentagonal ORCTs have five petals, while the plants with quadrilateral ORCTs have four petals; each angle of an ORCT corresponds to a petal (Figure 1and Figure 6). In flower sampling, we found only six specimens with quadrilateral petals, i.e., a ratio of 6:366 in the total sample of flowers collected. It is currently unclear what causes the formation of four-petal flowers in *V. major*, but given that it is a clonal species, it is unlikely that the variation of petal number is of genetic origin, although we cannot rule out that the sampled population was a mix of several clones. In fact, the situation is similar to *Syringa* spp. that typically have four-petal flowers, but occasionally form five-petal flowers, and sometimes even flowers with more than five petals in the same inflorescence.

Similar to flowers with pentagonal ORCTs, we fitted the coordinate data of the six quadrilateral ORCTs of *V. major* using Model 2. As with flowers with pentagonal ORCTs, all of the adjusted RMSE values for flowers with quadrilateral ORCTs were smaller than 0.05 (Table S3 in the online Supplementary Materials). Thus, Model 2 also well describes the shape of the ORCT of *V. major* with four petals (Figure 6b).

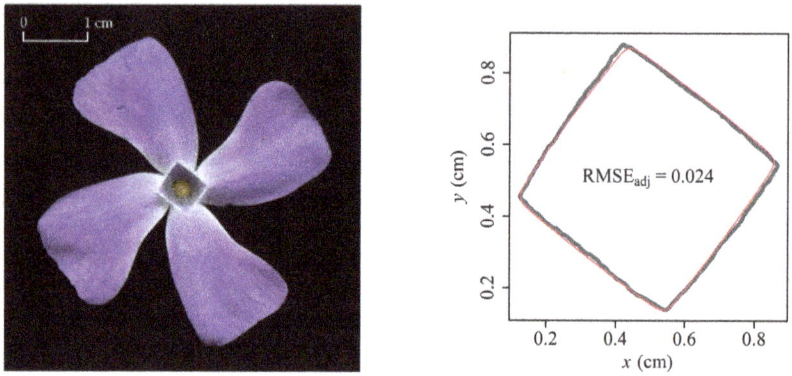

(a) **(b)**

Figure 6. The vertical projection of *V. major* flower with four petals (**a**), and the comparison between the observed and predicted ORCTs (**b**). In panel (**b**), the gray curve represents the observed ORCT and the red curve represents the predicted ORCT using Model 2. Presentation and symbols as in Figure 3.

5. Conclusions

The simplified Gielis equations with $m = 5$ or $m = 4$ are both valid in describing the outer rims of the corolla tubes (in top view) (ORCTs) of *V. major*. A further simplified Gielis model with $n_2 = n_3$ predicted a perfectly axial symmetrical ORCT. The use of the simpler model can avoid model overfitting and reduce the running time for completing parameter estimation. This work shows that the ORCTs of plants can be modeled by the Gielis equation and can provide a reference for future research on superelliptic shapes of plant organs or tissues, e.g., projections of calyxes and polygonal cross-sections of fruits, e.g., ref. [9]). Additionally, the small difference between the parameters n_2 and n_3 in Model 1 did not cause a large deviation for ORCTs from a perfectly axial symmetrical pattern. This work provides evidence that the natural morphology of plant organs and tissues follows the superformula proposed by Gielis [2], and the process of growth and development of corolla tubes exhibits a general but unknown biophysical mechanism that is commonly shared by other organs or tissues of many species, which at least can be reflected by the same mathematical model. Thus, it is worth further examining the validity of the Gielis equation in depicting more geometries found in nature to uncover the influence(s) of such a general biophysical mechanism on biological geometry and morphology.

Supplementary Materials: The following supporting information can be downloaded at: https://www.mdpi.com/article/10.3390/plants11151987/s1; Table S1: Fitted results of the planar coordinate data of ORCTs of *V. major* with five petals using Model 1; Table S2: Fitted results of the planar coordinate data of ORCTs of *V. major* with five petals using Model 2; and Table S3: Fitted results of the planar coordinate data of ORCTs of *V. major* with four petals using Model 2.

Author Contributions: Investigation, L.W. and Q.M.; formal analysis, L.W., Ü.N. and P.S.; writing—original draft preparation, L.W. and P.S.; writing—review and editing, Ü.N. and J.G.; supervision, Ü.N., J.G. and P.S.; L.W. and Q.M. contributed equally to this work. All authors have read and agreed to the published version of the manuscript.

Funding: This research received no external funding.

Institutional Review Board Statement: Not applicable.

Informed Consent Statement: Not applicable.

Data Availability Statement: The data can be found in Tables S1–S3 in the online Supplementary Materials.

Acknowledgments: The authors would like to thank Yabing Jiao and Long Chen for their valuable help in flower sampling and data acquisition. The authors are also thankful to three reviewers for their extremely constructive comments for improving the quality of this manuscript.

Conflicts of Interest: The authors declare no conflict of interest.

References

1. Lamé, G. *Examen des Différentes Méthodes Employées Pour Résoudre les Problèmes de Géométrie*; V. Courcier: Paris, France, 1818.
2. Gielis, J. A general geometric transformation that unifies a wide range of natural and abstract shapes. *Am. J. Bot.* **2003**, *90*, 333–338. [CrossRef] [PubMed]
3. Gielis, J. *The Geometrical Beauty of Plants*; Atlantis Press: Paris, France, 2017.
4. Shi, P.; Huang, J.; Hui, C.; Grissino-Mayer, H.D.; Tardif, J.C.; Zhai, L.; Wang, F.; Li, B. Capturing spiral radial growth of conifers using the superellipse to model tree-ring geometric shape. *Front. Plant Sci.* **2015**, *6*, 856. [CrossRef]
5. Shi, P.; Xu, Q.; Sandhu, H.S.; Gielis, J.; Ding, Y.; Li, H.; Dong, X. Comparison of dwarf bamboos (*Indocalamus* sp.) leaf parameters to determine relationship between spatial density of plants and total leaf area per plant. *Ecol. Evol.* **2015**, *5*, 4578–4589. [CrossRef] [PubMed]
6. Lin, S.; Zhang, L.; Reddy, G.V.P.; Hui, C.; Gielis, J.; Ding, Y.; Shi, P. A geometrical model for testing bilateral symmetry of bamboo leaf with a simplified Gielis equation. *Ecol. Evol.* **2016**, *6*, 6798–6806. [CrossRef] [PubMed]
7. Shi, P.; Ratkowsky, D.A.; Gielis, J. The generalized Gielis geometric equation and its application. *Symmetry* **2020**, *12*, 645. [CrossRef]
8. Tian, F.; Wang, Y.; Sandhu, H.S.; Gielis, J.; Shi, P. Comparison of seed morphology of two ginkgo cultivars. *J. For. Res.* **2020**, *31*, 751–758. [CrossRef]
9. Li, Y.; Quinn, B.K.; Gielis, J.; Li, Y.; Shi, P. Evidence that supertriangles exist in nature from the vertical projections of *Koelreuteria paniculata* fruit. *Symmetry* **2022**, *14*, 23. [CrossRef]
10. Shi, P.; Gielis, J.; Niklas, K.J. Comparison of a universal (but complex) model for avian egg shape with a simpler model. *Ann. N. Y. Acad. Sci.* 2022, *in press*. [CrossRef]
11. Weng, Y. Growth and form in biology: Generation of the plant morphology by spontaneous symmetry breaking based on a pressure field. *Chin. Phys. Lett.* **2004**, *21*, 211–214.
12. Gielis, J.; Caratelli, D.; Fougerolle, Y.; Ricci, P.E.; Gerats, T. A biogeometrical model for corolla fusion in asclepiad flowers. In *Modeling in Mathematics*; Gielis, J., Ricci, P.E., Tavkhelidze, I., Eds.; Atlantis Press: Paris, France, 2017; pp. 83–105.
13. Spíchal, L. Superelipsa a superformule. *Mat.–Fyzika–Inform.* **2020**, *29*, 54–69.
14. Angiosperm Phylogeny Group. An update of the Angiosperm Phylogeny Group classification for the orders and families of flowering plants: APG III. *Bot. J. Linn. Soc.* **2009**, *161*, 105–121. [CrossRef]
15. Endress, M.E.; Liede-Schumann, S.; Meve, U. An updated classification for Apocynaceae. *Phytotaxa* **2014**, *159*, 175–194. [CrossRef]
16. Ronse De Craene, L. *Floral Diagrams: An Aid to Understanding Flower Morphology and Evolution*; Cambridge University Press: Cambridge, UK, 2022.
17. Shi, P.; Ratkowsky, D.A.; Li, Y.; Zhang, L.; Lin, S.; Gielis, J. A general leaf-area geometric formula exists for plants—Evidence from the simplified Gielis equation. *Forests* **2018**, *9*, 714. [CrossRef]
18. Su, J.; Niklas, K.J.; Huang, W.; Yu, X.; Yang, Y.; Shi, P. Lamina shape does not correlate with lamina surface area: An analysis based on the simplified Gielis equation. *Glob. Ecol. Conserv.* **2019**, *19*, e00666. [CrossRef]
19. Shi, P.; Gielis, J.; Quinn, B.K.; Niklas, K.J.; Ratkowsky, D.A.; Schrader, J.; Ruan, H.; Wang, L.; Niinemets, Ü. 'biogeom': An R package for simulating and fitting natural shapes. *Ann. N. Y. Acad. Sci.* **2022**, *in press*. [CrossRef]
20. R Core Team. *R: A Language and Environment for Statistical Computing*; R Foundation for Statistical Computing: Vienna, Austria, 2022; Available online: https://www.R-project.org/ (accessed on 1 June 2022).
21. Nelder, J.A.; Mead, R. A simplex method for function minimization. *Comput. J.* **1965**, *7*, 308–313. [CrossRef]
22. Wei, H.; Li, X.; Huang, H. Leaf shape simulation of castor bean and its application in nondestructive leaf area estimation. *Int. J. Agric. Biol. Eng.* **2019**, *12*, 135–140. [CrossRef]
23. Milla, R.; Reich, P. The scaling of leaf area and mass: The cost of light interception increases with leaf size. *Proc. R. Soc. B Biol. Sci.* **2007**, *274*, 2109–2114. [CrossRef]
24. Niklas, K.; Cobb, E.; Niinemets, Ü.; Reich, P.; Sellin, A.; Shipley, B.; Wright, I. "Diminishing returns" in the scaling of functional leaf traits across and within species groups. *Proc. Natl. Acad. Sci. USA* **2007**, *104*, 8891–8896. [CrossRef]
25. Quinn, G.P.; Keough, M.J. *Experimental Design and Data Analysis for Biologists*; Cambridge University Press: Cambridge, UK, 2002; pp. 100–104.
26. Smith, R. Use and misuse of the reduced major axis for line-fitting. *Am. J. Phys. Anthropol.* **2009**, *140*, 476–486. [CrossRef]
27. Efron, B.; Tibshirani, R.J. *An Introduction to the Bootstrap*; Chapman and Hall/CRC: New York, NY, USA, 1993.
28. Sandhu, H.S.; Shi, P.; Kuang, X.; Xue, F.; Ge, F. Applications of the bootstrap to insect physiology. *Fla. Entomol.* **2011**, *94*, 1036–1041. [CrossRef]

29. Royston, P. Remark AS R94: A remark on Algorithm AS 181: The W test for normality. *Appl. Statist.* **1995**, *44*, 547–551. [CrossRef]
30. Liao, F.; Xie, Y.; Jiang, H. The effect of water stress on the physiology of Vinca major 'Variegata'. *Appl. Mech. Mater.* **2013**, *409–410*, 782–787. [CrossRef]
31. Shen, J. Research on the shade tolerance of *Vinca major*. *For. Sci. Technol.* **2012**, *37*, 15–17.
32. Wang, J.; Wang, J.; Zhou, J.; He, T.; Li, P. Recent progress of artificial lighting technique and effect of light on plant growth. *J. Nanjing For. Univer. (Nat. Sci. Ed.)* **2020**, *44*, 215–222.
33. Zhuang, Y. Comparative study on the water physiological characteristics of *Vinca major* and *Vinca major* 'Variegata'. *Anhui For. Sci. Technol.* **2013**, *39*, 19–24.
34. Niinemets, Ü. Leaf trait plasticity and evolution in different plant functional types. *Ann. Plant Rev.* **2020**, *3*, 473–522.

plants

MDPI

Article

The Modified Brière Equation and Its Applications

Jun Jin [1], Brady K. Quinn [2] and Peijian Shi [3,*]

1 Research Institute of Architecture, Southeast University, Nanjing 210096, China; 101010140@seu.edu.cn
2 Biological Effects Section, St. Andrews Biological Station, Fisheries and Oceans Canada, St. Andrews, NB E5B 0E4, Canada; brady.quinn@dfo-mpo.gc.ca
3 Bamboo Research Institute, College of Science, Nanjing Forestry University, Nanjing 210037, China
* Correspondence: pjshi@njfu.edu.cn; Tel.: +86-25-8542-7231

Abstract: The Brière equation (BE) is widely used to describe the effect of temperature on the development rate of insects, and it can produce both symmetrical and asymmetrical bell-shaped curves. Because of its elasticity in curve fitting, the integrated form of BE has been recommended for use as a sigmoid growth equation to describe the increase in plant biomass with time. However, the start time of growth predicted by the sigmoid growth equation based on the BE is not completely comparable to empirical crop growth data. In the present study, we modified the BE by adding an additional parameter to further increase its elasticity for data fitting. We termed this new equation the modified Brière equation (MBE). Data for the actual height and biomass of 15 species of plants (with two cultivars for one species) were fit with the sigmoid growth equations based on both the BE and MBE assuming that the growth start time was zero for both. The goodness of fit of the BE and MBE sigmoid growth equations were compared based on their root-mean-square errors and the corresponding absolute percentage error between them when fit to these data. For most species, we found that the MBE sigmoid growth equation achieved a better goodness of fit than the BE sigmoid growth equation. This work provides a useful tool for quantifying the ontogenetic or population growth of plants.

Keywords: axial symmetry; curve fitting; ontogenetic growth; sigmoid curve; symmetry

Citation: Jin, J.; Quinn, B.K.; Shi, P. The Modified Brière Equation and Its Applications. *Plants* **2022**, *11*, 1769. https://doi.org/10.3390/plants11131769

Academic Editor: Oleg Chertov

Received: 15 June 2022
Accepted: 1 July 2022
Published: 3 July 2022

1. Introduction

The ontogenetic growth trajectories of animals and plants usually exhibit a sigmoid pattern, and many mathematical equations have been proposed to describe the changes in growth data (e.g., biomass, diameter at breast height, height, length, etc.) with time [1–6]. Among these, the logistic equation is perhaps the most commonly used [5–8]. The three-parameter logistic equation assumes that the growth rate versus time curve is perfectly symmetrical but is not suited to all empirical datasets [9]. Thus, some equations predicting asymmetrical (skewed) growth rate curves have been used instead to reflect the growth trajectories of animals and plants [3,9,10].

In thermal biology, the development (or growth) rate (i.e., the proportion or amount of development or growth, respectively, completed per unit time) of organisms has been demonstrated to be an asymmetrical (usually left-skewed), bell-shaped function of temperature [11,12]. There are many different mathematical models that can produce skewed and symmetrical development (or growth) rate versus temperature curves [13,14]. Among these temperature-dependent development rate models, the equation proposed by Brière et al. [15] is relatively simple, yet it can produce an asymmetrical curve, includes biologically meaningful parameters, and is able to describe the effect of temperature on the development rate of many organisms:

$$r(T) = ax(T - T_{\min})(T_{\max} - T)^{1/m}, \quad (1)$$

where $r(T)$ represents the development rate at temperature T; T_{min} and T_{max} represent the lower and upper threshold temperatures of development, respectively, and $r(T) = 0$ when $T < T_{min}$ or $T > T_{max}$; and a and m are parameters to be estimated. Brière et al. [15] suggested that a simplified version of Equation (1) could be used by fixing $m = 2$; this simplified equation, referred to as the Brière-1 equation (versus the original Brière equation, or Brière-2 equation), is not considered further herein. All subsequent references to the Brière equation hereinafter refer to the original Brière equation, or Brière-2 equation. Up to 25 May 2022, the original paper by Brière et al. [15] has been cited 745 times according to Google Scholar.

Yin et al. [3] introduced a new sigmoid growth equation using the integrated form of the beta equation while replacing temperature with time as the independent variable, and it has been shown to provide a good fit to the actual growth data of plants [3,9,16,17]. Similarly, Cao et al. [18] presented a sigmoid growth equation based on the Brière equation. However, although the new growth equation introduced by Cao et al. [18] could fit the growth data of some crops very well, the start times of growth predicted by it were not reliable as the predicted start time seriously deviated from the actual sowing time [18]. Further, when the start time is manually set to zero, the prediction errors for the sigmoid equation proposed by Cao et al. [18] relative to empirical data can be quite large.

In this study, we propose a modification of the Brière equation by adding an extra parameter to render the new equation more flexible in curve fitting and reduce the dependency of the curve fitting on the start time of growth. We used empirical data from 15 species of plants (with two cultivars for one species) to test the validity of a new sigmoid growth equation based on this modified Brière equation (MBE) and compared the goodness of fit of the sigmoid growth equation based on the MBE to that based on the original Brière equation (BE).

2. Materials and Methods

2.1. Plant Materials

The seeds of 11 species of common agricultural crops (with two cultivars for one species, 12 crops in total), including sunflower (*Helianthus annuus*), peanut (*Arachis hypogaea*), black soybean (*Glycine max* 'Kuromame'), soybean (*Glycine max*), kidney bean (*Phaseolus vulgaris*), garden pea (*Pisum sativum*), adzuki bean (*Vigna angularis*), mung bean (*Vigna radiata*), cotton (*Gossypium herbaceum*), sweet sorghum (*Sorghum bicolor*), corn (*Zea mays*), and Mexican corn (*Zea mexicana*), in northern China were sown in a field at Jinan, China on 27 June 2011. We then randomly sampled 20 individuals from each species (or each cultivar) to measure their dry mass on each of 15 subsequent investigation dates from 11 July to 20 September 2011 once every five days on average. In total, 3000 individual samples were collected. All individuals were sampled between 7:00 a.m. to 8:00 a.m., and the roots were washed with running water to remove soil. For small crops, whole plants were dried for 24–48 h at 60 °C. However, for large crops, such as sunflower, sweet sorghum, corn, and Mexican corn, whole plants were dried for 72 h at 80 °C. Here, we defined the crops with mean whole-plant dry mass $\geq$ 150 g and mean aboveground height $\geq$ 1.5 m on the last investigation date (i.e., 85 days from the sowing time) as large crops and those with mean whole-plant dry mass < 150 g and mean aboveground height < 1.5 m at this time as small crops. Samples were weighed after drying to obtain their dry mass. Shi et al. [5] can be consulted for further details of the sampling procedures.

We also used the height data from four species of bamboo (*Phyllostachys iridescens*, *Phyllostachys mannii*, *Pleioblastus maculatus*, and *Sinobambusa tootsik*) previously collected at different investigation dates [9]. The bamboo shoots were grown at the Nanjing Forestry University campus in the spring of 2016. We defined the time when a bamboo shoot tip was first observed at ground level as time = 1, and the previous day as time = 0. We measured the height of each shoot at 12:00 p.m. every day at the early growth stage, and then measured the height at 12:00 p.m. once every three days when the height changed more slowly.

The above datasets were obtained from the 'crops' and 'shoots' datasets, respectively, included in the R (version 4.2.0) [19] package 'IPEC' (version 1.0.3) [20].

2.2. Methods

The original Brière equation was proposed to describe the effect of temperature on the development rate of many organisms, especially arthropods [13,15,21] (see Equation (1)). After replacing temperature T with time x, and the development rate $r(T)$ with the growth rate $f(x)$ in the original Brière equation, we have:

$$f(x) = ax(x - x_{\min})(x_{\max} - x)^{1/m}, \tag{2}$$

where $f(x)$ represents the growth rate at time x; $x_{\min}$ and $x_{\max}$ represent the start and end times of growth, respectively, and $f(x) = 0$ when $x < x_{\min}$ or $x > x_{\max}$; and a and m are parameters to be estimated.

Herein, we propose including an additional parameter to be estimated, δ, in Equation (2) to render it more elastic in curve fitting as follows:

$$f(x) = a\left|x(x - x_{\min})(x_{\max} - x)^{1/m}\right|^{\delta}. \tag{3}$$

We refer to Equations (2) and (3) as the BE and MBE for convenience hereinafter. The integrated forms of the BE and MBE were used to fit the dry mass versus time data of crops and those of height versus time data of bamboo shoots. Figure 1A illustrates the influence of the new δ parameter on the curve shapes plotted by the modified BE.

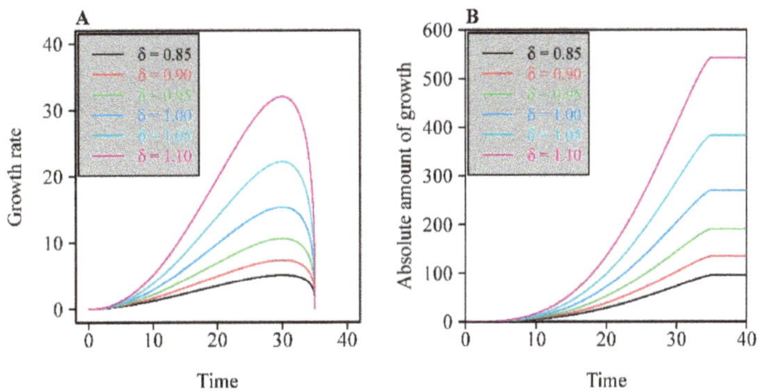

Figure 1. Influence of the new additional parameter, δ, on the shape of the curves plotting the modified Brière equation (**A**), and on those plotting the sigmoid equation based on the integrated form of the modified Brière equation (**B**). Here, $a = 0.01$, $m = 3$, $x_{\min} = 0$, and $x_{\max} = 35$. In this example, no specific units are used; the x-axis represents a generic time variable (hours, days, weeks, etc.) and the y-axes generic (**A**) growth rate (cm day^{-1}, g h^{-1}, etc.) and growth (cm, g, etc.) metrics.

After integrating the BE and MBE, two sigmoid growth equations are obtained, i.e.,

$$y = \begin{cases} 0 & x < x_{\min} \\ \int_{x_{\min}}^{x} f(x)\,dx & x \in [x_{\min}, x_{\max}] \\ 0 & x > x_{\max} \end{cases}. \tag{4}$$

Here, we obtained two sigmoid growth equations when using Equations (2) and (3) to represent $f(x)$, respectively. There is an explicit mathematical expression for the sigmoid equation based on the integrated form of BE [3,17], which is referred to as the BE sigmoid equation hereinafter for convenience. However, there is no explicit mathematical form for the integrated form of MBE. Thus, we calculated the numerical integral of MBE, which we refer to as the MBE sigmoid equation for convenience. Figure 1B illustrates the curves produced by the MBE sigmoid equation with different δ values.

The Nelder–Mead optimization method [22] was used to fit the above equations to each plant dataset by minimizing the residual sum of squares (RSS) between observed and predicted y values. For the empirical crop biomass and bamboo shoot height data we used, the start time of growth is actually known. Thus, we fixed the starting time to be zero, in which case Equation (3) had four parameters and Equation (2) had three parameters. We then calculated the root-mean-square error (RMSE) for each dataset:

$$\mathrm{RMSE} = \sqrt{\mathrm{RSS}/n}, \tag{5}$$

where n represents the number of data points. To examine whether the additional parameter in Equation (3) was warranted relative to using the simpler Equation (2), we used the percentage error of the absolute difference (APE, in %) between the RMSE of the BE sigmoid equation and that of the MBE sigmoid equation [23,24] as follows:

$$\mathrm{APE} = \frac{|\mathrm{RMSE_{BE}} - \mathrm{RMSE_{MBE}}|}{\mathrm{RMSE_{BE}}} \times 100\%. \tag{6}$$

As a rule of thumb, when APE > 5%, the introduction of the additional parameter (δ) in Equation (3) versus (2) is justified; otherwise, it is not worthwhile.

The whole-plant dry mass tends to be proportional to the whole-plant fresh mass [5,25] if there is no water loss. In that case, if the sigmoid equation based on the integrated form of MBE can fit the dry mass data well, it should also be applicable to fresh mass data. The difference here would only affect the numerical value of the parameter a in Equation (3). However, given the water loss during crop sampling, the dry mass is more reliable than the fresh mass. Thus, for the crops, we used the dry mass rather than the fresh mass to test the validities of the BE and MBE sigmoid equations.

The package 'biogeom' (version 1.0.5) [26,27] was used in R (version 4.2.0) [19] to estimate model parameters, and R (version 4.2.0) was also used to carry out all other calculations.

3. Results

For each of the crop species tested except kidney bean (11/12 datasets), the MBE sigmoid equation had a lower RMSE than the BE sigmoid equation, and the APE was greater than 5% (Figure 2). This confirmed the validity of the MBE sigmoid equation and its superiority relative to the BE sigmoid equation. Because of the lack of biomass data at the mature stages of sunflower and peanut, both sigmoid functions overestimated the end time of growth, leading the predicted curves to exhibit exponential growth at the early and middle growth stages (Figure 2A,B).

For each of the four species of bamboo considered, the MBE sigmoid equation also had a lower RMSE than the BE sigmoid equation, and the APE was larger than 5% for three of the four species (Figure 3). For *Pleioblastus maculatus*, the APE was equal to 4.68%, or approximately 5%. These results demonstrate that the BE and MBE sigmoid equations are both suited to representing data for growth in terms of height for bamboo shoots, at least provided that the initial time at which growth starts (i.e., $x = 0$) can be accurately known. Nevertheless, the MBE sigmoid equation is still better.

The estimated values of the parameters of the BE sigmoid equation and the MBE sigmoid equation for the 15 species of plants (with two cultivars for one species) tested are listed in Tables S1 and S2 in the online Supplementary Materials.

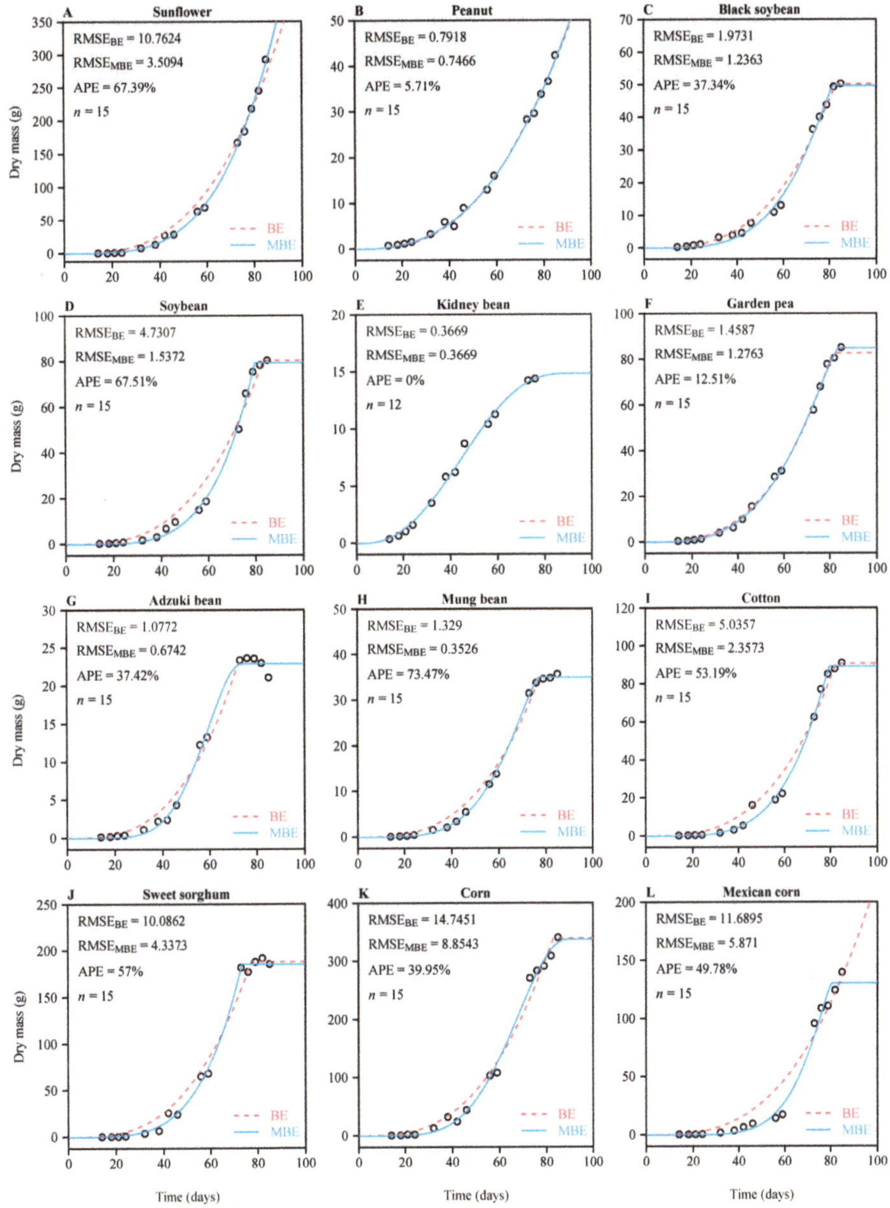

Figure 2. Results of fitting two sigmoid functions (based on the BE and MBE) to the whole-plant dry mass versus time data of 11 crop species (with two cultivars for one crop species). The small open circles represent the observed dry mass at different times after the sowing date; the red dashed curves represent the dry mass values predicted by the BE sigmoid equation; the blue solid curves represent the dry mass values predicted by the MBE sigmoid equation. RMSE represents the root-mean-square error between the observed and predicted y values; APE represents the percentage error of the absolute difference between the two equations' RMSE values; n represents the sample size. Panels (**A–L**) represent different crops.

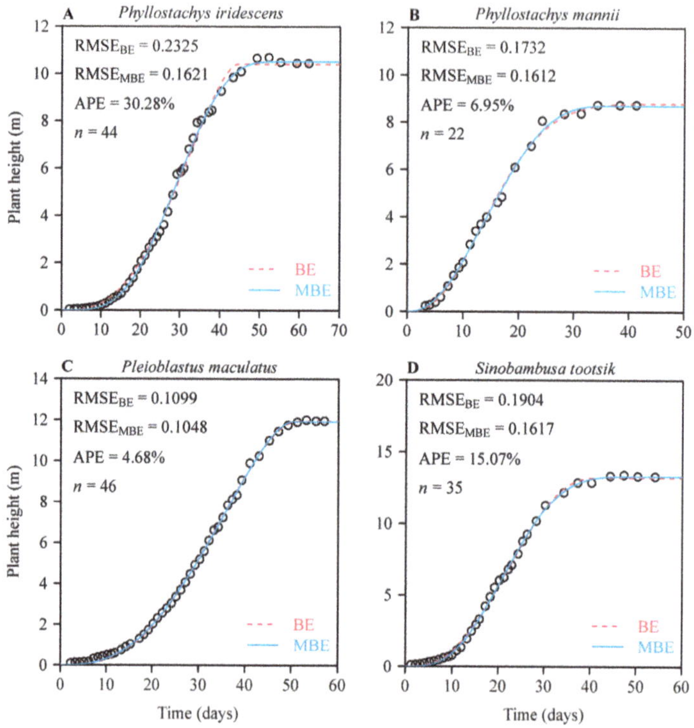

Figure 3. Results of fitting two sigmoid functions (based on the BE and MBE) to the shoot height versus time data of four species of bamboo. The small open circles represent the observed shoot height at different times (days) after the date when the shoot tip first emerged from the soil; the red dashed curves represent the height values predicted by the BE sigmoid equation; the blue solid curves represent the height values predicted by the MBE sigmoid equation. RMSE represents the root-mean-square error between the observed and predicted y values; APE represents the absolute percent error difference between the two equations' RMSE values; n represents the sample size. Panels (**A–D**) represent different bamboo species.

4. Discussion

4.1. Elasticity in Curve Fitting of the Two Sigmoid Equations

The inclusion of an additional parameter, δ, in the MBE greatly increased the elasticity of the MBE sigmoid equation in curve fitting; i.e., it increased the range of sigmoid curves with different curvatures that can be fit relative to the BE sigmoid equation (Figure 1). The larger the numerical value of δ is, the greater the curvature of the line will be that is generated by the MBE sigmoid equation. For 14 of the 16 tested datasets, the APE values were greater than 5%, which indicates that the increased elasticity in curve fitting achieved by the additional parameter in the MBE sigmoid equation outweighed the cost of the model's increased complexity [28]. In fact, the elasticities of the two sigmoid equations in curve fitting mainly depend on the abilities of the BE and MBE to accurately describe growth rates. However, it is more difficult to directly measure the growth rate per unit time in practice than to measure accumulated biomass or height after some time interval has elapsed. Although Cao et al. [18] showed that the BE sigmoid growth function achieved a good fit to the dry mass data of six of the twelve datasets of crops used herein, the estimated growth start times were later than the actual sowing time and the times seedings appeared (see Figure 1 in Ref. [18]). It is apparently reasonable to fix the start time to the sowing time given that all the seeds were planted on the same day. In our study, the MBE sigmoid function achieved a good fit to observations without estimating unreasonable start times, and it showed good elasticity in curve fitting through changes in the numerical value of the δ parameter. After all, the $x_{\min}$ parameter of the BE and MBE has an explicit geometrical and biological meaning, so it is unreasonable to over- or underestimate its numerical value. An over- or underestimated $x_{\min}$ value can easily mislead the user of such equations to predict inaccurate growth. On the other hand, the new δ parameter has no clear biological or geometrical meaning, and it only serves as a constant to be estimated to improve the flexibility of the equation in curve fitting.

4.2. Why We Did Not Compare the Two Sigmoid Equations with Other Equations

As stated in the Introduction section, there are many temperature-dependent development (or growth rate) equations that can produce a skewed curve [9,13,14,21]. Shi et al. [17] found that the sigmoid equation based on the integral of the beta equation [3] can achieve better goodness of fit than traditional growth equations, including the exponential and logistic equations [8], Gompertz equation [29], von Bertalanffy equation [1], and ontogenetic growth equation [2]. Shi et al. [9] demonstrated that the sigmoid equation based on the integral of the modified beta equation and that of the modified Lobry–Rosso–Flandrois (LRF) equation [30,31] can be better-suited to representing the ontogenetic growth data of animals and plants. However, the mathematical expressions of the BE and MBE are simpler than those of the beta equation, modified beta equation, LRF equation, and modified LRF equation (Equations (2) and (3) versus those published in Ref. [9]). In the present study, we did not compare the BE and MBE sigmoid equations with other sigmoid equations because we mainly wanted to examine the strengths and drawbacks of the BE and MBE sigmoid equations and to test whether the latter is better than the former for curve fitting. A systematic comparison among sigmoid equations is worth carrying out using more datasets in future studies.

4.3. Reliability of Estimated Parameters in the Two Sigmoid Equations

Although the estimated parameters, especially the start and end times of growth, are considered to be meaningful for predicting when these time points will occur during the growth of plants, the reliability of the prediction is largely constrained by whether the dataset includes a full range of growth data, especially those at the mature stage (i.e., the asymptotic values of plant biomass or height). If the asymptotic values of plant size at the mature stage are known, the predicted end time of growth tends to be reliable [6,9,18]; however, if such data are lacking, the predicted end time tends to be overestimated (see Figure 2A,B,L). To reduce the uncertainty of predictions and the complexity of the model, we suggest fixing the starting time and allowing the δ parameter to vary. It is more important to estimate the end time rather than the start time of growth for a crop product or forest resource management. If the estimated end time of growth is reliable, it can produce a reliable growth rate curve with which one can predict when plants reach their maximum growth rate.

4.4. Other Potential Applications of the Modified Brière Equation

Given that the MBE is flexible in fitting a skewed bell-shaped curve, it is potentially useful to model the profiles of ovate leaves [32]. Shi et al. [32] formed two axially symmetrical curves using the modified beta equation (also the modified LRF equation) to fit the ovate leaf shapes of *Neocinnamomum* plants (Lauraceae). Given that the MBE can generate similar skewed curves, the MBE should be similarly useful in describing ovate leaf shape. Relative to the BE, the MBE is more flexible due to its inclusion of an additional parameter, δ, as discussed above, and it can thus also fit the actual boundary coordinates of ovate leaves (Figure 4). Like the original BE, the MBE may also be useful in data-fitting and modelling of temperature-dependent development and growth rates in animals and plants [15,21]. While such additional applications of the MBE are beyond the scope of the present work (and, thus, we do not discuss them in detail here), it will be worthwhile for future studies to assess the validity of the MBE in fitting ovate and obovate leaf shapes and temperature-dependent development rate data.

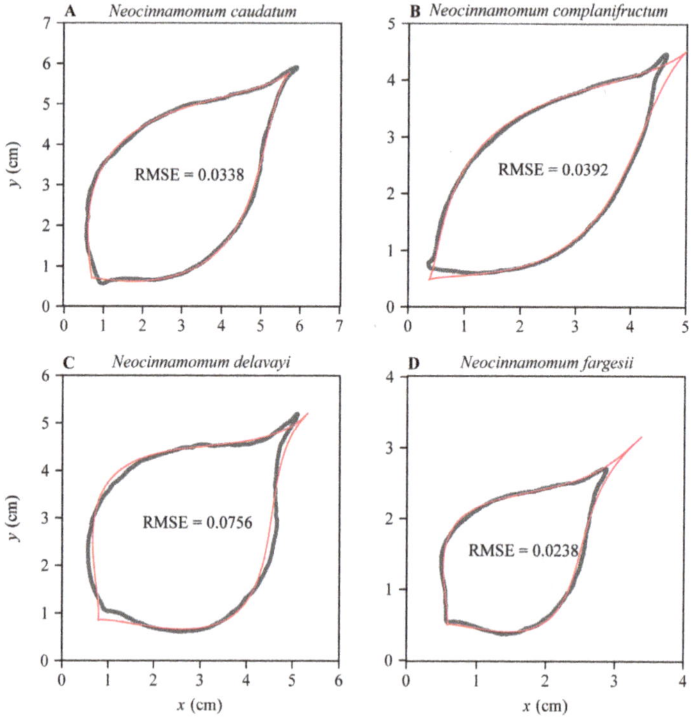

Figure 4. Results of fitting the MBE to the boundary coordinates of the leaves of four species of *Neocinnamomum*. The gray curves are scanned (actual) leaf perimeters, and the red curves are leaf perimeters predicted by the MBE. The boundary coordinate data came from the dataset '*Neocinnamomum*' in R package 'biogeom' (https://cran.r-project.org/web/packages/biogeom/index.html; accessed on 30 May 2022).

5. Conclusions

A better goodness of fit was obtained with the MBE sigmoid equation for 15 of the 16 datasets of plants than the BE sigmoid equation, and it had the same RMSE as that of the BE sigmoid equation for the one remaining species. For most datasets (14/16), the percentage errors of the absolute difference between the RMSE of the MBE sigmoid equation and that of the BE sigmoid equation were greater than 5%, which indicates that the addition of the δ parameter to the original Brière equation was worthwhile, improving the validity of the MBE sigmoid equation in reflecting the actual growth data of plants. In addition, by virtue of the estimated model parameters, whether the growth curve of a given species is left- or right-skewed can be evaluated (i.e., based on the derivative of the MBE sigmoid equation), and, thus, one can use this to predict the time associated with the maximum growth rate.

Supplementary Materials: The following supporting information can be downloaded at: https://www.mdpi.com/article/10.3390/plants11131769/s1, Table S1: The estimates of the model parameters for the whole-plant dry mass versus time data of 12 crops; Table S2: The estimates of the model parameters for the shoot height versus time data of four species of bamboo.

Author Contributions: Conceptualization, J.J., P.S. and B.K.Q.; methodology, P.S.; software, P.S. and B.K.Q.; formal analysis, J.J. and P.S.; investigation, J.J., P.S. and B.K.Q.; writing—original draft preparation, J.J. and P.S.; writing—review and editing, B.K.Q.; funding acquisition, J.J. All authors have read and agreed to the published version of the manuscript.

Funding: This research was funded by the National Key R&D Program of China (grant numbers 2020YFD1100405 and 2020YFD1100400).

Data Availability Statement: The data used in the present work have been packaged in package 'IPEC' (version 1.0.3; https://cran.r-project.org/web/packages/IPEC/index.html; accessed on 24 May 2022). The crop biomass growth data were tabulated in dataset 'crops', and the bamboo height growth data were tabulated in dataset 'shoots'.

Acknowledgments: We thank Meilin Fan, Shuiyuan Fang, Feng Ge, Xingyuan Men, Fang Ouyang, David A. Ratkowsky, and Chengkang Wang for their valuable help during the preparation of this work. We also thank the academic editor and four reviewers for their valuable comments.

Conflicts of Interest: The authors declare no conflict of interest.

References

1. von Bertalanffy, L. Quantitative laws in metabolism and growth. *Q. Rev. Biol.* **1957**, *32*, 217–231. [CrossRef] [PubMed]
2. West, G.B.; Brown, J.H.; Enquist, B.J. A general model for ontogenetic growth. *Nature* **2001**, *413*, 628–631. [CrossRef] [PubMed]
3. Yin, X.; Goudriaan, J.; Lantinga, E.A.; Vos, J.; Spiertz, H.J. A flexible sigmoid function of determinate growth. *Ann. Bot.* **2003**, *91*, 361–371. [CrossRef] [PubMed]
4. Makarieva, A.M.; Gorshkov, V.G.; Li, B.-L. Ontogenetic growth: Models and theory. *Ecol. Model.* **2004**, *176*, 15–26. [CrossRef]
5. Shi, P.; Men, X.; Sandhu, H.S.; Chakraborty, A.; Li, B.; Ouyang, F.; Sun, Y.; Ge, F. The "general" ontogenetic growth model is inapplicable to crop growth. *Ecol. Model.* **2013**, *266*, 1–9. [CrossRef]
6. Shi, P.; Ishikawa, T.; Sandhu, H.S.; Hui, C.; Chakraborty, A.; Jin, X.; Tachihara, K.; Li, B. On the 3/4-exponent von Bertalanffy equation for ontogenetic growth. *Ecol. Model.* **2014**, *276*, 23–28. [CrossRef]
7. Verhulst, P.-F. Recherches mathématiques sur la loi d'accroissement de la population. *Nouv. Mémoires L'académie R. Sci. Belles-Lett. Brux.* **1845**, *18*, 14–54.
8. Thornley, J.H.M.; Shepherd, J.J.; France, J. An open-ended logistic-based growth function: Analytical solutions and the power-law logistic model. *Ecol. Model.* **2007**, *204*, 531–534. [CrossRef]
9. Shi, P.; Fan, M.; Ratkowsky, D.A.; Huang, J.; Wu, H.; Chen, L.; Fang, S.; Zhang, C. Comparison of two ontogenetic growth equations for animals and plants. *Ecol. Model.* **2017**, *349*, 1–10. [CrossRef]
10. Paine, C.; Marthews, T.R.; Vogt, D.R.; Purves, D.; Rees, M.; Hector, A.; Turnbull, L.A. How to fit nonlinear plant growth models and calculate growth rates: An update for ecologists. *Methods Ecol. Evol.* **2012**, *3*, 245–256. [CrossRef]
11. Uvarov, B.P. Insects and climate. *Trans. Entomol. Soc.* **1931**, *79*, 1–232. [CrossRef]
12. Campbell, A.; Frazer, B.D.; Gilbert, N.; Gutierrez, A.P.; Mackauer, M. Temperature requirements of some aphids and their parasites. *J. Appl. Ecol.* **1974**, *11*, 431–438. [CrossRef]
13. Shi, P.; Reddy, G.V.P.; Chen, L.; Ge, F. Comparison of thermal performance equations in describing temperature-dependent developmental rates of insects: (I) empirical models. *Ann. Entomol. Soc. Am.* **2016**, *109*, 211–215. [CrossRef]
14. Ratkowsky, D.A.; Reddy, G.V.P. Empirical model with excellent statistical properties for describing temperature-dependent developmental rates of insects and mites. *Ann. Entomol. Soc. Am.* **2017**, *110*, 302–309. [CrossRef]
15. Briére, J.-F.; Pracros, P.; Le Roux, A.-Y.; Pierre, J.-S. A novel rate model of temperature-dependent development for arthropods. *Environ. Entomol.* **1999**, *28*, 22–29. [CrossRef]
16. Voorend, W.; Lootens, P.; Nelissen, H.; Roldán-Ruiz, I.; Inzé, D.; Muylle, H. LEAF-E: A tool to analyze grass leaf growth using function fitting. *Plant Methods* **2014**, *10*, 37. [CrossRef] [PubMed]
17. Shi, P.; Chen, L.; Hui, C.; Grissino-Mayer, H.D. Capture the time when plants reach their maximum body size by using the beta sigmoid growth equation. *Ecol. Model.* **2016**, *320*, 177–181. [CrossRef]
18. Cao, L.; Shi, P.; Li, L.; Chen, G. A new flexible sigmoidal growth model. *Symmetry* **2019**, *11*, 204. [CrossRef]
19. R Core Team. *R: A Language and Environment for Statistical Computing*; R Foundation for Statistical Computing: Vienna, Austria, 2022. Available online: https://www.R-project.org/ (accessed on 30 May 2022).
20. Shi, P.; Ridland, P.M.; Ratkowsky, D.A.; Li, Y. *IPEC: Root Mean Square Curvature Calculation*, R package version 1.0.3; CRAN: Vienna, Austria, 2022. Available online: https://cran.r-project.org/web/packages/IPEC/index.html (accessed on 24 May 2022).
21. Quinn, B.K. A critical review of the use and performance of different function types for modeling temperature-dependent development of arthropod larvae. *J. Therm. Biol.* **2017**, *63*, 65–77. [CrossRef]
22. Nelder, J.A.; Mead, R. A simplex method for function minimization. *Comput. J.* **1965**, *7*, 308–313. [CrossRef]
23. Yu, X.; Shi, P.; Schrader, J.; Niklas, K.J. Nondestructive estimation of leaf area for 15 species of vines with different leaf shapes. *Amer. J. Bot.* **2020**, *107*, 1481–1490. [CrossRef] [PubMed]
24. Yu, K.; Reddy, G.V.P.; Schrader, J.; Guo, X.; Li, Y.; Jiao, Y.; Shi, P. A nondestructive method of calculating the wing area of insects. *Ecol. Evol.* **2022**, *12*, e8792. [CrossRef] [PubMed]

25. Shipley, B.; Vu, T.-T. Dry matter content as a measure of dry matter concentration in plants and their parts. *New Phytol.* **2002**, *153*, 359–364. [CrossRef]
26. Shi, P.; Gielis, J.; Quinn, B.K. *'biogeom': Biological Geometries*, R package version 1.0.6.; CRAN: Vienna, Austria, 2022. Available online: https://cran.r-project.org/web/packages/biogeom/index.html (accessed on 30 May 2022).
27. Shi, P.; Gielis, J.; Quinn, B.K.; Niklas, K.J.; Ratkowsky, D.A.; Schrader, J.; Ruan, H.; Wang, L.; Niinemets, Ü. 'biogeom': An R package for simulating and fitting natural shapes. *Ann. N. Y. Acad. Sci.* 2022; *in press*. [CrossRef]
28. Spiess, A.-N.; Neumeyer, N. An evaluation of R squared as an inadequate measure for nonlinear models in pharmacological and biochemical research: A Monte Carlo approach. *BMC Pharmacol.* **2010**, *10*, 6. [CrossRef]
29. Gompertz, B. On the nature of the function expressive of the law of human mortality, and on a new mode of determining the value of life contingencies. *Philos. Trans. R. Soc. Lond.* **1825**, *115*, 513–583.
30. Lobry, J.R.; Rosso, L.; Flandrois, J.P. A FORTRAN subroutine for the determination of parameter confidence limits in non-linear models. *Binary* **1991**, *3*, 86–93.
31. Rosso, L.; Lobry, J.R.; Flandrois, J.P. An unexpected correlation between cardinal temperatures of microbial growth highlighted by a new model. *J. Theor. Biol.* **1993**, *162*, 447–463. [CrossRef]
32. Shi, P.; Yu, K.; Niklas, K.J.; Schrader, J.; Song, Y.; Zhu, R.; Li, Y.; Wei, H.; Ratkowsky, D.A. A general model for describing the ovate leaf shape. *Symmetry* **2021**, *13*, 1524. [CrossRef]

MDPI
St. Alban-Anlage 66
4052 Basel
Switzerland
www.mdpi.com

Plants Editorial Office
E-mail: plants@mdpi.com
www.mdpi.com/journal/plants

www.ingramcontent.com/pod-product-compliance
Lightning Source LLC
Chambersburg PA
CBHW040837250926
51850CB00007BA/1340
* 9 7 8 3 0 3 6 5 9 4 2 2 4 *